# ENERGY, ENVIRONMENT, AND SUSTAINABILITY

# ENERGY,
# ENVIRONMENT,
# AND SUSTAINABILITY

## Saeed Moaveni

MINNESOTA STATE UNIVERSITY, MANKATO

CENGAGE
Learning®

Australia • Brazil • Japan • Korea • Mexico • Singapore • Spain • United Kingdom • United States

***Energy, Environment, and Sustainability***
**Saeed Moaveni**

Product Director, Global Engineering:
  Timothy L. Anderson

Senior Content Developer: Mona Zeftel

Associate Media Content Developer:
  Ashley Kaupert

Product Assistant: Alex Sham

Marketing Manager: Kristin Stine

Director, Higher Education Production:
  Sharon L. Smith

Senior Content Project Manager:
  Martha Conway

Production Service: RPK Editorial Services, Inc.

Copyeditor: Shelly Gerger-Knechtl

Proofreader: Harlan James

Indexer: Shelly Gerger-Knechtl

Compositor: SPi Global

Senior Art Director: Michelle Kunkler

Cover and Internal Design: cmillerdesign

Cover Images:
  Seoul city lights: © Vincent St. Thomas/
            Shutterstock.com;
  Global warming: © kwest/Shutterstock.com

Intellectual Property
  Analyst: Christine Myaskovsky
  Project Manager: Sarah Shainwald

Text and Image Permissions Researcher:
  Kristiina Paul

Manufacturing Planner: Doug Wilke

Library of Congress Control Number: 2016952393

ISBN: 978-1-133-10509-1

**Cengage Learning**
20 Channel Center Street
Boston, MA 02210
USA

Cengage Learning is a leading provider of customized learning solutions with employees residing in nearly 40 different countries and sales in more than 125 countries around the world. Find your local representative at **www.cengage.com**.

Cengage Learning products are represented in Canada by Nelson Education Ltd.

To learn more about Cengage Learning Solutions, visit **www.cengage.com/engineering**.

Purchase any of our products at your local college store or at our preferred online store **www.cengagebrain.com**.

Unless otherwise noted, all items © Cengage Learning.

Printed in Canada
Print Number: 01          Print Year: 2016

# Contents

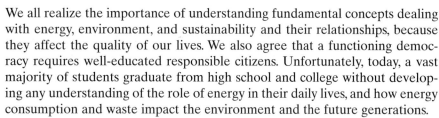

We all realize the importance of understanding fundamental concepts dealing with energy, environment, and sustainability and their relationships, because they affect the quality of our lives. We also agree that a functioning democracy requires well-educated responsible citizens. Unfortunately, today, a vast majority of students graduate from high school and college without developing any understanding of the role of energy in their daily lives, and how energy consumption and waste impact the environment and the future generations.

This book is an attempt to lay down the foundation for the development of responsible citizens with clear understanding of contemporary issues dealing with energy, environment, and sustainability. Great care has been exercised to use real-world examples to get important points across and foster critical thinking. The content is designed to develop the ability of students to go beyond mere understanding of the concepts but to also quantify their energy and environment footprints in order to determine whether their actions are sustainable. The content is also presented with a simple conversational tone with many visual aids to keep today's students engaged. The level of mathematical requirements is kept low so that the topics can be taught to all students. If students can add, subtract, and multiply, then they will be able to follow the examples presented in this textbook, solve the homework problems, and determine their environmental footprints. Moreover, in order to reach out to as many students as possible, the content is prepared as a general education course that can be taught at a community college or a university by instructors with various backgrounds including physics, science, or engineering. When it comes to energy, environment, and stainability, we must realize that *we are all in it together!*

## Organization

This book is organized into four parts and 12 chapters; each chapter begins by stating the **learning objectives (LO)** and concludes by summarizing what the student should have gained from studying that chapter. Relevant, everyday examples with which students can associate easily are provided throughout the book. Many **hands-on problems** conclude each chapter, asking the student to gather and analyze information. Moreover, these problems require students to make brief reports and presentations so that they learn the importance of good written and oral communication skills. To emphasize the significance of teamwork and to encourage group participation, many of the problems also require group work; some require the participation of the entire class. The main parts of the book are described below.

### Part One: Basic Concepts

In Part One, consisting of Chapters 1 and 2, we introduce the students to the importance of understanding basic concepts: human needs, energy, environment, sustainability, and everyday physical concepts.

**Chapter 1** provides an introduction to the current state of our world. It introduces the students to world population and its trends, basic human needs, and why it is important to understand concepts of energy, environment, and sustainability. We explain the traits of good global citizens, and the importance of developing good communication skills–all essential for a well-educated democratic society. We also emphasize that responsible citizens must have a good grasp of issues related to energy and environment and take active roles in their communities. We point out that although the activities of good citizens can be quite varied, there are some personality traits and involvement practices that typify a good citizen:

- Good citizens are well informed and have a firm grasp of current issues, particularly issues related to energy, environment, and sustainability.
- Good citizens have a desire to be life-long learners. For example, they are well read; they attend community meetings and presentations to stay abreast of new events and innovations in technologies and understand how new developments may affect their lives.
- Good citizens have good written and oral communication skills.
- Good citizens have time management skills that enable them to work productively, take good care of their families, and be active in their communities.
- Good citizens generally work in a team environment where they consult each other to solve complex problems that affect all of us.

**Chapter 2** explains the role and importance of fundamental dimensions (e.g., length, time, mass, temperature) and systems of units (e.g., feet, meter, seconds, pound, kilogram, degree Fahrenheit, Celsius) in our daily lives. We emphasize to the students that they have been using these concepts all their lives; we now define them in a formal way, so that students can understand and quantify more intelligently their own environmental impact, water and energy consumption rates and waste.

## Part Two: Energy

In Part Two, consisting of Chapters 3 through 7, we introduce students to the importance of understanding the basics of conventional and renewable energy, its sources and production, as well as consumption rates in homes, buildings, transportation, food production and manufacturing. **Chapter 3** explains the basic concepts related to energy and efficiency. These are concepts that every college graduate, regardless of his or her area of interest, should know. We need energy to build shelter, to cultivate and process food, to make goods, and to maintain our living places at comfortable settings. To quantify the requirements to build things, move or lift objects, or to heat or cool buildings, energy is defined and classified into different categories. We discuss what we mean by mechanical energy and thermal energy. The units of energy and power, including kilowatt-hour, Btu, kilowatt, and horsepower, are also discussed in this chapter. In **Chapter 4**, covers the basic concepts of electricity and electric power production. We also cover residential power consumption, particularly lighting systems, because lighting accounts for a major portion of electricity use in buildings; lighting systems have received much attention recently due to the energy and sustainability concerns. **Chapter 5** covers the fundamentals of heat transfer, and heat loss and gain in buildings. Space heating and air conditioning account for nearly fifty percent of energy use in homes in the United States. **Chapter 6** provides a comprehensive coverage of energy sources including gasoline, natural gas, coal, wood and their consumption rates. We detail how much energy we consume in our homes, buildings, transportation, and manufacturing sectors. **Chapter 7** explains renewable energy and its sources as well as the basic concepts related to solar energy, wind energy, and hydro-energy.

## Part Three: Environment

In Part Three, consisting of Chapters 8 through 11, we focus on the environment and introduce students to air and water and the anatomy of earth, its natural resources, and rates of consumption and waste. We emphasize that our earth has finite resources. **Chapter 8** provides general information about the atmosphere, weather and climate, along with outdoor and indoor air quality standards. **Chapter 9** covers water resources, quality standards, and consumption rates in our homes, in agriculture, and in the industrial and manufacturing sectors of our society. **Chapter 10** provides a detailed understanding of common materials that are used to make products and structures. **Chapter 11** discusses waste and recycling.

## Part Four: Sustainability

In Part Four, consisting of **Chapter 12**, we introduce key sustainability concepts, methods, and tools. Every college graduate must develop a keen understanding of the Earth's finite resources, environmental and socioeconomic issues related to sustainability, ethical aspects of sustainability, and the necessity for sustainable development. Students should also know about life-cycle based analysis, resource and waste management, and environmental impact analysis, and be familiar with sustainable-development indicators such as the U.S. Green Building Council (USGBC) and Leadership in Energy and Environmental Design (LEED) rating systems. Finally, in Chapter 12, we have included several personal and community-based projects to promote responsible citizenship and sustainability.

# Features

This book includes numerous features intended to promote active learning. These features include: (1) Learning Objectives (LO), (2) Discussion Starters, (3) Before You Go On, (4) Highlighted Key Concepts, (5) Summary, (6) Key Terms, (7) Apply What You Have Learned, and (8) Life-long Learning.

## Learning Objectives (LO)

Each chapter begins by stating the learning objectives (**LO**), enabling students to identify the most important concepts to take away from that chapter. These objectives are revisited throughout the chapter and are also highlighted within the chapter summary.

## Discussion Starters

Pertinent facts and articles serve as chapter openers to promote meaningful discussion and engage students. They provide a means to understanding the importance of what students are about to learn. A good way for the instructor to use a Discussion Starter is by giving students a few minutes to read it at the beginning of a class and then ask the students about their thoughts.

## Before You Go On

This feature encourages students to test their comprehension and understanding of the material discussed in a section by answering questions before they continue to the next section.

## Vocabulary

It is important for students to understand the importance of developing a complete vocabulary to persuade and converse about today's pressing issues. This feature promotes understanding of basic terminology by asking students to state the meaning of new terms that are covered in a section.

## Key Concepts

Key Concepts are highlighted and defined in special boxes throughout the book. The concepts are indexed at the end of each chapter so students can easily return to the definitions for review and study.

## Summary

Each chapter concludes by summarizing what the student should have gained from the chapter. These summaries are designed to help students comprehend and become proficient with the materials.

## Key Terms

At the end of each chapter the key concepts and terms are indexed so that students can use them for review or check back in the chapter for their meaning.

## Apply What You Have Learned

This feature, designed to highlight practical applications of course concepts, encourages students to apply what they have learned to an interesting problem or a situation. To emphasize the importance of teamwork and to encourage group participation, many of these problems require group work.

**Life-Long Learning Problems** that depict and apply concepts that are critical for life-long learning are clearly denoted by ☞ to draw attention to their importance.

## Supplements

An **Instructor's Solutions Manual**, containing detailed solutions to problems from the text, and **Lecture Note PowerPoint slides** are available via a secure, password-protected Instructor Resource Center at http://login.cengage.com.

## Acknowledgments

I am thankful to all the reviewers who offered general and specific comments including Loius D. Albright, Cornell University; Paul Dawson, Boise State University; John Gardner, Boise State University; Thomas Ortmeyer, Clarkson University; William Rauckhorst, Miami University; Malcolm M. Sanders; Chiang Shih, Florida State University; and Sesha Srinivasan, Tuskegee University.

I wish to acknowledge and thank the Global Engineering team at Cengage Learning for their dedication to this new book: Timothy Anderson, Product Director; Mona Zeftel, Senior Content Developer; Martha Conway, Senior Content Project Manager; Kristin Stine, Marketing Manager; Elizabeth Brown and Brittany Burden, Learning Solutions Specialists; Ashley Kaupert, Associate Media Content Developer; Teresa Versaggi and Alexander Sham, Product Assistants; and Rose Kernan of RPK Editorial Services, Inc. They have skillfully guided every aspect of this text's development and production to successful completion.

Thank you for considering this book, and I hope you enjoy it.

—*Saeed Moaveni*

# MindTap Online Course

*Energy, Environment, and Sustainability* is also available through **MindTap**, Cengage Learning's digital course platform. The carefully-crafted pedagogy and exercises in this trusted textbook are made even more effective by an interactive, customizable eBook, automatically graded assessments, and a full suite of study tools.

As an instructor using MindTap, you have at your fingertips the full text and a unique set of tools, all in an interface designed to save you time. MindTap makes it easy for instructors to build and customize their course, so you can focus on the most relevant material while also lowering costs for your students. Stay connected and informed through real-time student tracking that provides the opportunity to adjust your course as needed based on analytics of interactivity and performance. **End-of-chapter assessments** test students' knowledge of concepts in each chapter. **Videos** and **links to outside resources** help students learn about the global impacts of energy, environment, and sustainability.

## ‹ CHAPTER 5: THERMAL ENERGY: HEAT LOSS AND GAIN IN BUILDINGS

**Cost of Heating - Worksheet**
How to figure out what it costs to heat your house.

**Weather Data Depot: free downloads of heating & cooling degree days**
Your Source For Free Degree Day Reports Degree day data is useful to indicate how seasonal weather affec
impacts energy management, energy efficiency, and utility bill tracking.

**Calculating Degree Days**
Here are two different methods for calculating degree days from weather information on the internet.

📁 **Chapter 5: Thermal Energy Videos**

**Calculating a U-value for Heat Loss Calculation**
A guide to calculating a U-value for a room or building so that you can work out heat loss and how much he

**How to Calculate Heat Gain and Loss**
This video shows you how to calculate heat loss for a small building.

**Professors Seek Smarter Buildings, Develop Flexible Materials at NJIT**
Green Building Technologies at NJIT: Architects and Chemical Engineers are developing smart building mate
solar energy while limiting heat gain. Flexible building skins and batteries point to future efficiencies. Enviro
is focusing on batteries that are as thin and flexible as film - built from the carbon nanotube. "I think the fut
Architect Martina Decker focuses on flexible building materials, skins, and shades that respond to heat and
realize where the energy that we consume every day is coming from. We have to be very aware of what we

**Chapter 5 Quiz**
Take this quiz to see what you've learned in Chapter 5.

## How does MindTap benefit instructors?

- You can build and personalize your course by integrating your own content into the **MindTap Reader** (like lecture notes or problem sets to download) or pull from sources such as RSS feeds, YouTube videos, websites, and more. Control what content students see with a built-in learning path that can be customized to your syllabus.
- MindTap saves you time by providing you and your students with **automatically graded assignments and quizzes**. These problems include immediate, specific feedback, so students know exactly where they need more practice.

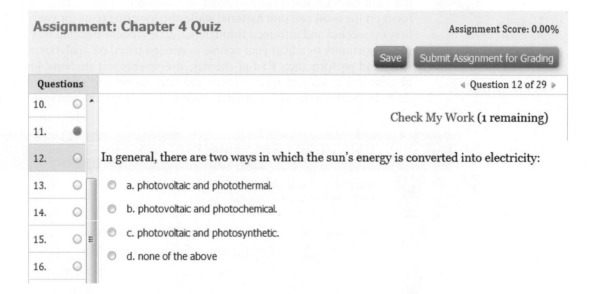

- The **Message Center** helps you to quickly and easily contact students directly from MindTap. Messages are communicated directly to each student via the communication medium (email, social media, or even text message) designated by the student.
- **StudyHub** is a valuable studying tool that allows you to deliver important information and empowers your students to personalize their experience. Instructors can choose to annotate the text with **notes** and **highlights**, share content from the MindTap Reader, and create **Flashcards** to help their students focus and succeed.
- The **Progress App** lets you know exactly how your students are doing (and where they might be struggling) with live analytics. You can see overall class engagement and drill down into individual student performance, enabling you to adjust your course to maximize student success.

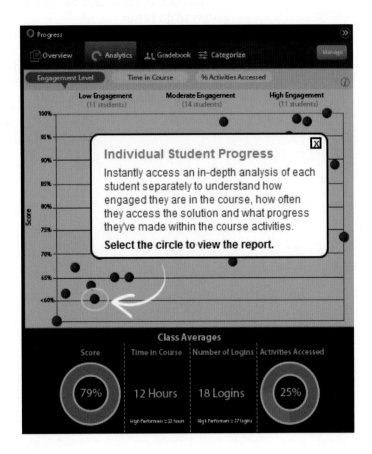

## How does MindTap benefit students?

- The **MindTap Reader** adds the abilities to have the content read aloud, to print from the reader, and to take notes and highlights while also capturing them within the linked **StudyHub App**.
- The **MindTap Mobile App** keeps students connected with alerts and notifications while also providing them with on-the-go study tools like Flashcards and quizzing, helping them manage their time efficiently.
- **Flashcards** are pre-populated to provide a jump start on studying, and students and instructors can also create customized cards as they move through the course.
- The **Progress App** allows students to monitor their individual grades, as well as their level compared to the class average. This not only helps them stay on track in the course but also motivates them to do more, and ultimately to do better.
- The unique **StudyHub** is a powerful single-destination studying tool that empowers students to personalize their experience. They can quickly and easily access all notes and highlights marked in the MindTap Reader, locate bookmarked pages, review notes and Flashcards shared by their instructor, and create custom study guides.

# Basic Concepts

In Part One of this book we introduce you to the importance of understanding basic concepts such as human needs, energy, environment, and sustainability. Good citizens are well informed and have a firm grasp of current issues, particularly those related to population trends, energy, environment, and sustainability; all essential for a well-educated democratic society. Responsible citizens also take active roles in their communities; have desire to be life-long learners; stay abreast of innovations in technologies and understand how new developments affect their lives; have time management skills that enable them to work productively, take good care of their families, and be active in their communities; and work in a team environment where they consult each other to solve problems that affect all of us.

In Part One, we also explain the role of fundamental dimensions such as length, time, mass, and temperature and systems of units such as foot (or meter), second, pound (kilogram), and degree Fahrenheit (Celsius) in our daily lives. You have been using these concepts all your lives; however, here we define them in a formal way, so that you can quantify more intelligently your own environmental impact, water and energy consumption rates, and waste.

# CHAPTER 1

# Introduction to Energy, Environment, and Sustainability

## LEARNING OBJECTIVES

**LO¹**   Basic Human Needs: understand the basic human needs, including clean air, clean water, food, and shelter

**LO²**   Energy: understand that it takes energy to address basic human needs and be familiar with energy consumption rates and sources in your daily life

**LO³**   Environment: explain what we mean by environment and be familiar with its main components

**LO⁴**   Sustainability: define sustainability and its role in your daily life

# Discussion Starter

We all want to make the world a better place, but how do we do it, and where do we start? Leo Tolstoy, a Russian novelist and philosopher, once said:

*"Everyone thinks of changing the world, but no one thinks of changing oneself."*

Increasingly, because of worldwide socio-economic trends, environmental concerns, and the Earth's finite resources, more is expected of all of us. As responsible global citizens, we are expected to consider the link among the Earth's finite resources and environmental, social, ethical, technical, and economical factors as we make decisions regarding the services that we use and the products we consume. This book is designed to introduce you—a college student— regardless of your area of study, personal interests, and future career path, to important issues such as energy, environment, and sustainability that affect all of us. A quote often attributed to Chief Seattle, Dkhw'Duw'Absh tribe (1786–1866) says it best:

*"What befalls the Earth befalls all the sons (and daughters) of the Earth. This we know: the Earth does not belong to man, man belongs to the Earth. All things are connected like the blood that unites us all. Man does not weave this web of life. He is merely a strand of it. Whatever he does to the web, he does to himself."*

**To the Students: What does all this mean to you? Have you thought about changing the world or yourself? Where do you start?**

## LO¹ 1.1 Basic Human Needs

During the past decades, much has been said about *energy, the environment,* and *sustainability.* What do they mean? Why is it important for you to understand these vital issues? Increasingly, because of worldwide socioeconomic trends, environmental concerns, and the Earth's finite resources, more is expected of all of us. As responsible global citizens, we are expected to consider the link among the Earth's finite resources and environmental, social, ethical, technical, and economical factors as we make decisions in our daily lives. In our decision-making process, we are expected to consider our energy and environment footprints and take into account factors such as the natural resources that were consumed to make a product. We also need to consider how much energy it takes to manufacture, transport, use, and finally dispose of the product.

This book is an attempt to introduce you to these very important concepts—issues that affect all of us. Presently, there is international competition for Earth's finite resources, because each nation works hard to address its own energy, water, and food security. A human body is made of many interacting parts that work well together and share resources effectively. Furthermore, when a part of our body—even as small as our little finger—is in pain, the body as a whole is uncomfortable until the pain is gone. We should develop a similar, holistic view of our societies: one that increases commonality of human purpose, and one that gives a greater meaning to life beyond the walls of our homes, beyond the boundaries of our cities, and beyond our own countries. It is imperative that we all understand that *we are all in it together*, and in order to address our energy, clean air and water, and food security intelligently, we need to work together and be well educated in topics such as energy, environment, and sustainability. *It is only then that we consume resources in such a way that meets our present needs without compromising the ability of future generations to meet their needs.*

We as people, regardless of where we live, need the following things: clean air, clean water, food, and shelter. In our modern society, we also need various modes of transportation to get to different places, because we live and work in different locations or may wish to visit friends and relatives. We also like to have some sense of security, to be able to relax, and to be entertained. We need to be liked and appreciated by our friends and family, as well. As you also know, some of us have a good standard of living, while others who live in developing countries do not. You will probably agree that our world would be a better place if every one of us had clean air and water, enough food to eat, a comfortable and safe place to live, meaningful work to do, and some time for relaxation, family, and friends.

At the turn of the 20th century, there were approximately six billion of us inhabiting the Earth; as a means of comparison, the world population about 100 years ago, at the turn of the 19th century, was one billion. Think about this! It took us since the beginning of human existence to year 1900 to reach a population of one billion. Then it only took 100 years to increase the population fivefold.

According to the latest estimates and projections of the U.S. Census Bureau, the world population will reach 9.3 billion people by the year 2050. Not only will the number of people inhabiting the Earth continue to rise, but the age structure of the world population will also change. The world's elderly population—people at least 65 years of age—will more than double in the next 25 years (see Figure 1.1).

How is this information relevant? Well, let's start with our most essential need, clean air; without it we cannot live.

Rawpixel.com/Shutterstock.com

It is expected that the world population will reach 9.3 billion people by the year 2050.

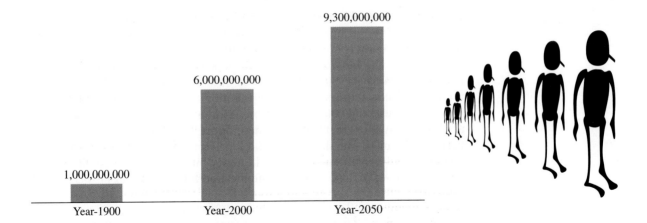

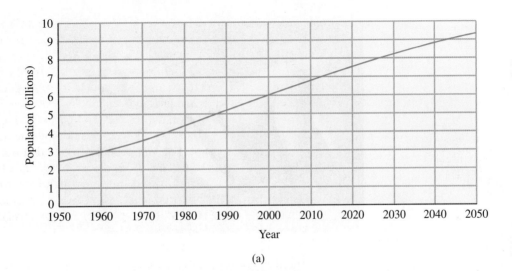

(a)

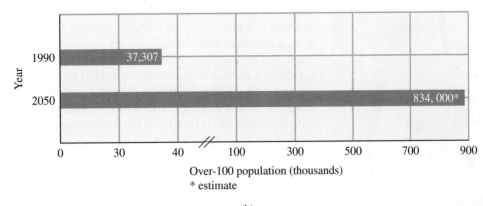

(b)

**FIGURE 1.1**  (a) The latest projection of world population growth. (b) The latest estimate of U.S. elderly population growth.
*Source:* Data from the U.S. Census Bureau

**We Need Clean Air** Every day, human activities through *stationary* and *mobile* sources contribute to pollution of the outdoor air. Power plants, factories, and dry cleaners are examples of stationary sources that create outdoor air pollution. The *mobile* sources of air pollution such as cars, buses, trucks, planes, and trains also add to the level of outdoor air pollution. In addition to these man-made sources, *natural* air pollution also occurs due to forest fires, windblown dust, and volcanic eruptions. Moreover, because most of us spend approximately 90 percent of our time indoors, the indoor air quality is also very important to our short-term and long-term health. In recent years, we have used more synthetic materials in newly built homes that could give off harmful vapors. We also use more chemical pollutants, such as pesticides and household cleaners.

Liukov/Shutterstock.com

**We Need Clean Water** Our next essential need is water. Droughts are good reminders of how significant water is to our daily lives. In addition to quantity, quality is also a concern. As you would expect, human activities and naturally occurring microorganisms contribute to the contaminant level in our water supply. For example, in agriculture, fertilizers, pesticides, or animal waste from large cattle, pig, or poultry farms contribute to water pollution. Other human activities such as mining, construction, manufacturing goods, landfills, and waste water treatment plants are also major contributors.

**We Need Food** To lead a normal active life, we need to consume a certain number of calories that come from eating beef, lamb, pork, poultry, fish, eggs, dairy products, fruits, bread, vegetables, and the like. In the American diet, carbohydrates, protein, and fat are the main sources of calories.

The total number of food calories a person needs each day to lead an active and healthy life depends on factors such as gender, age, height, weight, and level of physical activity. Moreover, in order to maintain a healthy body weight, calories consumed from food and drinks must equal calories expended through daily activities.

Ilya Andriyanov/Shutterstock.com

monticello/Shutterstock.com

> To maintain a healthy body weight, calories consumed from food and drinks must equal calories expended through daily activities.

Therefore, if you consume more calories than you expend, you will gain weight. As we later explain in Chapter 3, the energy content of food is typically expressed in Calories (with an uppercase C). For example, a banana has about 100 Calories, whereas a medium serving of French fries has around 400 Calories. One Calorie is equal to 1,000 calories (with a lowercase c), and one calorie is formally defined as the amount of energy required to raise the temperature of one gram (1 g) of water by one degree Celsius (1°C). For now, don't worry if you don't fully understand what one calorie represents; this and other concepts will be explained in greater detail in Chapters 2 and 3.

In the United States, by law, dietary guidelines for Americans are reviewed and published every five years by the U.S. Department of Agriculture (USDA) and the U.S. Department of Health and Human Services (HHS).

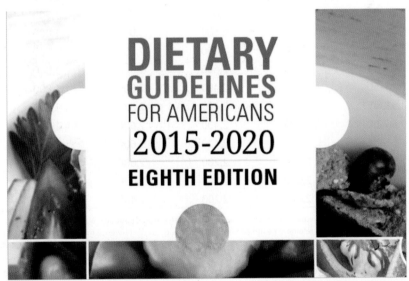

ODPHP, 2015–2020 Dietary Guidelines for Americans, http://health.gov/dietaryguidelines/2015/

Table 1.1 shows the estimated total calorie needs for weight maintenance based on age, gender, and physical activity level. This data is from the *Dietary Guidelines for Americans 2015* USDA and HHS report. As shown in Table 1.1,

**TABLE 1.1**     **Estimated Calorie Needs per Day by Age, Gender, and Physical Activity Level**

Estimated amounts of Calories[a] needed to maintain Calorie balance for various gender and age groups at three different levels of physical activity. The estimates are rounded to the nearest 200 Calories for assignment to a USDA food pattern. An individual's Calorie needs may be higher or lower than these average estimates.

| Activity Level[b] (Age) | Male | | | Female[c] | | |
|---|---|---|---|---|---|---|
| | Sedentary | Moderately Active | Active | Sedentary | Moderately Active | Active |
| 2 | 1,000 | 1,000 | 1,000 | 1,000 | 1,000 | 1,000 |
| 3 | 1,200 | 1,400 | 1,400 | 1,000 | 1,200 | 1,400 |
| 4 | 1,200 | 1,400 | 1,600 | 1,200 | 1,400 | 1,400 |
| 5 | 1,200 | 1,400 | 1,600 | 1,200 | 1,400 | 1,600 |
| 6 | 1,400 | 1,600 | 1,800 | 1,200 | 1,400 | 1,600 |
| 7 | 1,400 | 1,600 | 1,800 | 1,200 | 1,600 | 1,800 |
| 8 | 1,400 | 1,600 | 2,000 | 1,400 | 1,600 | 1,800 |
| 9 | 1,600 | 1,800 | 2,000 | 1,400 | 1,600 | 1,800 |
| 10 | 1,600 | 1,800 | 2,200 | 1,400 | 1,800 | 2,000 |
| 11 | 1,800 | 2,000 | 2,200 | 1,600 | 1,800 | 2,000 |
| 12 | 1,800 | 2,200 | 2,400 | 1,600 | 2,000 | 2,200 |
| 13 | 2,000 | 2,200 | 2,600 | 1,600 | 2,000 | 2,200 |
| 14 | 2,000 | 2,400 | 2,800 | 1,800 | 2,000 | 2,400 |
| 15 | 2,200 | 2,600 | 3,000 | 1,800 | 2,000 | 2,400 |
| 16 | 2,400 | 2,800 | 3,200 | 1,800 | 2,000 | 2,400 |
| 17 | 2,400 | 2,800 | 3,200 | 1,800 | 2,000 | 2,400 |
| 18 | 2,400 | 2,800 | 3,200 | 1,800 | 2,000 | 2,400 |
| 19–20 | 2,600 | 2,800 | 3,000 | 2,000 | 2,200 | 2,400 |
| 21–25 | 2,400 | 2,800 | 3,000 | 2,000 | 2,200 | 2,400 |
| 26–30 | 2,400 | 2,600 | 3,000 | 1,800 | 2,000 | 2,400 |
| 31–35 | 2,400 | 2,600 | 3,000 | 1,800 | 2,000 | 2,200 |
| 36–40 | 2,400 | 2,600 | 2,800 | 1,800 | 2,000 | 2,200 |
| 41–45 | 2,200 | 2,600 | 2,800 | 1,800 | 2,000 | 2,200 |
| 46–50 | 2,200 | 2,400 | 2,800 | 1,800 | 2,000 | 2,200 |
| 51–55 | 2,200 | 2,400 | 2,800 | 1,600 | 1,800 | 2,200 |
| 56–60 | 2,200 | 2,400 | 2,600 | 1,600 | 1,800 | 2,200 |
| 61–65 | 2,000 | 2,400 | 2,600 | 1,600 | 1,800 | 2,000 |
| 66–70 | 2,000 | 2,200 | 2,600 | 1,600 | 1,800 | 2,000 |
| 71–75 | 2,000 | 2,200 | 2,600 | 1,600 | 1,800 | 2,000 |
| 76+ | 2,000 | 2,200 | 2,400 | 1,600 | 1,800 | 2,000 |

[a]Based on Estimated Energy Requirements (EER) equations, using reference heights (average) and reference weights (healthy) for each age-gender group. For children and adolescents, reference height and weight vary. For adults, the reference man is 5 feet 10 inches tall and weighs 154 pounds. The reference woman is 5 feet 4 inches tall and weighs 126 pounds. EER equations are from the Institute of Medicine, *Dietary Reference Intakes for Energy, Carbohydrate, Fiber, Fat, Fatty Acids, Cholesterol, Protein, and Amino Acids*, Washington D.C.: The National Academies Press, 2002.

[b]Sedentary means a lifestyle that includes only the light physical activity associated with typical day-to-day life. Moderately active means a lifestyle that includes physical activity equivalent to walking about 1.5 to 3 miles per day at 3 to 4 miles per hour, in addition to the light physical activity associated with typical day-to-day life. Active means a lifestyle that includes physical activity equivalent to walking more than 3 miles per day at 3 to 4 miles per hour, in addition to the light physical activity associated with typical day-to-day life.

[c]Estimates for females do not include women who are pregnant or breastfeeding.

*Source:* Based on the Institute of Medicine, *Dietary Reference Intakes for Energy, Carbohydrate, Fiber, Fat, Fatty Acids, Cholesterol, Protein, and Amino Acids*, Washington D.C.: The National Academies Press, 2002.

adult women need to consume between 1,800 and 2,400 Calories per day, while adult men may require 2,400 to 3,200 Calories. The low values represent caloric intake for sedentary conditions, whereas the higher values are for active individuals.

Not all Americans are able to follow the dietary guidelines. Here are some facts about American calorie imbalance that are worth noting:

- According to the USDA Economic Research Service, in recent years, nearly 15 percent of American households have been unable to get enough food to meet their daily Calorie needs.
- At the other end of the spectrum, many Americans (among all subgroups of the population) are overweight or obese because their daily Calorie intake exceeds their activity level needs.
- In an article entitled "U.S. Lets 141 Trillion Calories of Food Go to Waste Each Year," Eliza Braclay writes that "The sheer volume of food wasted in the U.S. each year should cause us some shame, given how many people are hungry both in our own backyard and abroad." This is happening in America, while 1 in 9 people in the world (data from World Food Programme Organization) do not have enough food to lead a normal life. The 141 trillion Calories that represent approximately 1,250 Calories per person per day in the United States is the result of nearly 130 billion pounds of food that was lost. Moreover, it is important to understand that the wasted food is worth over \$100 billion. According to USDA, the top three food groups lost (in a recent year) were dairy products (25 billion pounds, or 19 percent of all the lost food); vegetables (25 billion pounds, or 19 percent); and grain products (18.5 billion pounds, or 14 percent). The USDA's Economic Research Service also points out that, if we were to reduce this waste, the price of food worldwide might go down. In addition, most of us do not realize that a vast amount of energy is spent in the food supply chain, and when food is wasted, valuable resources such as water and fossil fuels (that go into growing, processing, and transporting the food) are also wasted.

Have you ever thought about how much energy it takes to feed you every day? Let's start with a simple example and assume that you had some cereal for breakfast this morning. Now think about what it takes to grow a cereal crop such as corn and wheat (incidentally, corn, wheat, and soybeans make up the majority of field crop inputs to the U.S. food supply). Think about the energy that needs to be spent to plant the seeds, make and apply fertilizers, irrigate the field, harvest the crop, and finally transport it to a processing plant. Next consider how much energy it takes to process the corn into the cereal, make plastic bags and attractive boxes to contain it, and deliver the cereal boxes to the supermarkets. Moreover, we like to have some milk with our cereal. Well, that is going to require additional energy to make the cattle feed, run the milking machines, produce milk containers, and build and operate refrigerated trucks. And after the milk gets to the supermarket, it needs to be placed in cold storage, requiring energy to maintain its low temperature. After you bring the milk home, you need to store it in a refrigerator, which also consumes energy. You get the picture! Now, think about all of the other food and drinks that you consume in a single day. In Chapters 2 through 7, we explain important concepts related to energy and power that every good citizen should understand.

Now let's consider what happens to a dollar spent on food. According to the USDA, for a typical dollar spent (in a recent year) by U.S. consumers on domestically produced food (including both grocery store and eating-out purchases): 12.1 cents went to agribusiness and farm production, 15.8 cents to food processing, 2.7 cents to packaging, 3.3 cents to transportation, 9.3 cents to a wholesale trader, 13 cents to the food retailer, 31.1 cents to services provided by food service establishments, 5.6 cents to energy costs, 3.3 cents to finance and insurance costs, and 3.8 cents to pay for activities such as advertising, legal, and accounting services (see Figure 1.2).

As you can see, in the United States, if you were to buy a loaf of bread, say for $1, approximately 12 cents goes to the actual cost of the flour and the remaining 88 cents goes to paying for processing, packaging, transportation, advertisement, and so on. But if the price of wheat doubles from 12 cents to 24 cents, assuming no changes in other costs, the bread will cost only an additional 12 cents, which is an increase of 12 percent in total cost (incidentally, in the United States, we spend approximately 12.9 percent of our income on food). The share of U.S. household consumer expenditures by major categories for 2013 is shown in Figure 1.3. In contrast, in developing countries, poor people may spend as much as 80 percent of their income on food.

Consequently, the poor cannot afford to buy processed, cooked, or packaged food. For example, instead of buying already-baked bread, they buy the flour and make the bread themselves. So for these people, when the price of wheat doubles, the cost of bread is also doubled—an increase of 100 percent! The 2015 world hunger map is shown in Figure 1.4. Next time, when you are about to waste food, think carefully!

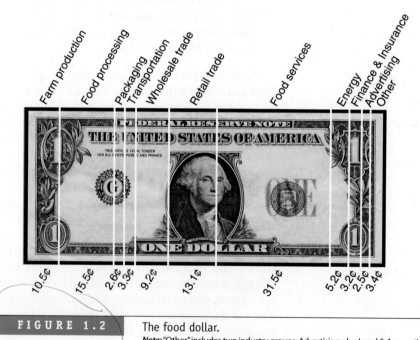

Farm production · Food processing · Packaging · Transportation · Wholesale trade · Retail trade · Food services · Energy · Finance & Insurance · Advertising · Other

10.5¢   15.5¢   2.6¢   3.3¢   9.2¢   13.1¢   31.5¢   5.2¢   3.2¢   2.5¢   3.4¢

**FIGURE 1.2**     The food dollar.
*Note:* "Other" includes two industry groups: Advertising plus Legal & Accounting.
*Source:* USDA, Economic Research Service.

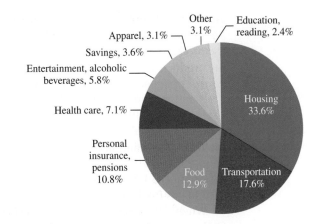

| FIGURE 1.4 | The 2015 World Hunger Map. |

*Source:* Juliann / Shutterstock.com, data from World Food Programme, World Hunger Map.

**We Need Natural Resources to Make Goods and Provide Services** In addition to our need for clean air, water, and food, we as a society create and use products and services that make our lives better (Figure 1.5). Think about all the products that we use in our everyday lives, such as cars, computers, clothing, home appliances, heating and cooling equipment, healthcare devices, and the tools and machines used to make these products. Also think about our infrastructure; for example, homes, malls, commercial buildings, highways, airports, communication systems, mass-transit systems, and the power plants that supply the power to maintain this framework. Often, we forget that there are many people behind the scenes who are responsible for finding suitable ways and designing the necessary equipment to extract raw materials, petroleum, and natural gas from the Earth.

When we use a product such as a smart phone, an electronic tablet, a car, a clothes washing machine, an oven, or a refrigerator, we need to be mindful of what type of materials went into making the product, where the materials came from, how much energy it took to produce the product, and eventually, what it would take to recycle or dispose of it. We will discuss common materials used in making products and infrastructure in Chapter 10.

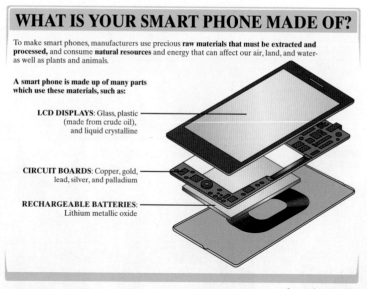

## WHAT IS YOUR SMART PHONE MADE OF?

To make smart phones, manufacturers use precious **raw materials that must be extracted and processed,** and consume **natural resources** and energy that can affect our air, land, and water- as well as plants and animals.

A smart phone is made up of many parts which use these materials, such as:

**LCD DISPLAYS**: Glass, plastic (made from crude oil), and liquid crystalline

**CIRCUIT BOARDS**: Copper, gold, lead, silver, and palladium

**RECHARGEABLE BATTERIES**: Lithium metallic oxide

Source: Based on EPA.

arka38/Shutterstock.com     vovan/Shutterstock.com   Alexandru Nika/Shutterstock.com     You can more/Shutterstock.com

**FIGURE 1.5**   Examples of products used in our daily life.

**What Happens to Products When Disposed?** As a good global citizen, it is also important to understand what happens to products when we discard them. Each year, the U.S. Environmental Protection Agency collects and reports data on the generation and disposal of waste in the United States. According to the latest available data, in 2012, people in the United States generated 251 million tons of trash (approximately 4.4 pounds per person per day) of which only 86.6 million tons were recycled or composted. We will discuss municipal and industrial waste and recycling in Chapter 11.

kanvag/Shutterstock.com

Evan Lorne/Shutterstock.com

## Before You Go On

Answer the following questions to test your understanding of the preceding section:

1. What are the basic human needs?

2. In your own words, describe energy, environment, and sustainability.

3. What are some of the consequences of increasing world population?

## LO² 1.2 Energy

*Without energy, we cannot do anything!* Therefore, *energy* should be the starting point for a better understanding of our environmental footprint. We need energy to keep our homes comfortable, to make goods, and to provide services that allow us to enjoy a high standard of living. We use energy in our homes for space heating and cooling, hot water, lighting, appliances, and electronics. We also use energy in our cars for personal and business travel. In addition to our personal energy requirements, we need energy for businesses and industry to make and transport all kinds of products and food; to make materials for and to erect buildings, and to build and maintain our infrastructure (roads, bridges, railroad systems, airports, etc.). To understand your daily energy needs, tomorrow morning when you get up, just look around you and think carefully. During the night, your bedroom was kept at the right temperature thanks to the heating or cooling system in your place of residence. When you turn on the lights or your TV, be assured that thousands of people at power plants around the country are making certain the flow of electricity remains uninterrupted.

hans engbers/Shutterstock.com

When you are getting ready to take your morning shower, think about the clean water you are about to use: Where did it come from and how is it heated? That water could be coming to your place thanks to a network of piping systems and water treatment facilities. Moreover, the water could be heated by natural gas, electricity, or fuel oil that is brought to your home thanks to the work and effort of many people behind the scenes. When you get ready to dry yourself with a towel, think about what types of machines and how much energy was consumed to produce the towel. Think about the machines used to plant and pick the cotton, transport it to a factory, clean it, and dye it a color that is pleasing to your eyes. Think about other machines that were used to weave the fabric and send it to sewing machines. Also, think about where the towel was made and all of the energy consumed to transport it to the store from which you purchased it. The same is true of the clothing you are about to put on. Next, let's say you are about to have some cereal. As mentioned previously, the milk was kept fresh in your refrigerator and the cereal was made available due to the efforts of farmers and people in a food processing plant; each requires energy to produce and transport to grocery stores. Now you are ready to get into your car, take a bus, or ride the subway. Think about the amount of materials and energy needed to make your transportation system and to move it along. So, you see *there is nothing that you do in your daily life that does not involve energy.* As we have been emphasizing, there are certain concepts that every citizen, regardless of his or her area of interest, should know. As a good global citizen, you need to have a firm grasp of energy: its sources, generation, and consumption rates.

The world energy consumption by fuel type is shown in Figure 1.6. In 2011 (the most recent available data), 519 quadrillion Btu of energy was consumed worldwide, and as shown in Figure 1.6, petroleum, coal, and natural gas made up nearly 86 percent of all the fuel used to generate energy. One quadrillion is equal to $10^{15}$ or 1,000,000,000,000,000, and Btu denotes British thermal units. One Btu represents the amount of energy needed to raise the temperature of one pound of water by 1 degree Fahrenheit (1°F). For example, to take a nice long shower, you need to raise the temperature of 20 gallons of water (approximately 170 pounds of water) from 70 to 120°F (a temperature rise of 50°F). The amount of energy required to achieve this task is about 8,500 Btu.

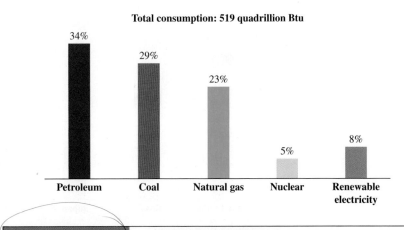

**FIGURE 1.6** World energy consumption by fuel type (the most recent available data).

And if you were to multiply this number by 365 days, you would obtain the annual amount, which is nearly 3.13 million Btu. This example gives you a sense of how much energy it takes for just one of your daily activities (hot shower) and a better understanding of the relative magnitude of the Btu value shown in Figure 1.6. As is the case with any new concepts you learn, energy has its own terminology, and as a good citizen, you need to become aware of certain terms. We will explain energy and power-related concepts in detail in Chapter 3.

In 2011, the five countries with the largest energy consumptions were China, the United States, Russia, India, and Japan, as shown in Figure 1.7. The per capita consumption for these countries is shown in Figure 1.8. Note that the United States has the largest per capita energy consumption in the world with a value of 313 billion Btu.

In the United States, to keep track of how we consume energy in our society, the Energy Information Administration (EIA) classifies the energy consumption rates by major sectors of our economy. These sectors are organized into *industrial*, *transportation*, *residential*, and *commercial*. The percentage of energy consumed by major sectors of the economy is depicted in Figure 1.9.

> The United States Energy Information Administration (EIA) classifies the energy consumption rates by major sectors of our economy: Industrial, Transportation, Residential, Commercial.

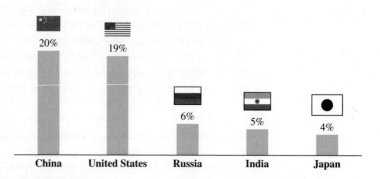

**FIGURE 1.7**    World energy consumption by top five countries.

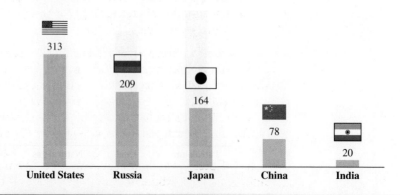

**FIGURE 1.8**    Per capita consumption of selected countries.
Unit: million Btu.

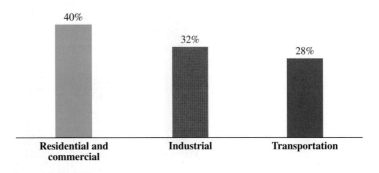

| FIGURE 1.9 | Share of energy consumed by major sectors of the economy. |

- The *residential sector* accounts for energy use in homes and apartments. Think about the space heating and cooling equipment, lighting systems, all electronic devices, and appliances such as refrigerators, freezers, ovens, washers, and dryers that are used in our homes every day.
- The *commercial sector* keeps track of energy use in schools, municipal buildings, hospitals, hotels, shopping malls, restaurants, police stations, places of worship, and warehouses. As shown in Figure 1.9, in 2013, the residential and commercial sectors accounted for 40 percent of total energy used in the United States.
- The *industrial sector* accounted for 32 percent of our total energy used in the United States. This value represents the share of total energy needed for all of the facilities and equipment for construction, mining, agriculture, and manufacturing.
- The *transportation sector*, which includes energy use by all types of vehicles (motorcycles, cars, trucks, buses, trains, subways, aircraft, boats, barges, ships, etc.) to transport people and goods, accounted for 28 percent of the total energy used in 2013. Think about all of the cars, buses, trains, planes, and subway systems that are used to transport people. Also, consider all of the trucks, trains, barges, and planes that are used every day to carry goods. According to the EIA, most of our transportation energy is consumed by automobiles and light trucks; gasoline and diesel fuel account for nearly 85 percent of energy consumed by vehicles.

Now that you have a good idea about the share of energy use in industrial, transportation, residential, and commercial sectors of our economy, let us look at the types of fuel, such as petroleum, natural gas, and coal, that are used to generate energy. Figure 1.10 shows the total energy consumption in the United States by fuel/energy source. As shown in Figure 1.10, fossil fuels (petroleum, natural gas, and coal) make up 82 percent of the total fuel/energy source.

In Chapter 6, we discuss in greater detail energy sources such as gasoline, natural gas, coal, wood, and their consumption rates. We will also explain in more detail how much energy we consume in our homes and buildings and for transportation and manufacturing.

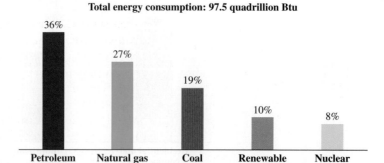

**Total energy consumption: 97.5 quadrillion Btu**

| | |
|---|---|
| FIGURE 1.10 | Total energy consumption in the U.S. by fuel/energy source (the most recent available data). |

# Before You Go On

Answer the following questions to test your understanding of the preceding section:

1.  Why do we need energy?

2.  According to the Energy Information Administration, what are the major sectors of the economy that consume energy?

3.  How do we consume energy at home?

4.  What types of fuel are used in the residential sector?

*Vocabulary—It is important for you as a good global citizen to understand that you need to develop a comprehensive vocabulary to communicate effectively. Throughout this book, we ask you to define the meaning of new words. This feature promotes your vocabulary growth. State the meaning of the following words:*

Btu

Quadrillion

# LO³  1.3  Environment

*Environment* is one of those terms that mean different things to different people. For example, as a computer user, we may talk about the *desktop environment*, which means the user interface with the computer: icons, windows, folders, toolbars, etc. On the other hand, to a civil engineer or a construction manager, the *built environment* refers to man-made structures such as roads, water piping networks, fuel distribution piping networks, buildings, or electric power networks. Often when we talk about **environment**, we mean the *natural environment*, which includes all living (plants, animals) and non-living things (air, water, rocks) that exist on or within the Earth. It is also important

robert_s/Shutterstock.com

to realize that each one of these categories could be sub-divided further. For example, water could be grouped as above ground (rivers, ponds, lakes, seas, oceans) or below ground (aquifers).

As you learned in school, our home, the Earth, is the third planet from the Sun. As you also know, over 70 percent of the Earth's surface is covered with bodies of water: the oceans, seas, lakes, and rivers. Oceans play an important role in moderating the Earth's surface temperature. You also know that, because of the abundance of water on its surface, the Earth appears blue when viewed from space, and hence the name: the blue planet. Moreover, to better represent the Earth's structure, it is divided into major layers that are located above and below its surface. *Atmosphere* represents the air that covers the surface of the Earth. The air extends approximately 90 miles from the surface of the Earth to a point called "the edge of space." The solid portion of the Earth itself is made up of different layers with different characteristics. Its mass is composed mostly of iron, oxygen, and silicon (approximately 32 percent iron, 30 percent oxygen, and 15 percent silicon). The Earth also contains other elements such as sulfur, nickel, magnesium, and aluminum. The structure below the Earth's surface is generally grouped into four layers: *crust*, *mantle*, *outer core*, and *inner core* (see Figure 1.11). This classification is based on the properties

Above the surface of the Earth

Atmosphere: 0 – 90 miles (0 – 140 km)

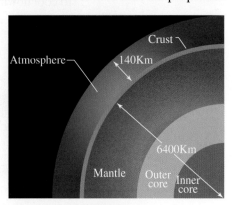

Thickness of different layers (approximate values)

Crust: 0 – 25 miles (0 – 40 km)
Mantle: 25 – 1800 miles (40 – 2900 km)
Outer core: 1800 – 3200 miles (2900 – 5200 km)
Inner core: 3200 – 4000 miles (5200 – 6400 km)

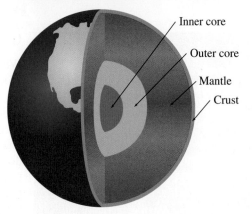

**FIGURE 1.11**          The structure of the Earth.

of materials and the manner by which the materials move or flow in each layer. We explain the Earth's structure in greater detail in Chapter 10. However, at this point, it is important for you to understand that the raw materials that make up the products that we use in our daily lives come from the Earth's crust. The crust makes up about 0.5 percent of the Earth's total mass and 1 percent of its volume. We also discuss common materials used in making products and building our infrastructure in Chapter 10.

## Air

We all need air to sustain life. The Earth's atmosphere, which we refer to as *air*, is a mixture of approximately 78 percent nitrogen, 21 percent oxygen, and a very small amount of argon and other gases such as carbon dioxide, sulfur dioxide, and nitrogen oxide. The atmosphere also contains water vapor. The water vapors in the atmosphere in the form of clouds allow for the transport of water from the oceans to land in the form of rain and snow. At higher altitudes, the Earth's atmosphere also contains ozone.

Even though gases such as carbon dioxide make up a small percentage of the Earth's atmosphere, they play a significant role in maintaining a thermally comfortable environment for us and other living species. For example, the ozone absorbs most of the ultraviolet radiation arriving from the sun that could harm us. Carbon dioxide plays an important role in sustaining plant life; however, if the atmosphere contains too much carbon dioxide, it will not allow the Earth to cool down effectively by radiation.

Ciprian Stremtan/Shutterstock.com

Air is a mixture of mostly nitrogen, oxygen, and small amounts of other gases such as argon, carbon dioxide, sulfur dioxide, and nitrogen oxide. Carbon dioxide plays an important role in sustaining plant life; however, if the atmosphere contains too much carbon dioxide, it will not allow the Earth to cool down effectively by radiation.

## Greenhouse Gases

When solar energy passes through the Earth's atmosphere, some of it is absorbed, some of it is scattered, and some of it is reflected by clouds, dust, pollutants, and different types of gases or water vapor in the atmosphere. The solar energy that reaches the Earth's surface warms the Earth, and eventually, some of the absorbed energy is radiated back toward space, as the Earth's surface cools down in the evenings. Many gases present in the atmosphere trap some of this heat and consequently prevent the Earth's surface and its atmosphere from cooling (Figure 1.12). The gradual warming of the Earth's atmosphere is commonly referred to as the *greenhouse effect*, and the gases that cause the warming are called *greenhouse gases* (Figure 1.12).

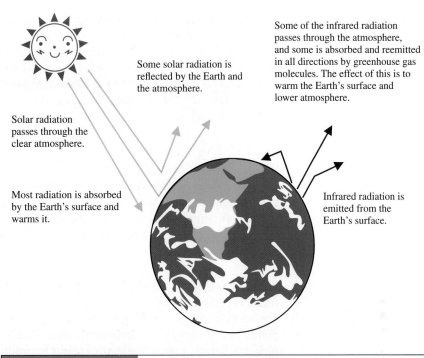

FIGURE 1.12    The greenhouse gas effect.
*Source:* Based on Energy Information Administration

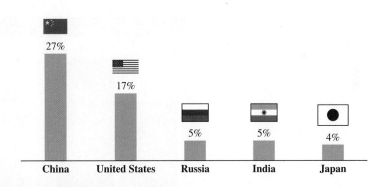

FIGURE 1.13    Energy-related carbon dioxide emissions by top five emitters.

The five countries with the largest energy-related carbon dioxide emissions are shown in Figure 1.13. Note that China is the largest emitter followed by the United States. We discuss air, air quality standards, and our individual roles in contributing to indoor and outdoor air pollution in Chapter 8.

## Water

Every living thing also needs water to sustain life. In addition to drinking water, we need water for many of our daily activities, including cooking, grooming, washing, and for fire protection. Water is not only transported to homes for our

Dmitry Naumov/Shutterstock.com

domestic use, but it also has many other applications. We need water to grow fruits, vegetables, nuts, cotton, trees, and so on. Water is commonly used in the mining industry, as a cooling or cleaning agent in a number of food processing plants, and in many other industrial operations. Water also is used in all steam power-generating (thermoelectric) plants to produce electricity. Here are some important data related to water, agriculture, and food security as reported by the United Nations.

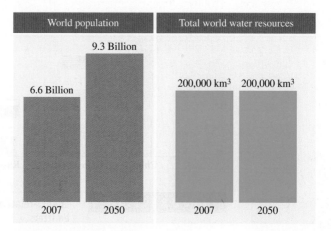

- The daily drinking water requirement per person is 2 to 4 liters, but it takes 2,000 to 5,000 liters of water to produce one person's daily supply of food.
- It takes 1,000 to 3,000 liters of water to produce just one kilogram of rice and 13,000 to 15,000 liters to produce one kilogram of grain-fed beef.
- In 2007, the estimated number of undernourished people worldwide was 923 million.
- From now to 2050, the world's water will have to support the agricultural systems that will feed and create livelihoods for an additional 2.7 billion people.

- The extent of land under irrigation in the world is 277 million hectares, about 20 percent of which is cropland. Rain-fed agriculture is practiced on the remaining 80 percent of the arable land.
- The Intergovernmental Panel on Climate Change predicts yields from rain-dependent agriculture could be down by 50 percent by 2020.

In this mentioned data, one liter is approximately equal to a quarter of a gallon (1 liter ≈ ¼ gallon), and one kilogram is equal to 2.2 pounds. We will explain systems of units in greater detail in Chapter 2.

To better understand the ***water cycle***, see Figure 1.14. Radiation from the sun evaporates water; water vapors form into clouds, and eventually, under favorable conditions, water vapor turns into liquid water or snow and falls back on the land and into the ocean. On land, depending on the amount of precipitation, part of the water infiltrates the soil, part of it may be absorbed by vegetation, and part runs as streams or rivers and collects into natural reservoirs called lakes. ***Surface water*** refers to water in reservoirs, lakes, rivers, and streams. ***Groundwater***, on the other hand, refers to the water that has infiltrated the ground; surface water and groundwater eventually return to the ocean, and the water cycle is completed. In addition to understanding the water cycle, *it is also*

> The total amount of water on the Earth is constant—we don't lose or gain water on the Earth.

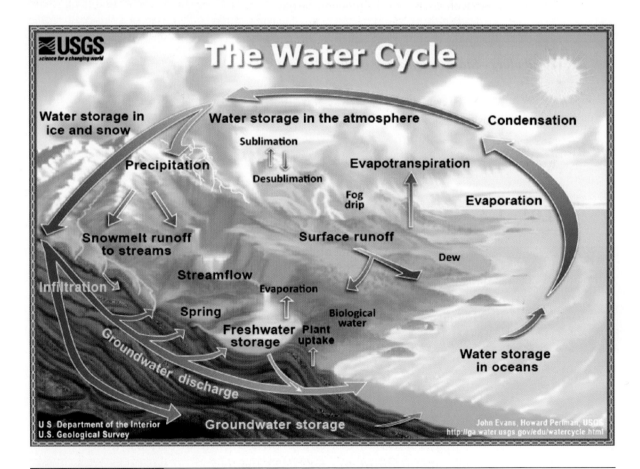

The water cycle.
*Source:* USGS

*important to realize that the amount of water that is available to us on the Earth is constant.* Even though water can change phase from liquid to vapor or from liquid to ice, the total amount remains constant—we don't lose or gain water on the Earth. For example, when you take a shower, the water you used could end up elsewhere and be used for an entirely different purpose, such as cooking (of course, after it has been treated and filtered). We discuss water resources, consumption rates, and quality standards in Chapter 9.

## Before You Go On

Answer the following questions to test your understanding of the preceding section:

1.  What does the word environment mean to you?

2.  What are the major layers of the Earth?

3.  What are the main gases that make up air?

4.  In your own words, describe the greenhouse gas effect.

5.  In your own words, describe the water cycle.

*Vocabulary—State the meaning of the following words:*

Air

Greenhouse gases

## LO⁴  1.4  Sustainability

Sustainability could be defined as consuming resources in such a way that meets our present needs without compromising the ability of future generations to meet their needs.

As we mentioned previously, much has been written or said about "sustainability." But what does it mean to you and why is it important for you, as a good global citizen, to have a good grasp of this concept? To start with, it is important to know that there is no universal definition for *sustainability*. It means different things to different professions. However, one of the generally accepted definitions is

*"design and development that meets the needs of the present without compromising the ability of future generations to meet their needs."*

As you know by now, we need to design and produce goods and services to enjoy a high standard of living. We also need to address our infrastructure needs, energy, and food security. As a society, we are expected to design and provide goods and services that increase the standard of living and advance health care, while also addressing serious environmental and sustainability concerns. We must consider the link among the Earth's finite resources and our environmental, social, ethical, technical, and economical factors. Therefore, we need global citizens who can come up with solutions that address energy and food security and simultaneously address the sustainability issues. The shortage of citizens who understand the concept of sustainability—people who can apply the sustainability concepts, methods, and tools to their problem-solving and decision-making processes—could have serious negative consequences for our future. Because of this fact, in recent years, many institutions of higher education and organizations have come out in support of sustainability education. As you study this book, you gradually will learn more details about these concepts, methods, and tools. Hopefully, you will apply them to your decision-making process and way of life in order to make the world a better place for all of us!

## Attributes of Good Global Citizens

Now that you have a general sense of why you need to know about energy, environment, and sustainability, you may be wondering about how to get involved and how to become a good global citizen. *As good global citizens, we need to realize that the choices that we make in our everyday lives affect all of us.* We need to change our behavior, especially with respect to the way we consume energy and use the finite resources available to us. Computers, smart electronic devices, and computer-controlled machines are continuously reshaping our way of life. Such tools influence the way we do things and help provide us with the necessities of our lives—clean water, food, and shelter. We need to become

wavebreakmedia/Shutterstock.com

lifelong learners so that we can make informed decisions and anticipate and react to the global changes caused by technological innovations as well as population and environmental changes. Although the activities of good citizens may be varied, there are some personality traits and practices that typify most of today's good global citizens.

- Good citizens are well informed and have a firm grasp of basic concepts and current issues, particularly issues related to energy, the environment, and sustainability.
- Good citizens have a desire to be life-long learners. For example, they are well read, attend community and town meetings to stay abreast of new events, and they learn about how innovations and new technologies may affect their lives.
- Good citizens have good written and oral communication skills.
- Good citizens have time management skills that enable them to work productively, take good care of their families, and be active in their communities.
- Good citizens generally work well in a team environment where they consult each other to solve complex problems that affect all of us.

## Communication

As good global citizens, you need to develop good written and oral communication skills in order to express your thoughts, present a concept, provide an analysis of a problem and its solution, or show your findings from a research project that you have done. Starting right now, it is important to understand that *the ability to communicate your solution to a problem is as important as the solution itself.* You may spend weeks on a project, but if you cannot effectively communicate to others, the results of all your efforts may not be understood and appreciated. *In this book, to emphasize that a good global citizen should*

Rido/Shutterstock.com

*have good communication skills, we ask you to write reports and give presentations*. These reports might be lengthy and contain charts and graphs, or they may take the form of a brief memorandum. These forms of communication are explained next.

**Written Reports** One type of written report is a progress report. *Progress reports* are a means of communicating to others in an organization or to the sponsors of a project how much progress has been made and which of the main objectives of the project have been achieved to date. Based on the total time period required for a project, progress reports may be written for a period of a week, a month, several months, or a year. The format of the progress report may be dictated by a manager in an organization or by the project's sponsors. In your case, your instructor will specify how often you need to write a progress report for say, a term project.

*Short memos* are yet another way of conveying information in a brief way to interested individuals. Generally, short memos are under two pages in length. A general format for a short memo header is shown below. The header of the memo contains information such as the date, who the memo is from, to whom it is being sent, and a subject line. This is followed by the main body of the memo.

> **Date:** *May 3, 2015*
> **From:** *Mr. John Doe*
> **To:** *Members of Project X*
> **Re:** *Proposed Wind Energy Farm*

As the name implies, *detailed reports* are comprehensive and provide a great deal of information. These reports generally contain the following items: title, abstract, objectives, analysis, data and results, a discussion of results, conclusions and recommendations, and references. Whenever you write a report, you must include a list of references that show the reader where you obtained some of the information. For the references, you may want to use the following format styles:

> *For Books:* Author, *title* (italicized), publisher, place of publication, date (year), and page(s).
> *For Journal Articles:* Author, "title of article" (enclosed in quotation marks), name of journal, volume number, issue number, year, and page(s).
> *For Internet Materials:* Author (or company), title (or page tab), date accessed, and URL address.

**Oral Presentations** Some of the problems at the end of the chapters in this book require you to give oral presentations. You already communicate orally with others all the time. Informal communication is part of our daily life. We may talk about sports, weather, what is happening around the world, or a homework assignment. However, when it comes to formal presentations, there are certain rules and strategies that you need to follow. Your oral presentation may show the results of all your efforts regarding a project that you may have spent weeks or months to develop. If the listener cannot follow

lightpoet/Shutterstock.com

you, then all of your efforts become insignificant. It is very important that all information be conveyed in a manner easily understood by the listener.

An oral technical presentation in many ways is similar to a written one. You need to be well organized and have an outline of your presentation ready, similar to the format for a written report. It may be a good idea to write down what you are planning to present. Remember, it is harder to erase or correct what you say after you have said it than to write it down on a piece of paper and correct it before you say it. You want to make every effort to ensure that what is said (or sent) is what is understood (or received) by the listener.

*Rehearse* your presentation before you deliver it formally. You may want to ask a friend to listen and provide helpful suggestions about your style of presentation, delivery, content of the talk, and so on. Present the information in a way that will be understood easily by your audience. Avoid using terminology or phrases that may be unfamiliar to listeners. If you have to give a longer talk, then you may want to mix your presentation with some humor or tell some interesting related story to keep your audience's attention. Maintain eye contact with everyone in your audience, not just one or two people. Use good visual aids. When possible, incorporate charts, graphs, animated drawings, short videos, or a model. You may also want to have copies of the outline, along with notes on the important concepts and findings, ready to hand out to interested audience members. In summary, when giving an oral presentation, be organized, be well prepared, get right to the point, and consider the needs and expectations of your listeners.

## Teamwork

Some of the projects in this book require teamwork. Therefore, it is important to say a few words about teamwork and conflict resolution. A *team* may be defined as a group of individuals with complementary expertise, problem-solving skills, and talents who are working together to solve a problem or achieve a common goal. A good team is one that gets the best out of each other. The individuals making up a good team know when to compromise for the good of the team and its common goal. Communication is an essential part of successful teamwork. The individuals making up the team need to clearly understand the role of each team member and how each task fits together.

**Common Traits of Good Teams** More and more, employers are looking for individuals who not only have a good grasp of contemporary issues but who can also work well with others in a team environment. Successful teams have the following components:

- The project that is assigned to a team must have clear and realistic goals. These goals must be understood and accepted by all members of the team.
- The team should be made up of individuals with complementary expertise, problem-solving skills, backgrounds, and talents.
- The team must have a good leader.
- The team leadership and the environment in which discussions take place should promote openness, respect, and honesty.
- Team needs and goals should come before individual needs and goals.

**Conflict Resolution** When a group of people work together, conflicts sometimes arise. Conflicts can be the result of miscommunication, personality differences, or the way events and actions are interpreted by a member of a team. Managing conflicts is an important part of a team dynamic. When it comes to managing conflicts, a person's response may be categorized in one of the following ways.

- There are those in a team environment who try to avoid conflicts. Although this may seem like a good approach, it demonstrates low assertiveness and a low level of cooperation. Under these conditions, the person who is most assertive will dominate the team, making progress as a whole difficult. *Accommodating team members* are highly cooperative, but their low assertiveness could result in poor team decisions. This is because the ideas of the most assertive person in the group may not necessarily reflect the best solution.
- *Compromising team members* demonstrate a moderate level of assertiveness and cooperation. Compromised solutions should be considered as a last resort. Again, by compromising, the team may have sacrificed the best solution for the sake of group unity.
- A better approach is the *collaborative* "conflict resolution" approach, which demonstrates a high level of assertiveness and cooperation by the team. With this approach, instead of pointing a finger at someone and blaming an individual for the problem, the conflict is treated as a problem to be solved by the team. The team proposes solutions, means of evaluation, and perhaps combines solutions to reach an ideal solution. Furthermore, in order to reach a resolution to a problem, a plan with clear steps must be laid out.

Good communication is an integral part of any conflict resolution. One of the most important rules in communication is to make sure that the message sent is the message received—without misunderstanding. *Team members must listen to each other.* Good listeners do not interrupt; they allow the speaker to feel at ease, do not get angry, and do not criticize. You may want to ask relevant questions to let the speaker know that you really are listening.

## Before You Go On

Answer the following questions to test your understanding of the preceding section:

1. Why is it important for all of us to understand and apply sustainable practices to our everyday lives?

2. What are common traits of good global citizens?

3. State the differences between these written communications: progress report, executive summary, short memo, and detailed technical report.

4. What do we mean by team and teamwork?

5. Explain how you would resolve a conflict that may arise when working in a team environment.

# SUMMARY

## LO¹  Basic Human Needs

You should have a good understanding of basic human needs and the growing world population. At the turn of the 20th century, there were approximately six billion of us inhabiting the Earth. According to the latest estimates, the world population will reach 9.3 billion people by the year 2050. We need clean air, clean water, food, and shelter. As a society, we create and consume many different products and services. Think about all the products and services that you used yesterday.

We need energy to address our needs, such as building structures, growing food, and having access to clean water. The energy use per capita in the world has been increasing steadily as the economies of the world grow. Added to these concerns is the expected rise in the population of the world from the current 6.6 billion to about 9.3 billion people by the mid-21st century! Stationary, mobile, and natural sources contribute to outdoor air pollution. Human activities, such as mining, construction, manufacturing goods, and agriculture, contribute to water pollution. In order to address our needs and maintain a good standard of living, we as a society are faced with the problems of finding energy sources and reducing pollution and waste.

## LO²  Energy

You should have a good understanding of the significant role energy plays in your daily life and realize that without energy we cannot keep our homes warm and well lit, move our cars, make products and structures, grow food, or have easy access to water, shelter, and other essential needs. You should be familiar with energy consumption rates in our society. You should know that coal, natural gas, and petroleum provide the majority of our energy needs, and realize that the majority of the coal mined in the United States is used for generating electricity. So don't waste electricity!

## LO³  Environment

By environment we mean *our natural environment,* which includes all living (plants, animals) and non-living (air, water, rocks) things that exist on or within the Earth. It is also important to realize that each one of these categories could be subdivided further. For example, water could be grouped as above ground (rivers, ponds, lakes, seas, oceans) or below ground (aquifers). You should understand what we mean by greenhouse gases and how you can reduce the amount of greenhouse gases that are produced due to your everyday activities.

## LO⁴ Sustainability

It is very important for us to understand that, because of worldwide socioeconomic trends, environmental concerns, and Earth's finite resources, more is expected of each one of us. As a society, we are expected to design and provide goods and services that increase the standard of living and advance health care while considering the links between the Earth's finite resources and environmental, social, ethical, technical, and economical factors. One of the generally accepted definitions of sustainability is *"design and development that meets the needs of the present without compromising the ability of future generations to meet their own needs."*

## KEY TERMS

Air  22
Atmosphere  21
Commercial Sector  19
Earth's Crust  21
Earth's Inner Core  21
Earth's Outer Core  21

Energy  16
Environment  20
Greenhouse Gases  22
Groundwater  25
Industrial Sector  19
Mantle  21

Residential Sector  19
Surface Water  25
Sustainability  26
Team  30
Transportation Sector  19
Water Cycle  25

## Apply What You Have Learned

This is a possible term project for your class. Prepare a website for waste reduction, greenhouse gas reduction, and energy-saving measures that could be used on your campus. Elect a group leader, and then divide up the tasks among yourselves. Think about ways to measure the success of the project. As you work on the project, also take note of both the pleasures and problems that arise from working in a team environment. Write a brief report about your experiences working as a team on this project. What are your recommendations for others who may work collaboratively on similar projects?

## PROBLEMS

*Problems that promote life-long learning are denoted by* 🔑

1.1 Each of you is to ask an older adult (for example, your grandparents) to think back to when they graduated from high school or college and to create a list of products and services that are available in their everyday lives now that were not available to them then. Ask them if they ever imagined that these products and services would be available today. To get your conversation started, here are few examples: cellular phones, ATM cards, personal computers, airbags in cars, price scanners at the supermarket, E-Z Passes for tolls, and so on. Ask them to explain how these products have made their lives better (or worse).

wavebreakmedia/Shutterstock.com

Dmitry Kalinovsky/Shutterstock.com

**1.2** Use your imagination to compile a list of products and services that are not available now that you think will be readily available in the next 20 years. Present the list to the class and explain which products and services you most look forward to using.

**1.3** Record how much trash you generate each week. For a period of one week, maintain a daily logbook to keep track of what and how much you throw away and recycle each day. Suggest ways you could reduce waste and increase your own recycling. Compile your findings into a report and present it to the class.

**1.4** Estimate how many cans of soda or other beverages you drink each year and calculate the amount of aluminum (in pounds or kilogram) that was used to make the cans. A 12-ounce (355 milliliters) empty aluminum can has a mass of 0.0445 pounds (20.18 grams). State your assumptions, and explain your calculations.

**1.5** Estimate the amount of copy or printing paper that you use every year. A 500-sheet ream of copy paper has an approximate mass of 5 pounds (2.27 kg). How much of this consumption is truly necessary, and how much of your own paper consumption could be avoided? State your assumptions.

**1.6** Estimate how much water you consume each year when showering. To determine your shower water consumption:

a. Obtain a container of a known volume (for example, an empty gallon-size water or milk container) and then time how long it takes to fill the container.

b. Calculate the volumetric flow rate in gallons or liters per minute.

c. Measure the time that you spend on average when taking showers. Calculate the volume of the water consumed taking a shower on a daily basis.

d. Multiply the daily value by 365 to get the yearly value.

**1.7** Estimate how much gasoline you consume each year for driving around town, doing errands, going to school, travelling to and from work, or just traveling from place to place.

**1.8** Estimate how much food you consume annually for breakfast, lunch, dinner, and snacks. For a period of one week, maintain a daily log to keep track of what and how much you eat each day. Based on this analysis, estimate your annual food consumption. State all your assumptions.

**1.9** Electric motors, which are found in many appliances and devices around your home, consume lots of energy. Identify at least five products at home that use electric motors. Could you get by without using any of them as often as you do now?

**1.10** Identify at least five different energy-consuming products or practices at home and suggest ways to reduce consumption, such as turning off the light when you leave a room.

**1.11** Electronic communication is becoming increasingly important. In your own words, identify the various situations under which you should write a letter, send an e-mail, send a text message, make a telephone call, or talk to someone in person. Explain why one particular form of communication is preferable to the others available.

**1.12** In a brief report, discuss why we need various modes of transportation. How did they evolve? Discuss the roles of public transportation, water transportation, highway transportation, railroad transportation, and air transportation.

**1.13** Visit the U.S. Department of Energy website and collect energy consumption data for each sector of the economy for the most recent year. Prepare a brief report discussing your findings.

**1.14** To increase public awareness about the importance of energy, the environment, and sustainability and to promote global citizenship education among the younger generation, prepare and give a 15-minute presentation at a mall near your town. You need to do some planning ahead of time and ask permission from the proper authorities.

**1.15** If this class has a term project, present your final work, on the date set by your instructor, at your school dining hall or during half-time of a sporting event. If the project has a competitive component, hold the design competition at the suggested locations as well.

**1.16** Prepare a 15-minute oral presentation about energy and its use in our everyday lives. The next time you go home, present it to the juniors in your high school.

**1.17** Investigate how much trash is generated on your campus each week. Suggest ways to reduce waste and increase recycling. Compile your findings in a brief report and present it to the class. State all your assumptions.

**1.18** A gallon (3.8 liters) of gasoline that weighs 6.3 pounds (2.85 kg) can produce 20 pounds (9.1 kg) of carbon dioxide. Yes, 20 pounds (9.1 kg) of carbon dioxide! Assume 100 million people with cars (with 20 miles/gallon (10.6 km/liters) gasoline

consumption rates) decide to walk 3 miles (4.8 km) a day (approximately an hour) instead of driving their cars. What would be the reduction in pounds of carbon dioxide released into the atmosphere on a yearly basis?

**1.19** Make a list of clothing, shoes, and accessories that you purchased last year. List the materials that you think were used to make these items. Discuss the origin of the materials.

**1.20** Look around your home and estimate how many feet (or meters) of visible copper wire are in use for extension and power cords for common items such as a hairdryer, TV, cell phone charger, laptop computer, or a lamp. Write a brief report and discuss your findings.

Fred Stein Archive/Contributor/Archive Photos/Getty Images

*"Education is what remains after one has forgotten everything he learned in school."* —Albert Einstein (1879–1955)

# Fundamental Dimensions and Systems of Units

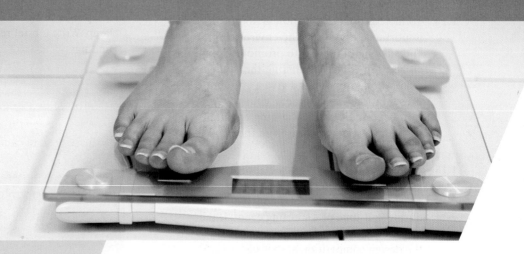

## LEARNING OBJECTIVES

**LO¹** Fundamental Dimensions and Units: explain what they mean and give examples

**LO²** Systems of Units: know what systems of units represent and give examples of International (metric) and U.S. Customary units for length, time, mass, force, and temperature

**LO³** Dimensional Homogeneity and Unit Conversion: know how to check for dimensional homogeneity and convert data from SI or the metric system of units to U.S. Customary units and vice versa

**LO⁴** Components and Systems: explain what they mean and give examples of their use

# Discussion Starter

Yuganov Konstantin / Shutterstock.com

HAPPY BIRTHDAY!

Africa Studio / Shutterstock.com

Fundamental dimensions play important roles in our everyday lives. Throughout history, human beings have realized that we need only a few physical dimensions or quantities to describe our surroundings and daily events. For example, we need a *length dimension* to describe how tall, how long, or how wide something is. We also use two or more length dimensions to calculate and describe the area and volume of something. *Time* is another physical dimension that we need to answer questions such as: "How old are you?", "When are you coming over?", or When is the next test?" Based on what we know about our world, we use seven fundamental dimensions to correctly express ourselves in our surroundings today. They are *length, mass, time, temperature, electric current, amount of substance,* and *luminous intensity.*

The other important concept that we have learned is that not only do we need to define these physical dimensions to describe our surroundings, but we also need some way to scale or divide them into units. For example, the dimension *time* can be divided into both small and large portions, such as seconds, minutes, hours, days, months, years, and so on.

**To the Students:** Take a few minutes and think about the previously mentioned dimensions and consider how frequently you used them in your daily life activities during the past week. Did you use the *length dimension* or its related quantities such as area or volume? Did you use the *time dimension*? Did you use the *mass dimension*? Did you use the *temperature dimension*? How did you use these dimensions? To get you started, during the past seven days, did you purchase any food, drink, or gasoline? If so, what amounts (express units) did you buy? Did you discuss the weather conditions with anyone?

# LO¹  2.1  Fundamental Dimensions and Units

In this section, we introduce you to the concepts of dimensions and units. You have been using these concepts all your life; here we define them in a formal way. For example, when asked, "How tall are you?" you may respond with, "I am 6 feet (183 centimeters) tall." When asked, "What is the temperature outside today?" you could answer with something like, "Today is going to be hot and might reach 100 degrees Fahrenheit (38°Celsius)."

luminaimages / Shutterstock.com        TerryM / Shutterstock.com

In this chapter, you are introduced to some very important concepts that you need as a foundation to understand other materials in the subsequent chapters. Read this chapter carefully, and remind yourself that a good understanding of fundamental dimensions and units is very important in understanding our energy and environmental footprints. We all want to be good global citizens, so we need to learn these concepts well. Don't be intimidated by definitions and numbers. You can do it!

The evolution of the human intellect has taken shape over a period of thousands of years. Men and women all over the world observed and learned from their surroundings. They used the knowledge gained from their observations of nature to design, develop, test, and fabricate tools, shelter, weapons, water transportation, and the means to cultivate and produce more food. Moreover, people realized that they needed only a few physical quantities called *dimensions* to fully describe natural events and their surroundings.

A *dimension* is a physical quantity, such as *length, mass, time*, or *temperature*, that makes it possible for us to communicate

StockLite / Shutterstock.com

iakov Filimonov / Shutterstock.com

rangizzz / Shutterstock.com

Vitaliy Netiaga / Shutterstock.com

about and analyze our surroundings and events. For example, the *length* dimension describes how tall, how long, or how wide something is. People also realize that some things are heavier than other things, so there is a need for another physical quantity (dimension) to describe that observation: the concept of mass and weight. Early humans did not fully understand the concept of gravity; consequently, the distinction between mass and weight, which is a force, was made later in history.

*Mass* represents the amount of matter that makes up all things. Then what is force? The simplest form of a *force* that represents the interaction of two objects is a push or a pull. When you push or pull on a vacuum cleaner, that interaction between your hand and the vacuum cleaner is called force. In this example, the force is exerted by one object (your hand) on another object (the vacuum cleaner) by direct contact.

Not all forces result from direct contact. For example, gravitational force is not exerted by direct contact. If you hold this book, say, 3 feet above the ground and let it go, what happens? It falls! This is due to the gravitational force that is exerted by the Earth on the book. Gravitational forces act at a distance. The weight of an object is the force that is exerted on the object by the Earth's gravity.

*Time* is another physical dimension that humans need to understand in order to be able to explain their surroundings and answer questions such as, "How old are you?" or "How long does it take to go from here to there?" The response to these questions in the past may have been something like, "I am two hundred Moons old," or "It takes two days to go from our village to the village on the other side of the mountains."

To describe how cold or hot something is, humans need yet another physical quantity, or physical dimension, that we now refer to as *temperature*. Think about the important role of temperature in your everyday life to describe the various states of things. Do you know the answer to some of these questions?

What is your deep body temperature?
What is the room air temperature?
What is the temperature of the water that you used this morning to take a shower?
What is the temperature of the air inside your refrigerator that kept the milk cold overnight?
What is the temperature inside the freezer section of your refrigerator?
What is the temperature of the air coming out of your hair dryer?

Once you start thinking about the role temperature plays in quantifying what goes on in our surroundings, you realize that you could ask hundreds of similar questions.

Let us now examine what we mean by temperature more closely. Temperature provides a measure of the molecular activity of an object. All objects and living things are made of matter, and matter itself is made up of atoms, or chemical elements. Moreover, atoms are combined naturally or in a laboratory setting to create molecules. For example, as you already know, water molecules are made of two atoms of hydrogen and one atom of oxygen. Temperature represents the level of molecular activity of a substance. The molecules of a substance at a high temperature are more active than at a lower temperature.

Early humans relied on the sense of touch or vision to measure how cold or how warm something was. In fact, we still rely on touch today. When you are planning to take a bath, you first turn the hot and cold water on and let the bathtub fill with water. Before you enter the tub, however, you first touch the water to feel how warm it is. Basically, you are using your sense of touch to get an indication of the temperature. Of course, using touch alone, you can't quantify the temperature of water accurately. You cannot say, for example, that the water is at 104°F(40°C). Another example of how people rely on their senses to quantify temperature is the way blacksmiths use their eyes to estimate how hot a fire is. They judge the temperature by the color of the burning fuel before they place an iron piece in the fire. From these examples, you see that our senses are useful in judging how cold or how hot something is, but they are limited in accuracy and cannot quantify a value for a temperature. Thus, we need a measuring device that can provide information about the temperature of something more accurately and effectively. This need led to the development of thermometers, which are based on thermal expansion or contraction of a fluid, such as alcohol, or a liquid metal, such as mercury.

Today, based on what we know about our physical world, we use *seven fundamental dimensions* to correctly express ourselves in our surroundings. They are *length, mass, time, temperature, electric current, amount of substance*, and *luminous intensity*. With the help of these base dimensions, we can present all other necessary physical quantities that describe how things work.

> Dimension is a physical quantity, such as length, time, mass, and temperature, that makes it possible for us to describe our surroundings and events.

By now, you understand why we need to formally define physical variables using fundamental dimensions. The other important fact you need to realize is that early humans needed not only physical dimensions to describe their surroundings, but also some way to scale or divide these physical dimensions. As such, we have developed standard *units* to scale and measure these dimensions. For example, time is considered a physical dimension, but it can be divided into both small and large portions

bikeriderlondon / Shutterstock.com                    McCarthy's PhotoWorks / Shutterstock.com (units), such as seconds, minutes, hours, days, months, years, decades, centuries, millennia, and so on. Today, when someone asks you how old you are, you reply by saying, "I am 19 years old." You don't say that you are approximately 6,939 days, or 170,000 hours old, even though these statements may very well be true at that instant! Or to describe the distance between two cities, we may say that they are 100 miles (161 kilometers) apart; we don't say the cities are 528,000 feet (161,000 meters) apart. The point of these examples is that we use appropriate divisions of physical dimensions to keep numbers manageable. We have learned to create an appropriate scale for these fundamental dimensions and divide them properly so that we can describe particular events, the size of an object, the thermal state of an object, or its interaction with its surroundings correctly, and do so without much difficulty.

## Physical Laws

As we mentioned earlier, men and women all over the world observed and learned from their surroundings. They used the knowledge gained from their observations of nature to design, develop, test, and fabricate all kind of products to address their needs. Let us now say a few words about how significant observations are formulated into ***physical laws***, so that they can be used to help design products that we use in our daily lives. Having had a high school education, you have a pretty good idea of what we mean by mathematics. But what do we mean by a *physical* law? Well, the universe, including the Earth that we live on, was created a certain way. There are differing opinions as to the origin of the universe. Was it put together by God, or did it start with a big bang? We won't get into that discussion here. But what is important is that we have learned through our own observation and by the collective effort of those before us that things work a certain way in nature. For example, if you let go of something that you are holding in your hand, it will fall to the ground. That is an observation that we all agree upon. We can use words to explain our observations or use another language, such as mathematics and formulas, to express our findings. Sir Isaac Newton (1642–1727) formulated that observation into a useful mathematical expression that we know today as the *universal law of gravitational attraction*. Our understanding of this and other physical laws has allowed us to design things such as escalators, elevators, parachutes, planes, and satellites.

Physical laws are based on observation and experimentation and are expressed using mathematical formulae.

Another important law that all of you have heard about is *Newton's second law of motion*. If you place a book on a smooth table and push it hard enough, it will move. This is simply the way things work. Newton observed this and formulated his observation into what we call Newton's second law of motion. This is not to say that other people had not made this simple observation before, but Newton took it a few steps further. He noticed that, as he increased the mass of the object being pushed while keeping the magnitude of the force constant (pushing with the same effort), the object did not move as quickly. Moreover, he noticed that there was a direct relationship between the magnitude of the push, the mass of the object being pushed, and the acceleration of the object. He also noticed that there was a direct relationship between the direction of the force and the direction of the acceleration. Newton's second law of motion now forms the basis for a discipline called mechanics, which is used to design and predict the behavior of all types of moving things such as the atmosphere, rivers, cars, and planes.

Again, an important point to remember is that the *physical laws are based on observations*. Moreover, we use mathematics and basic physical quantities to express our observations in the form of a law. Even so, to this day we may not fully understand why nature works the way it does. We just know it works. There are physicists who spend their lives trying to understand on a more fundamental basis why nature behaves the way it does. As another example, when you place some hot object in contact with a cold object, the hot object cools down, while the cold object warms up, until they both reach an equilibrium temperature somewhere between the two initial temperatures. From your everyday experience, you know that the cold object does not get colder while the hot object gets hotter! Why is that? Well, it is just the way things work in nature! The second law of thermodynamics, which is based on this observation, simply states that heat flows spontaneously from a high-temperature region to a low-temperature region. The object with the higher temperature (more energetic molecules) transfers some of its energy to the low-temperature (less energetic molucules) object. When you put some ice

Joshua Resnick / Shutterstock.com

cubes in a glass of warm soda, the soda cools down, while the ice warms up and eventually melts away. You may call this "sharing resources." Unfortunately, most people do not follow this law closely when it comes to social issues. Again, it is important to note that the second law of thermodynamics, which is based on observation, is used to design all kinds of everyday products such as refrigerators and heating and cooling systems.

Take the time to understand these concepts and other upcoming topics in this book, as they are meant to encourage you to think, evaluate, and analyze. A good understanding of these topics will allow you to make good decisions at home or work and in the way you interact with the environment. If you don't take the time to understand the basics, you are likely to make poor decisions that indirectly will affect all of us!

Educated global citizens are also good bookkeepers. What do we mean by this? Any of us with a checking account knows the importance of accurate record keeping. In order to avoid problems, most of us keep track of the transactions in terms of payments (e.g., credit card debits, rent, loan payments) and deposits (paychecks, received loans). Good bookkeepers can tell you instantly what the balance in their account is. They know they need to add to the recorded balance whenever they deposit some money and subtract from the balance with every withdrawal from the account. In a similar way, as good global citizens, it is important to keep track of how much energy and materials we consume and how much waste we produce *every day*. Remember, the way we treat our environment is judged by Mother Nature—a judge who is unforgiving of our errors.

## Before You Go On

Answer the following questions to test your understanding of the preceding section:

1. Name at least four fundamental dimensions.

2. What is the difference between dimension and unit?

3. Name at least two units that you use every day.

4. What is the difference between mass and weight?

5. What do we mean by a physical law and what are such laws based on?

6. Give two examples of physical laws.

*Vocabulary—State the meaning of the following words:*

Dimension

Unit

Mass

Weight

## LO² 2.2 **Systems of Units**

In the previous section, we explained that a dimension or physical quantity such as time can be divided into small and large units or portions such as seconds, hours, and days. Throughout the world, there are several systems of units in use today. The most common are the *International System* (*abbreviated as SI, from French Systéme International d' Unites* or sometimes called *metric units*) and the *U.S. Customary System of units*. Let us now examine the systems of units in greater detail.

### International System (SI) of Units

> Meter, kilogram, second, Kelvin (or degree Celsius), ampere, mole, and candela are units of length, mass, time, temperature, electric current, amount of substance, and luminous intensity in the SI System.

We begin our discussion of systems of units with the International System (SI) of units, because SI is the most common system of units used in the world. The origin of the present day International System of units can be traced back to 1799 with **meter** and **kilogram** used as the first two *base* or *fundamental units*. It was not until 1946 that the proposal for the **ampere** as a *base unit* for electric current (explained in more detail in Chapter 4) was approved by the General Conference on Weights and Measures (CGPM). In 1954, the CGPM included units of *degree Kelvin* (for absolute temperature) and **candela** (for luminous intensity), and the **mole** for amount of substance was added as a *base unit* in 1971. A list of SI basic units is given in Table 2.1.

You need not memorize the formal definitions of base units as provided by the Bureau International des Poids et Mesures (BIPM). From your everyday life experiences you have a pretty good idea about some of them. For example, you know how short a time period a second is, or how long a period a year is. However, you may need to develop a "feel" for some of the other base units. For example, how long is a meter? How tall are you? Are you under 2 meters or perhaps above 2 meters? In general, most adults are between 1.6 and 2 meters tall. What is your body mass in kilograms?

Developing a "feel" for units will make you a better informed global citizen. Also, when you travel abroad, the knowledge of these units could be quite useful to you. In 1960, the first series of *prefixes* and symbols of *decimal multiples* of SI units were adopted. Over the years, the list has been extended to include those listed in Table 2.2. When studying Table 2.2, note that **nano** ($10^{-9}$), **micro** ($10^{-6}$), centi ($10^{-2}$), kilo ($10^3$), **mega** ($10^6$), **giga** ($10^9$), and **tera** ($10^{12}$) are examples of decimal multiples and prefixes used with SI units. You already use some of these multiples and prefixes in your daily conversations. For example, when describing an electronic file or a hard drive size, you may say megabytes, gigabytes, or terabytes.

The units for other physical quantities that we use in our lives can be derived from the base units. For example, the *unit for force is the* **newton (N)**, which is derived from Newton's second law of motion. One newton is defined *as a magnitude of a force that, when applied to 1 kilogram of mass, will accelerate the mass (change its speed) at a rate of 1 meter per second squared* ($m/s^2$). That is, $1 \text{ N} = (1 \text{ kg})(1 \text{ m/s}^2)$. As well-educated global citizens, it is also important to know the difference between mass and weight. As mentioned previously, the weight of an object is the force that is exerted on the mass of the object by the Earth's gravity and is based on the universal law of gravitational attraction. The following mathematical

> Nano ($10^{-9}$), *micro* ($10^{-6}$), *centi* ($10^{-2}$), *kilo* ($10^3$), *mega* ($10^6$), *giga* ($10^9$), and *tera* ($10^{12}$) are examples of decimal multiples and prefixes used with SI units.

| TABLE 2.1 | The SI Base Units | | |
|---|---|---|---|
| **Physical Quantity (Dimension)** | | **Name of SI Base Unit** | **SI Symbol** |
| **Length** | 1.6 m–2.0 m  Range of height for most adults | Meter | m |
| **Mass** | 50 kg–120 kg  Range of mass for most adults | Kilogram | kg |
| **Time** | Fastest person can run 100 meters in approximately 10 seconds | Second | s |
| **Thermodynamic temperature** | Ice water: 0°C or 273 K  Comfortable room temperature: 22°C or 295 K | Kelvin | °C or K |
| **Electric current** | 120 volts  120 watts  Electrical device  1 amp | Ampere | A |
| **Amount of substance** | Uranium 238 ← One of the heaviest atoms known  Gold 197  Silver 108  Copper 64  Calcium 40  Aluminum 27  Carbon 12 ← Common carbon is used as a standard  Helium 4  Hydrogen 1 ← Lightest atom | Mole | mol |
| **Luminous intensity** | A candle has luminous intensity of approximately 1 candela | Candela | cd |

relationship shows the relationship among the weight of an object, its mass, and the acceleration due to gravity.

$$\text{weight} = (\text{mass})(\text{acceleration due to gravity}) \qquad 2.1$$

For example, an apple with a mass of 100 grams or 0.1 kilograms has an approximate weight of 1 newton. Or, a one-liter bottle of water has a mass of one kilogram and an approximate weight of 10 newtons, as shown in Figure 2.1.

| TABLE 2.2 | The List of Decimal Multiples and Prefixes Used with SI Base Units | |
|---|---|---|
| **Multiplication Factors** | **Prefix** | **SI Symbol** |
| $1{,}000{,}000{,}000{,}000{,}000{,}000{,}000{,}000 = 10^{24}$ | yotta | Y |
| $1{,}000{,}000{,}000{,}000{,}000{,}000{,}000 = 10^{21}$ | zetta | Z |
| $1{,}000{,}000{,}000{,}000{,}000{,}000 = 10^{18}$ | exa | E |
| $1{,}000{,}000{,}000{,}000{,}000 = 10^{15}$ | peta | P |
| $1{,}000{,}000{,}000{,}000 = 10^{12}$ | tera | T |
| $1{,}000{,}000{,}000 = 10^{9}$ | giga | G |
| $1{,}000{,}000 = 10^{6}$ | mega | M |
| $1000 = 10^{3}$ | kilo | k |
| $100 = 10^{2}$ | hecto | h |
| $10 = 10^{1}$ | deka | da |
| $0.1 = 10^{-1}$ | deci | d |
| $0.01 = 10^{-2}$ | centi | c |
| $0.001 = 10^{-3}$ | milli | m |
| $0.000{,}001 = 10^{-6}$ | micro | μ |
| $0.000{,}000{,}001 = 10^{-9}$ | nano | n |
| $0.000{,}000{,}000{,}001 = 10^{-12}$ | pico | p |
| $0.000{,}000{,}000{,}000{,}001 = 10^{-15}$ | femto | f |
| $0.000{,}000{,}000{,}000{,}000{,}001 = 10^{-18}$ | atto | a |
| $0.000{,}000{,}000{,}000{,}000{,}000{,}001 = 10^{-21}$ | zepto | z |
| $0.000{,}000{,}000{,}000{,}000{,}000{,}000{,}001 = 10^{-24}$ | yocto | y |

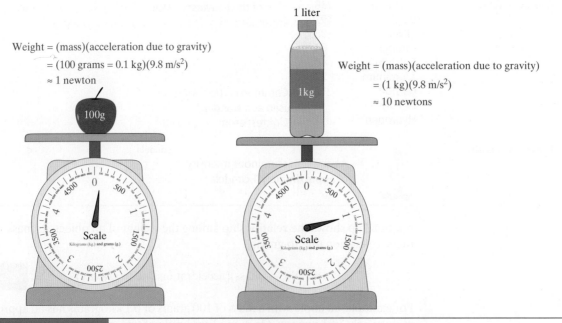

Weight = (mass)(acceleration due to gravity)
   = (100 grams = 0.1 kg)(9.8 m/s²)
   ≈ 1 newton

Weight = (mass)(acceleration due to gravity)
   = (1 kg)(9.8 m/s²)
   ≈ 10 newtons

**FIGURE 2.1**   The relationship between mass and weight.

Africa Studio / Shutterstock.com

This is a good place to say a few words about acceleration due to the Earth's gravity, which has an approximate value of 9.8 m/s² (or 9.81 m/s² to be more exact). To better understand what this value represents, consider a situation where you let go of something from the rooftop of a tall building (see Figure 2.2). If you were to express your observation, you will note the following. At the instant the object is released, it has zero speed. The speed of the object will then increase by 9.8 m/s each second after you release it, resulting in speeds of 9.8 m/s after 1 second, 19.6 m/s after 2 seconds, 29.4 m/s after 3 seconds, and so on. Moreover, when an object changes speed, we say it is accelerating. Weight represents an equivalent force that we must exert to prevent the object from falling or accelerating toward the ground. For example, when you are holding on to a suitcase above the ground, you feel the force that your hand has to apply to prevent the suitcase from accelerating and falling to ground. To better understand the difference between mass and weight, let us next consider the following example.

**EXAMPLE 2.1**

Michael Ransburg / Shutterstock.com

Consider a situation where an exploration vehicle having a mass of 250 kilograms on the Earth (gravity$_{Earth}$ = 9.8 m/s²) is sent to the Moon and planet Mars to explore their surfaces. What is the mass of the vehicle on the Moon where acceleration due to gravity is 1.6 m/s² and on Mars where gravity$_{Mars}$ = 3.7 m/s²? What is the weight of the vehicle on the Earth, on the Moon, and on Mars?

The mass of the vehicle is 250 kg on the Moon and on Mars as well. The mass of the vehicle is always 250 kg, regardless of where it is located. The mass represents the matter that makes up the vehicle; since that does not change, the mass remains constant.

However, the weight of the vehicle varies depending on the gravitational pull of the location. On the Earth, the vehicle will have a weight of

$$\text{weight}_{\text{on Earth}} = (250 \text{ kg})(9.8 \frac{\text{m}}{\text{s}^2}) = 2,450 \text{ N}$$

Whereas on the Moon and Mars, the weight of the vehicle on each is, respectively,

$$\text{weight}_{\text{on Moon}} = (250 \text{ kg})(1.6 \frac{\text{m}}{\text{s}^2}) = 400 \text{ N}$$

$$\text{weight}_{\text{on Mars}} = (250 \text{ kg})(3.7 \frac{\text{m}}{\text{s}^2}) = 925 \text{ N}$$

So as you can see, the vehicle will weigh the least on the surface of the moon, and it would require the least amount of effort to lift it off the moon's surface.

**FIGURE 2.2**    The change in the speed of a falling object as a function of time.

Let us now turn our attention to the SI units for temperature: Celsius and Kelvin. As we explained earlier, thermometers, which are based on thermal expansion or contraction of a fluid, such as alcohol, or a liquid metal, such as mercury, provide a quantitative measure of temperature. As you probably know almost everything will expand and its length increase when you increase its temperature, and it will contract and its length decrease when you decrease its temperature. Most of you have seen a thermometer with a graduated glass rod that is filled with mercury or alcohol. On the *Celsius* scale, under standard atmospheric conditions, the value of zero was arbitrarily assigned to the temperature at which water freezes, and the value of 100 was assigned to the temperature at which water boils. It is important to understand that the numbers were assigned *arbitrarily*. If someone had decided to assign a value of 100 to the ice water temperature and a value of 1000 to boiling water, we would have had a very different type of temperature scale today! In fact, as you will see in the next section, in the U.S. Customary system of units, on a *Fahrenheit* temperature scale, under standard atmospheric conditions, the temperature at which water freezes is assigned a value of 32, and the temperature at which the water boils is assigned a value of 212.

Because both the Celsius and the Fahrenheit scales are arbitrarily defined, scientists recognized a need for a better temperature scale. This need led to the definition of an absolute scale, the *Kelvin* and *Rankine scales*, which are based on the behavior of a perfect gas, where at zero absolute temperature, all molecular activities of the gas will stop.

In SI, the unit of temperature is expressed in degree Celsius (°C) or in terms of absolute temperature Kelvin (K), and the relationship between Celsius and Kelvin is given by

$$\text{temperature (K)} = \text{temperature (°C)} + 273 \qquad \textbf{2.2}$$

In Chapters 3, 4, and 5, we discuss the physical meaning, significance, and relevance of additional SI units, including those used to quantify energy, power, electricity, and heat flow in our daily activities.

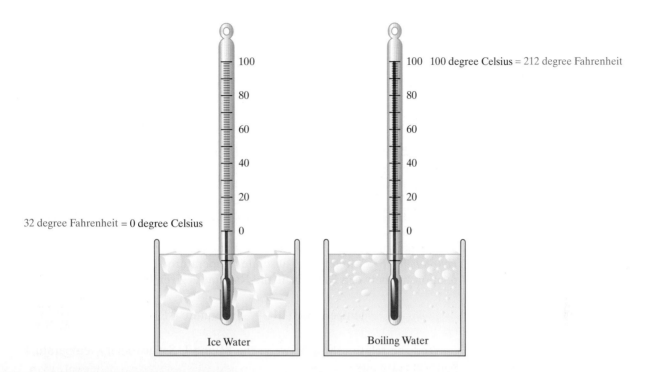

32 degree Fahrenheit = 0 degree Celsius

100 degree Celsius = 212 degree Fahrenheit

Ice Water

Boiling Water

## U.S. Customary System of Units

Feet, pound mass, second, Rankine or Fahrenheit, ampere, mole, and candela are units of length, mass, time, temperature, electric current, amount of substance, and luminous intensity in the U.S. Customary system.

In the United States, a system of units called U.S. Customary is used. In the U.S. Customary system, the unit of length is a *foot* (ft), which is equal to 0.3048 meters; the unit of mass is a *pound mass* (lbm), which is equal to 0.4535 kg; and the unit of time is a *second* (s). The unit of force is a *pound force* (lbf), and 1 lbf is defined as the weight of an object having a mass of 1 lbm at a location where acceleration due to gravity is 32.2 ft/s². One pound force is equal to 4.448 newtons (N).

Because the pound force is not defined using Newton's second law and Equation (2.1), the correction factor $32.2 \dfrac{\text{lbm} \cdot \text{ft}}{\text{lbf} \cdot \text{s}^2}$ must be used with mass in many formulas. The reason is, as mentioned earlier, in the U.S. Customary system, one pound mass is defined as having a weight of one pound force; so if you were to use Equation (2.1) to calculate the weight of one pound mass, you would get

$$\text{weight} = (\text{mass})(\text{acceleration due to gravity})$$

$$\text{weight} = (1 \text{ lbm}) \left( 32.2 \frac{\text{ft}}{\text{s}^2} \right) = 32.2 \frac{\text{lbm} \cdot \text{ft}}{\text{s}^2}$$

According to this equation, one pound mass has a weight of $32.2 \dfrac{\text{lbm} \cdot \text{ft}}{\text{s}^2}$, which numerically is not equal to 1 pound force as initially defined! So to avoid this problem, we must first divide the value of the mass by the correction factor $32.2 \dfrac{\text{lbm} \cdot \text{ft}}{\text{lbf} \cdot \text{s}^2}$ as shown here.

$$\text{weight} = (\text{mass})(\text{acceleration due to gravity})$$

$$\text{weight} = \left( \overbrace{\frac{1 \text{ lbm}}{32.2 \dfrac{\text{lbm} \cdot \text{ft}}{\text{lbf} \cdot \text{s}^2}}}^{\text{mass}} \right) \left( 32.2 \frac{\text{ft}}{\text{s}^2} \right) = 1 \text{ lbf}$$

Now because of the correction factor, the initial definition of one pound mass having a weight of one pound force is upheld, and the distinction between mass (pound mass) and weight (pound force) is also made. Many people find the need for the correction factor confusing. Don't worry too much! Just remember, when using U.S. Customary units, the correction factor given for the mass must be used in many formulas.

The unit of temperature in the U.S. Customary system is expressed in *degree Fahrenheit* (°F) or in terms of absolute temperature *degree Rankine* (°R). The relationship between Fahrenheit and Rankine is given by

$$\text{temperature } (°R) = \text{temperature } (°F) + 460 \qquad \boxed{2.3}$$

And the relationships between the SI and U.S. Customary temperature scales are given by

$$\text{temperature } (°C) = \frac{5}{9}[\text{temperature } (°F) - 32] \qquad \boxed{2.4}$$

or

$$\text{temperature } (°F) = \frac{9}{5}\text{temperature } (°C) + 32 \qquad \boxed{2.5}$$

Next, we look at an example that demonstrates how to use these temperature relationships.

---

**EXAMPLE 2.2**

What is the equivalent value of $T = 50°C$ in degrees Fahrenheit, Rankine, and Kelvin?

To convert the value of temperature ($T$) from degrees Celsius to Fahrenheit, we use Equation (2.5) and substitute the value of 50 for the temperature (°C) variable as shown.

$$\text{temperature } (°F) = \frac{9}{5}\text{temperature } (°C) + 32 = \frac{9}{5}(50) + 32 = 122°F$$

And to convert the result to degree Rankine, we use Equation (2.3):

$$\text{temperature } (°R) = \text{temperature } (°F) + 460 = 122 + 460 = 582°R$$

Finally, to covert the value of $T = 50°C$ to Kelvin, we use Equation (2.2):

$$T(K) = T(°C) + 273 = (50) + 273 = 323 \text{ K}$$

A list of U.S. Customary basic units is given in Table 2.3.

The relationships among magnitudes of length, mass, and temperature in SI and U.S. Customary units are depicted in Figure 2.3. When examining Figure 2.3, note that *1 meter is slightly larger than 3 feet, and 1 kilogram is slightly larger than 2 pounds,* and *every 10 degrees Celsius difference is equal to 18 degrees Fahrenheit difference.* Examples of both SI and U.S. Customary units used in our everyday lives are shown in Table 2.4.

**TABLE 2.3**  **The U.S. Customary Base Units**

| Physical Quantity (Dimension) | | Name of U.S. Customary Base Unit | U.S. Customary Symbol |
|---|---|---|---|
| **Length** | 5 ft – 6.5 ft<br>Range of height for most adults | Foot | ft |
| **Mass** | 110 lbm – 265 lbm<br>Range of mass for most adults | Pound mass | lbm |
| **Time** | Fastest person can run 100 meters in approximately 10 seconds | Second | s |
| **Thermodynamic temperature** | Ice water: 32°F or 492°R<br>Comfortable room temperature: 72°F or 532°R | Rankine | °F or °R |
| **Electric current** | 120 volts  120 watts  Electrical device  1 amp | Ampere | A |
| **Amount of substance** | Uranium 238 ← One of the heaviest atoms known<br>Gold 197<br>Silver 108<br>Copper 64<br>Calcium 40<br>Aluminum 27<br>Carbon 12 ← Common Carbon is used as a standard<br>Helium 4<br>Hydrogen 1 ← Lightest atom | Mole | mol |
| **Luminous intensity** | A candle has luminous intensity of approximately 1 candela | Candela | cd |

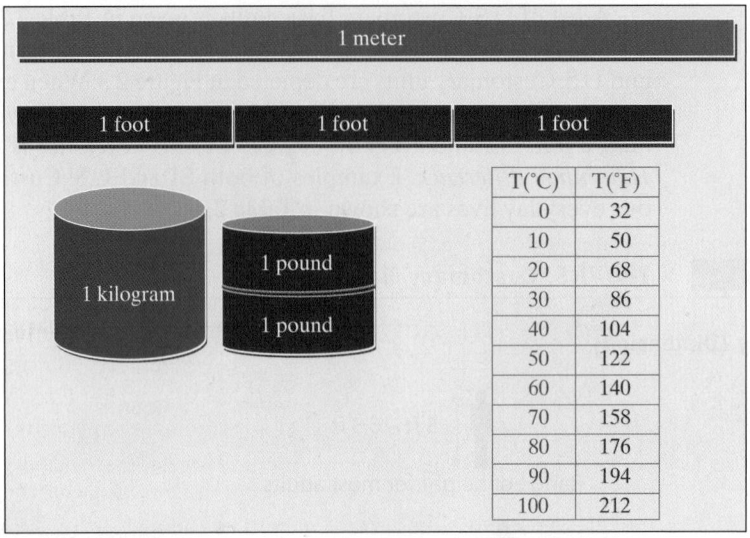

1 meter
1 foot
1 foot
1 foot
1 kilogram
1 pound
1 pound
T(°C) T(°F)
0 32
10 50
20 68
30 86
40 104
50 122
60 140
70 158
80 176
90 194
100 212

**FIGURE 2.3** The relationships among magnitudes of various SI and U.S. Customary units. Note that 1 m is slightly larger than 3 ft, 1 kg is slightly larger than 2 lbm, and every 10°C difference is equal to an 18°F difference.

**TABLE 2.4 Examples of SI and U.S. Customary Units in Everyday Use**

| **Examples of Usage** | **SI Units Used** |
|---|---|
| Medication dose such as pills | 100 mg, 250 mg, or 500 mg |
| Sports: | |
| Swimming | 100 m breaststroke or butterfly stroke |
| Running | 100 m, 200 m, 400 m, 5,000 m, and so on |
| Automobile engine capacity | 2.2 L (liter), 3.8 L (1 liter = 1,000 $cm^3$) |
| Light bulbs | 60 W, 100 W, or 150 W |
| Electric consumption | Kilowatt-hour (kWh) |
| Radio broadcasting: | |
| Signal frequencies | 88–108 MHz (FM broadcast band)<br>0.54–1.6 MHz (AM broadcast band) |
| Police, fire frequencies | 153–159 MHz |
| Global positioning system signals | 1,575.42 MHz and 1,227.60 MHz |
| **Examples of Usage** | **U.S. Customary Units Used** |
| Fuel tank capacity of an automobile | 20 gallons or 2.67 $ft^3$ (1 $ft^3$ = 7.48 gallons) |
| Sports (length of a football field) | 100 yd (1 yard = 3 feet) |
| Power capacity of an automobile | 150 hp (1 hp = 550 lb·ft/s) |
| Distance between two cities | 100 miles (1 mile = 5,280 ft) |

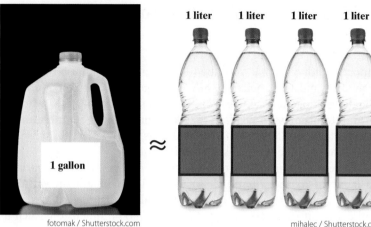

1 liter    1 liter    1 liter    1 liter

≈

1 gallon

fotomak / Shutterstock.com

mihalec / Shutterstock.com

As shown in Table 2.4, a common SI unit for volume is the *liter*, which is equal to 1,000 cm³ (cubic centimeters), and 1,000 liters is equal to 1 cubic meter (i.e., 1,000 liters = 1 m³). Also, note that 1 cubic foot is 7.48 gallons (1 ft³ ≈ 7.5 gallons). Good numbers to remember: a liter of water has a mass of 1 kilogram, and a gallon of water has a mass of approximately 8.3 pounds.

The watt (W) and horsepower (hp) are units of power in SI and U.S. Customary units, respectively, and kilowatt-hour (kWh) is an SI unit of energy. We discuss these units in greater detail in Chapter 3 after we explain the different forms of energy and power. The units of frequency are commonly expressed in kilohertz (kHz), megahertz (MHz), or gigahertz (GHz). Frequency represents the number of cycles per seconds. For example, the alternating electric current in your home is 60 cycles per second or a hertz (Hz). *Alternating current* (ac) is the flow of electric charge that periodically reverses. We discuss this concept in Chapter 4 when we discuss electricity.

## Before You Go On

Answer the following questions to test your understanding of the preceding section:

1. What are the two most common systems of units?

2. What are the base SI units?

3. Name at least three prefixes and symbols of decimal multiples of SI units.

4. What are the units of mass and weight in U.S. Customary units?

5. What do we mean by absolute zero temperature?

*Vocabulary—State the meaning of the following words:*

Absolute zero temperature

Rankine temperature scale

Kelvin temperature scale

newton

## LO³  2.3  Dimensional Homogeneity and Unit Conversion

Another important concept that you need to understand is that all formulas used in any analysis must be *dimensionally homogeneous*. What do we mean by dimensionally homogeneous? Can you, say, add someone's height who is 6 feet (183 cm) tall to his mass of 285 lbm (129 kg) and his body temperature of 98°F (36.7°C); that is, $6 + 285 + 98 = 389$ (or in SI units, $183 + 129 + 36.7 = 348.7$)? Of course not! What are the units of such a calculation?

Therefore, if we were to use the formula $L = a + b + c$ in which the variable $L$ on the left-hand side of the equation has a dimension of length, the variables $a$, $b$, and $c$ on the right-hand side of the equation also must have dimensions of length. Otherwise, if variables $a$, $b$, and $c$ had *different* dimensions such as length, mass, and temperature, respectively, the given formula would be inhomogeneous, which would be like adding someone's height to his mass and body temperature!

Now you know why it is important to check for dimensional homogeneity in a formula when performing an analysis. It is equally important to check for the consistency of units in an analysis. Some of you may recall that not too long ago NASA lost a spacecraft called the Mars Climate Orbiter because two groups of engineers working on the project neglected to communicate correctly their calculations with appropriate units. According to an internal review conducted by NASA's Jet Propulsion Laboratory, "a failure to recognize and correct an error in a transfer of information between the Mars Climate Orbiter spacecraft team in Colorado and the mission navigation team in California led to the loss of the spacecraft." The peer review findings indicated that one team used U.S. Customary units (e.g., foot and pound) while the other used SI units (e.g., meter and kilogram) for a key spacecraft operation. According to NASA, the information exchanged between the teams was critical to the maneuvers required to place the spacecraft in the proper Mars orbit. The waste of our taxes could have been averted easily. As you can see, when performing analysis, a need to convert from one system of units to another arises. You don't have to become an engineer to learn to convert information from one system of units to another correctly. Given today's global economy, you may end up working in Europe or Asia and as a result find yourself in a situation where you need to convert some data from,

+            +            = ?

| TABLE 2.5 | Systems of Units and Conversion Factors | | |
|---|---|---|---|
| **Dimension** | **System of Units** | | **Conversion Factors** |
| | **SI** | **U.S. Customary** | |
| Length | meter (m) | foot (ft) | 1 m ≈ 3.3 ft |
| | | | 1 ft ≈ 0.3 m |
| Time | second (s) | second (s) | none |
| Mass | kilogram (kg) | pound mass (lbm) | 1 kg ≈ 2.2 lbm |
| | | | 1 lbm ≈ 0.45 kg |
| Force | newton (N) $1\,N \approx (1\,kg)\left(1\,\dfrac{m}{s^2}\right)$ | One pound mass* weighs one pound force at sea level | 1 N ≈ 0.225 lbf 1 lbf ≈ 4.45 N |
| Temperature | degree Celsius (°C) or Kelvin (K) $K = °C + 273$ | degree Fahrenheit (°F) or degree Rankine (°R) $°R = °F + 460$ | |

*The relationship between pound mass and pound force is not defined using Newton's second law.

say, feet to meters or pound mass to kilograms. The conversion factors for units commonly encountered in daily life are shown in Table 2.5. Example 2.3 shows the steps that you need to take to convert from one system of units to another.

**EXAMPLE 2.3**

As we mentioned earlier, with today's global economy, you could travel to or end up working in Europe or Asia and find yourself in a situation where you need to convert some data from, say, feet to meters or pound mass to kilograms. In this example, we show the steps that you need to take to convert information from U.S. Customary to metric units. Make sure to study them carefully, as these steps become important in subsequent chapters.

Now, consider a person who is 6 feet and 3 inches tall and weighs 185 pounds (lbf) driving a car at a speed of 65 miles per hour over a distance of 25 miles between two cities. The outside air temperature is 80°F. Let us now convert all of the values given in this example from U.S. Customary to SI units.

The steps to convert the person's height from feet and inches to meters and centimeters are explained below.

Karramba Production / Shutterstock.com

$$\text{height} = \underbrace{\left(6\text{ ft} + \underbrace{(3\text{ in.})\underbrace{\left(\frac{1\text{ ft}}{12\text{ in.}}\right)}_{\text{step 1}}}_{\text{step 2}}\right)\left(\frac{0.3048\text{ m}}{1\text{ ft}}\right)}_{\text{step 3}} = 1.905\text{ m}$$

or

$$\overbrace{\text{height} = (1.905 \text{ m})\left(\frac{100 \text{ cm}}{1 \text{ m}}\right)}^{\text{step 4}} = 190.5 \text{ cm}$$

**Step 1.** Start with converting the 3 inch value into feet by realizing that 1 foot is equal to 12 inches. The expression $\left(\dfrac{1 \text{ ft}}{12 \text{ in.}}\right)$ conveys the same fact, except when you write it in fraction form and multiply it by the 3 inch value as

$$(3 \cancel{\text{ in.}})\left(\frac{1 \text{ ft}}{12 \cancel{\text{ in.}}}\right)$$

the inch units in the numerator and denominator cancel out, and the 3 inch value is now represented in feet.

**Step 2.** Add the results of step 1 to 6 feet.

**Step 3.** Multiply the results of step 2 by $\left(\dfrac{0.3048 \text{ m}}{1 \text{ ft}}\right)$, because 1 foot (ft) is equal to 0.3048 meters, and the foot units in the numerator and denominator also cancel out. This step leads to the person's height being expressed in meters as

$$(6 \cancel{\text{ ft}}) + (1 \cancel{\text{ in.}})\left(\frac{1 \cancel{\text{ ft}}}{12 \cancel{\text{ in.}}}\right)\left(\frac{0.3048 \text{ m}}{1 \cancel{\text{ ft}}}\right)$$

**Step 4.** To convert the result of step 3 from meters to centimeters, we multiply 1.854 $\cancel{\text{m}}$ by $\left(\dfrac{100 \text{ cm}}{1 \cancel{\text{ m}}}\right)$, because 1 meter is equal to 100 cm, and this step cancels out the meter in the numerator and denominator.

The step to convert the person's weight from pound force to newtons is shown below.

$$\overbrace{\text{weight} = (185 \text{ lbf})\left(\frac{4.448 \text{ N}}{1 \text{ lbf}}\right)}^{\text{step 5}} = 822.8 \text{ N}$$

**Step 5.** To convert the person's weight, multiply the 185 pound force value by $\left(\dfrac{4.448 \text{ N}}{1 \text{ lbf}}\right)$, because 1 pound force (lbf) is equal to 4.448 newtons (N). This leads to pound force units in the numerator and denominator canceling out and the person's weight being expressed in newtons as $(185 \cancel{\text{ lbf}})\left(\dfrac{4.448 \text{ N}}{1 \cancel{\text{ lbf}}}\right)$.

The steps to convert the speed of the car from miles per hour to kilometers per hour are

$$
\text{speed} = \overbrace{\overbrace{\overbrace{\left(65\,\frac{\text{miles}}{\text{h}}\right)\left(\frac{5,280\ \text{ft}}{1\ \text{mile}}\right)}^{\text{step 6}}\left(\frac{0.3048\ \text{m}}{1\ \text{ft}}\right)}^{\text{step 7}}\left(\frac{1\ \text{km}}{1,000\ \text{m}}\right)}^{\text{step 8}} = 104.6\,\frac{\text{km}}{\text{h}}
$$

**Step 6.** To convert the speed of the car from 65 miles per hour to kilometers per hour, start by converting the 65 miles value to feet; Since 1 mile is equal to 5,280 feet, multiply the 65 miles by 5,280. $\left(65\,\frac{\cancel{\text{miles}}}{\text{h}}\right)\left(\frac{5,280\ \text{ft}}{1\ \cancel{\text{mile}}}\right) = \left((65)(5,280)\,\frac{\text{ft}}{\text{h}}\right)$. This step cancels out the miles units in the numerator and denominator and results in the speed value being represented in feet per hour (ft/h).

**Step 7.** Next, multiply the results of step 6 by $\left(\frac{0.3048\ \text{m}}{1\ \text{ft}}\right)$, because 1 foot is equal to 0.3048 meters. This step cancels out the foot units in the numerator and denominator and leads to

$$
\left(65\,\frac{\cancel{\text{miles}}}{\text{h}}\right)\left(\frac{5,280\ \cancel{\text{ft}}}{1\ \cancel{\text{miles}}}\right)\left(\frac{0.3048\ \text{m}}{1\ \cancel{\text{ft}}}\right) = 104,607\ \text{m/h}
$$

**Step 8.** To convert the result of step 7 from meters per hour (m/h) to kilometers per hour (km/h), note that 1 kilometer is equal to 1,000 meters, and multiply $\left(104,607\,\frac{\cancel{\text{m}}}{\text{h}}\right)$ by $\left(\frac{1\ \text{km}}{1,000\ \cancel{\text{m}}}\right)$ to cancel out the meter unit in the numerator and denominator. The speed of the car is now expressed in kilometers per hour (km/h).

The steps to convert the distance traveled between two cities from miles to kilometers are similar to steps discussed previously.

$$
\text{distance} = \overbrace{\overbrace{\overbrace{(25\ \text{miles})\left(\frac{5,280\ \text{ft}}{1\ \text{mile}}\right)}^{\text{step 9}}\left(\frac{0.3048\ \text{m}}{1\ \text{ft}}\right)}^{\text{step 10}}\left(\frac{1\ \text{km}}{1,000\ \text{m}}\right)}^{\text{step 11}} = 40.2\ \text{km}
$$

**Step 9.** Convert the miles to feet by multiplying $(25\ \cancel{\text{miles}})\left(\frac{5,280\ \text{ft}}{1\ \cancel{\text{mile}}}\right)$.

**Step 10.** Convert the feet to meters by $(25\ \cancel{\text{miles}})\left(\frac{5,280\ \cancel{\text{ft}}}{1\ \cancel{\text{mile}}}\right)\left(\frac{0.3048\ \text{m}}{1\ \cancel{\text{ft}}}\right)$

**Step 11.** Convert the meters to kilometers by

$$
(25\ \cancel{\text{miles}})\left(\frac{5,280\ \cancel{\text{ft}}}{1\ \cancel{\text{mile}}}\right)\left(\frac{0.3048\ \cancel{\text{m}}}{1\ \cancel{\text{ft}}}\right)\left(\frac{1\ \text{km}}{1,000\ \cancel{\text{m}}}\right)
$$

To convert the air temperature from degrees Fahrenheit to Celsius, we substitute for $T(°F)$ the value 80 in Equation (2.3).

$$T(°C) = \frac{5}{9}[T(°F) - 32]$$

$$T(°C) = \frac{5}{9}[80 - 32] = 26.7°C$$

---

**EXAMPLE 2.4**

*"I'm not overweight. I'm just nine inches too short."*

—SHELLEY WINTERS

You don't have to lie about your mass! For those of us who might be slightly massive (or as commonly said, overweight), it might be wiser to express our mass in kilograms rather than in pound mass.

For example, a person who has a body mass of 150 pound mass (lbm) would sound skinny if he/she were instead to convert this value and express his/her body mass in kilograms (kg).

$$(150 \text{ lbm})\left(\frac{1 \text{ kg}}{2.2 \text{ lbm}}\right) = (150 \text{ lbm})\left(\frac{1 \text{ kg}}{2.2 \text{ lbm}}\right) = 68 \text{ kg}$$

To convert the mass from pound mass (lbm) to kilograms (kg), we note that 1 kg is equal to 2.2 lbm, and to obtain the result in kilograms, multiply the 150 lbm by the conversion factor of $\frac{1 \text{ kg}}{2.2 \text{ lbm}}$, which reads 1 kg is equal to 2.2 lbm. This step cancels out the pound mass units in the numerator and denominator as shown.

As you can see from the result, 150 lbm is equal to 68 kg, and therefore, he/she is telling the truth about his/her body mass. So you don't have to lie about your body mass; knowledge of units can bring about instant results without any exercise or diet.

---

## Before You Go On

Answer the following questions to test your understanding of the preceding section:

1. Why is it important to know how to convert from one system of units to another?

2. What do we mean by dimensional homogeneity? Give an example.

3. Show the steps that you would take to convert your height from feet and inches to meters and centimeters.

4. Show the steps that you would take to convert your weight from pound force to newtons.

***Vocabulary—State the meaning of the following words:***

Dimensional homogeneity

Unit conversion

# LO⁴ 2.4 Components and Systems

Every product is considered a system that serves a purpose. A system is made up of smaller parts called components.

As good global citizens, when purchasing a product, you should be mindful of the entire life cycle of the product. Think about the natural resources that were used to make the product and how much energy it took to produce, transport, use, and eventually dispose of it. As is the case with any new areas you explore, the concepts of energy, environment, and sustainability have their own terminologies. Make sure you spend a little time familiarizing yourself with these terms so you can follow the concepts later. Let us now focus on some terms that you may have heard but do not fully understand. For example, what do we mean by a system and its components?

Every product that you own or will purchase someday is considered a *system* and is made of *components*. Let us start with a simple example to demonstrate what we mean by a system and its components. Most of us own a winter coat, which can be looked at as a system. First, note that the coat serves a purpose. Its primary function is to offer additional insulation so that our body heat does not escape as quickly and as freely as it would without protective covering.

The coat may be divided into smaller components: the fabric comprising the main body of the coat, insulating material, a liner, threads, zipper(s), and buttons. Moreover, each component may be further subdivided into smaller components. For example, the main body of the jacket may be divided into sleeves, a collar, pockets, the chest section, and the back section (see Figure 2.4). Each component serves a purpose: The pockets were designed to hold things, the sleeves cover our arms, and so on. The main function of the zipper is to allow us to open and close the front of the jacket freely. It too consists of smaller components. Hence, a well-designed coat not only looks appealing to the eyes, but also has functional components, keeps us warm during the winter, and is made of sustainable materials.

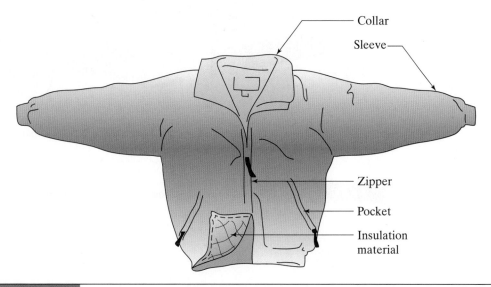

**FIGURE 2.4** A simple system and its components.

Other systems are similar to a winter coat. Any given product (system) can be divided into smaller, manageable subsystems, and each subsystem can be further divided into smaller and smaller components. The components of a well-designed system should function and fit well together so that the primary purpose of the product is attained. Let us consider another example. The primary function of a car is to move us from one place to another in a reasonable amount of time. The car must provide a comfortable area for us to sit in. Furthermore, it must provide some protection from the outside elements, such as harsh weather and harmful objects.

The automobile consists of thousands of parts. When viewed in its entirety, it is a complicated system; however, it may be divided into major subsystems, such as an electrical system, body, chassis, power train, and a heating and/or air conditioning unit (see Figure 2.5). Each major subsystem can be further subdivided into smaller components. For example, the main body of the car consists of doors, hinges, locks, windows, and so on; the electrical system of a car consists of a battery, a starter, an alternator, wiring, lights, switches, radio, microprocessors, and so on.

The next time you purchase a product, think of it in terms of a system and its components. Again, be mindful of the entire life cycle of the product. Ask yourself, "What natural resources were used to make the product?", "How much energy did it take to produce and transport the product?", and "How much energy would it take to use the product and eventually dispose of it?" We discuss the concepts of sustainability and life-cycle analysis in greater detail in Chapter 12.

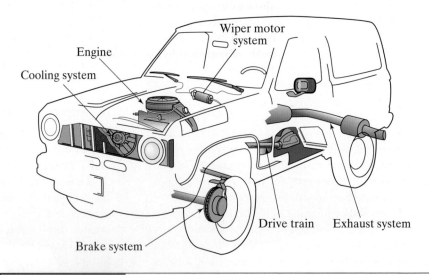

**FIGURE 2.5**     A system and some of its main components.

## Before You Go On

Answer the following questions to test your understanding of the preceding section:

1.  What is the difference between a component and a system?
2.  What are the major components of a building?
3.  How would you define the major components of a supermarket?
4.  How would you define the major components in your college?

*Vocabulary — State the meaning of the following words:*

Component

System

## SUMMARY

### LO¹ Fundamental Dimensions and Units

By now, you should understand the importance of fundamental dimensions in everyday life, and why—as a good global citizen—you should develop a good grasp of them. As people, we have realized that we need only a few physical dimensions or quantities to describe our surroundings and daily events. For example, we need a length dimension to describe how tall, how long, or how wide something is. Time is another physical dimension that we need to answer questions such as, "How old are you?" or "How long does it take to go from here to there?" You should also know that today, based on what we know about our world, we need seven fundamental dimensions to correctly express ourselves in our surroundings. They are length, mass, time, temperature, electric current, amount of substance, and luminous intensity. The other important concept that you should know is that not only do we need to define these physical dimensions to describe our surroundings, but we also need some way to scale or divide them into units. For example, the time dimension can be divided into both small and large portions, such as seconds, minutes, hours, days, months, years, etc.

You should also realize that physical laws are based on observation and experimentation. We have learned through observation and by the collective effort of those before us that things work a certain way in nature. For example, if you let go of something that you are holding in your hand, it will fall to the ground. This is an observation that we all agree upon. We can use words to explain our observations or use another language, such as mathematics and formulas, to express our findings. Sir Isaac Newton (1642–1727) formulated that observation into a useful mathematical expression that we know as the universal law of gravitational attraction.

### LO² System of Units

The SI system of units (*from French: Systéme International d' Unites*) is the most common system of units used in the world. You should be familiar with these units of length (meter), time (second), mass (kilogram), temperature (Kelvin or degree Celsius), electric current (ampere), amount of substance (mole), and luminous intensity (candela). You also should have a good feeling for what these units represent (for example, how much a kilogram is) and appreciate their importance in your daily life. SI units also make use of a series of prefixes and symbols of decimal multiples such as mega, giga, kilo, etc., to expand on their representation.

The U.S. Customary System of units is used only in the United States. You should also be familiar with the U.S. Customary units of length (feet), time (second), mass (pound mass), temperature (degree Rankine or degree Fahrenheit), electric current (ampere), amount of substance (mole), and luminous intensity (candela). You also should have a good feeling for what these units represent. The U.S. Customary units also make use of some of the SI prefixes and symbols of decimal multiples such as mega, giga, etc.

### LO³ Dimensional Homogeneity and Unit Conversion

You should know what we mean when we say an equation must be dimensionally homogeneous. For example, you already know that you cannot add someone's height of 6 feet (183 cm) to his mass of 285 lbm (129 kg) and his body temperature of 98°F (36.7°C); that is, 6 + 285 + 98 = 389 (or in SI units, 183 + 129 + 36.7 = 348.7). What would be the result of such a calculation? Therefore, if you were to use the formula $L = a + b + c$, in which the variable $L$ on the left-hand side of the equation has a dimension of length, then the variables $a$, $b$, and $c$ on the right-hand side of the equation must also have dimensions of length. Common sense! It is also important (and useful) to know how to convert values from one system of units to another. For example, you should be able to convert SI data given in meters or kilograms to U.S. Customary units of feet and pound mass and vice versa.

### LO⁴ Components and Systems

Every product that you own or will purchase some day is considered a system and is made of components. The next time you purchase a product, think of it in terms of a system and its components, and be mindful of the entire life cycle of the product. Could the components of the system be recycled and used for another purpose?

## KEY TERMS

| | | |
|---|---|---|
| Ampere 44 | Giga 44 | Nano 44 |
| Candela 44 | Kelvin 48 | Physical Law 41 |
| Celsius 48 | Kilogram 44 | Pound Force 49 |
| Component 59 | Mass 39 | Pound Mass 49 |
| Dimension 38 | Mega 44 | Rankine 48 |
| Fahrenheit 48 | Meter 44 | System 59 |
| Feet 49 | Micro 44 | Tera 44 |
| Force 39 | Mole 44 | Unit 40 |

## Apply *What You Have Learned*

You are planning a trip to Europe. In order to prepare yourself for your visit, you need to convert the following data from U.S. Customary units to SI units: your height from feet and inches to meters and centimeters; your mass from pound mass to kilograms; your desired room-temperature thermostat setting from Fahrenheit to Celsius; one-half gallon of drinking water to liters; fifteen gallons of gasoline to liters; and speed limits of 30, 40, 50, and 60 from miles per hour to kilometers per hour. If you reside outside the U.S. and are planning a trip to the U.S., convert your data from SI to U.S. Customary units.

Oleksiy Mark / Shutterstock.com

# PROBLEMS

*Problems that promote life-long learning are denoted by* 🔑

**2.1** Convert the information given in the accompanying table from SI units to U.S. Customary units. Show all steps of your solutions. See Example 2.3.

| Convert from SI Units | To U.S. Customary Units |
|---|---|
| 120 km/h | miles/h and ft/s |
| 100 m³ | ft³ |
| 80 kg | lbm |
| 900 N | lbf |
| 9.81 m/s² | ft/s² |

**2.2** Convert the information given in the accompanying table from U.S. Customary to SI units. Show all steps of your solutions. See Example 2.3.

| Convert from U.S. Customary Units | To SI Units |
|---|---|
| 65 miles/h | km/hr and m/s |
| 120 lbm/ft³ | kg/m³ |
| 200 lbm | kg |
| 200 lbf | N |

**2.3** Convert your age from years, months, weeks, and days to hours. How old are you in hours?

**2.4** A house has a given floor space of 2,000 ft². Convert this area to m².

**2.5** Calculate the volume of water in a large swimming pool with dimensions of 50 m × 25 m × 2 m. Express your answer in liters, m³, gallons, and ft³.

**2.6** A 500 sheet ream of copy paper has thickness of 2.25 in. What is the average thickness of each sheet in mm?

**2.7** A barrel can hold 42 gallons of oil. How many liters of oil are in the barrel?

**2.8** Express the kinetic energy [½(mass)(speed)²] of a car with a mass of 1,200 kg moving at a speed of 100 km/h. First, you need to convert the speed from km/h to the fundamental units of m/s. Show the conversion steps. (*Note*: We explain the concept of kinetic energy in Chapter 3.)

**2.9** A machine shop has a rectangular floor shape with dimensions of 30 ft by 50 ft. Express the area of the floor in ft², m², in², and cm². Show the conversion steps.

**2.10** A trunk of a car has a listed luggage capacity of 18 ft³. Express the capacity in in³, m³, and cm³. Show the conversion steps.

**2.11** An automobile has a 3.5 liter engine. Express the engine size in in³. Show the conversion steps. Note that 1 liter is equal to 1,000 cm³.

**2.12** The density of air that we breathe at standard room conditions is 1.2 kg/m³. Express the density in U.S. Customary units. Show the conversion steps.

**2.13** On a summer day in Phoenix, Arizona, the inside room temperature is maintained at 68°F while the outdoor air temperature is a sizzling 110°F. What is the outdoor–indoor temperature difference in degrees (a) Fahrenheit or (b) Celsius?

**2.14** A person who is 180 cm tall and weighs 750 newtons is driving a car at a speed of 90 kilometers per hour over a distance of 80 kilometers. The outside air temperature is 30°C and has a density of 1.2 kg/m³. Convert all of the values given from SI to U.S. Customary units.

**2.15** Convert the given values: (a) area $A = 16$ in² to ft² and (b) volume $V = 64$ in³ to ft³.

**2.16** The acceleration due to gravity $g$ is 9.81 m/s². Express the value of $g$ in U.S. Customary units. Show all conversion steps.

**2.17**  Atmospheric pressure is the weight of the column of air over an area. For example, under standard conditions, the atmospheric pressure is 14.7 lbf/in². This value means that the column of air in the atmosphere above a surface with an area of 1 in² will exert a force of 14.7 lbf. Convert the atmospheric pressure in the given units to the requested units:

(a) 14.7 lbf/in² to lbf/ft²,

(b) 14.7 lbf/in² to Pa,

(c) 14.7 lbf/in² to kPa, and

(d) 14.7 lbf/in² to bars.

Show all of the conversion steps. [*Note*: One Pascal (1 Pa) is equal to one newton per meter squared (1 Pa = 1 N/m²) and 1 bar = 100 kPa.]

**2.18**  The density of water is 1,000 kg/m³. Express the density of water in lbm/ft³ and lbm/gallon. (*Note*: 7.48 gallons = 1 ft³.)

**2.19**  Look up the given U.S. Customary specifications (body size, trunk size, engine size, and gas consumption) for a car of your choice and convert your findings to SI units. If you live outside of the United States, convert the data from SI to U.S. Customary units.

**2.20**  Look up the U.S. specifications (capacity and dimensions) for a home appliance such as a clothes washer. Convert your findings to SI units. If you live outside of the United States, convert the data from SI to U.S. Customary units.

Everett Historical / Scutterstock.com

*"If I have seen farther than others, it is because I have stood on the shoulders of giants."*

— SIR ISAAC NEWTON (1642–1727)

# Energy

# 2

In Part Two, we introduce you to the basics of conventional and renewable energy, its sources and production, as well as energy consumption rates in homes, buildings, transportation, food production, and manufacturing. We need energy to build shelter, cultivate and process food, make goods, and to maintain our living places at comfortable temperatures. To quantify the requirements to build things, to move or lift objects, or to heat or cool buildings, energy is defined and classified into different categories. In this part of the book, you will learn what we mean by energy and power and their common units, including Joule, pound-foot, Btu, kilowatt-hour, kilowatt, and horsepower. We also discuss the basic concepts of electricity and electric power production. We cover residential power consumption, particularly lighting systems, because lighting accounts for a major portion of electricity use in buildings; lighting systems have received much attention recently due to energy and sustainability concerns. We also explain the fundamentals of heat transfer, as well as heat loss and gain in buildings. Space heating and air conditioning account for nearly fifty percent of energy use in homes in the United States. A comprehensive coverage of energy sources such as gasoline, natural gas, coal, and wood, along with their consumption rates, is also provided. We discuss in detail how much energy we consume in homes, buildings, transportation, and manufacturing sectors. We also explain the basic concepts related to solar energy, wind energy, and hydro-energy.

# Energy and Power

## LEARNING OBJECTIVES

**LO¹**   Energy and Work: describe what we mean by energy and work

**LO²**   Forms of Energy: describe how we quantify what it takes to move things (kinetic energy), to lift things (potential energy), and to heat or cool things (thermal energy)

**LO³**   Difference between Energy and Power: explain the difference between energy and power

**LO⁴**   Energy Contents (Heating Values) of Fuels: explain what energy contents of fuel represent

# *Discussion Starter*

We quantify and express our energy consumption rates using different units, including the kilowatt-hour, Btu, and Calorie. For example, you will learn in this chapter that one kilowatt-hour represents the amount of energy consumed during 1 hour by a device that uses 1,000 watts (W) or one kilowatt (kW) of electric power and that one British thermal unit (Btu) represents the amount of thermal energy needed to raise the temperature of one pound mass (lbm) of water by one degree Fahrenheit (°F). You may also recall from our discussion in Chapter 1 that the energy content of food is typically expressed in Calories (with an uppercase C). For example, a banana has about 100 Calories, whereas a medium serving of French fries has around 400 Calories. One Calorie is equal to 1,000 calories (with a lowercase c); and one calorie is defined as the amount of energy required to raise the temperature of one gram of water by one degree Celsius (°C).

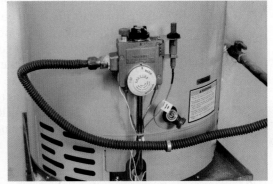

Jo Ann Snover / Shutterstock.com

Jim Barber / Shutterstock.com

monticello / Shutterstock.com

**To the Students:** What is energy? What is power? What is the difference between energy and power? What do you think your annual energy consumption is in terms of kilowatt-hours, Btu, and Calories?

## LO¹ 3.1 **Energy**

Energy, because of its importance in our daily lives, is a concept that every good citizen, regardless of his or her area of interest, should know. Think about your everyday activities and what it takes to maintain your lifestyle today. For example, can you use your phone indefinitely without charging it? Can you run your car forever without putting some form of fuel in it? Or, can you keep your house warm in the winter months with a heating unit that won't require some form of energy input? During the past few decades, much has been said about energy consumption rates, greenhouse gases, and climate change and why we should adopt policies that promote energy savings and make use of more green energy. But what is energy?

*Energy* is one of those abstract terms that you already have a good feel for. We need energy to do things. For instance, you already know that we need energy to make various products and to operate them. We need energy to cultivate and process food. We need energy to construct buildings such as houses, apartments, malls, supermarkets, schools, and hospitals. We also need energy to maintain our living and working places at comfortable temperatures. You already know this much; however, what you may not know is that energy has different forms. In order to better explain the requirements to make things like a smart phone, to build structures like a house, to move things like a car, to lift things like an elevator, or to heat or cool a home, we define and classify energy into different categories such as *kinetic energy, potential energy*, and *thermal energy*. For example, in order to quantify how much energy it takes to move something, we make use of the concept called kinetic energy; to determine what it takes to lift an elevator or an escalator, we use potential energy; or to calculate how much fuel such as natural gas it takes to keep our homes warm during the winter months, we use the concept of thermal energy.

**We need energy to do things.**

Nataliya Hora / Shutterstock.com

CHAIYA / Shutterstock.com

Orientaly / Shutterstock.com

Lee Prince / Shutterstock.com

Another idea that you are familiar with is that in order perform a task, you have to do the *work*! But, what do we mean by the term *work*? Work could mean different things to different people. For example, it could mean effort, labor, job, occupation, or employment. However, in a physics context, ***work*** is performed when *a force moves an object through a distance*. That is,

$$\text{work} = (\text{force})(\text{distance})$$

**3.1**

We explained the concept of force in the previous chapter. Recall that the simplest form of a force that represents the interaction of two objects is a push or a pull. When you push on a lawnmower, that interaction between your hand and the lawnmower is called *force*. When an automobile pulls a trailer, a force is exerted by the bumper hitch on the trailer. Also, note that, symbolically, a force is represented by an arrow. Therefore, when you push on a lawnmower and move it through a distance, you perform work, or when an automobile pulls a trailer and moves it through a distance, work is performed. The SI unit for work is called ***joule*** (J), and one joule represents the work done by a force with a magnitude of one newton (1 N) acting through a distance of one meter (1 m); then from Equation (3.1): 1 J = (1 N)(1 m). The U.S. Customary unit for work is pound-force-ft (lbf · ft) and is obtained by substituting pound force (lbf) for the unit of force and foot (ft) for the unit of distance in Equation (3.1); that is, 1 lbf · ft = (1 lbf)(1 ft). Moreover, the relationship between the magnitude of joule (J) and pound-force-ft (lbf · ft) is given by: 1 J ≈ 0.74 lbf · ft or 1 lbf · ft ≈ 1.4 J.

An additional concept that you need to clearly understand is that we need to *spend energy* to do the work. For example, if you don't eat, you won't have the energy to push the lawnmower. You may recall from your high school education that metabolism is the biochemical process by which our bodies convert food and drinks (fuel) into energy. You may also recall that, even if we are resting and watching TV or sleeping, we still need energy for basic body functions such as blood circulation and breathing. For the example of an automobile pulling a trailer, if you do not have gasoline (fuel) in the car to burn to convert it into energy and subsequently to do work and move the car, you cannot pull the trailer. As such, we need to expend energy to do the work.

We need to spend energy to do the work.

ARENA Creative / Shutterstock.com

Robert J. Beyers II / Shutterstock.com

## Before You Go On

Answer the following questions to test your understanding of the preceding section:

1. Name three forms of energy.

2. In your own words, explain what is meant by force.

3. In your own words, explain what is meant by work.

4. What is the difference between work and energy?

***Vocabulary—State the meaning of the following words:***

Force

Work

Energy

# LO² 3.2 Forms of Energy

As mentioned previously, to quantify the requirements to move objects such as our cars, to lift things like an elevator, or to heat or cool our homes, energy is classified into different categories such as *kinetic energy, potential energy,* and *thermal energy.* Let's now look at each form of energy in more detail.

## Kinetic Energy—What It Takes to Move Things

***Kinetic energy*** is a way by which we quantify how much energy is required to move something. At one time or another in your life, you have pushed on an object to change its position. Therefore, intuitively, you understand that the amount of work (or energy required) to move something depends on two things: (1) the mass of the object and (2) how fast you want to move it. The bigger the mass of the object, the harder it is to move the object. An object having a known mass $m$ and moving with a given speed $V$ has *kinetic energy*, which is equal to:

Kinetic energy quantifies the amount of energy required to move something.

$$\text{Kinetic energy} = \left(\frac{1}{2}\right)(\text{mass})(\text{speed})^2 = \left(\frac{1}{2}\right)mV^2 \qquad \text{3.2}$$

The SI unit for kinetic energy is the joule (J), which is obtained by substituting kilogram (kg) for the unit of *mass* and meter per second (m/s) for the units of *speed*, as shown next. Also, note that the one-half factor in the kinetic energy equation has no units.

$$\text{kinetic energy} = \left(\frac{1}{2}\right)(\text{kg})(\text{m/s})^2 = \left(\frac{1}{2}\right)(\text{kg})(\text{m}^2/\text{s}^2)$$

recall from Chapter 2 that $1 \text{ N} = (1 \text{kg})\left(1\frac{\text{m}}{\text{s}^2}\right)$

$$= \underbrace{\left(\frac{1}{2}\right)}_{\text{no units}} \overbrace{(\text{kg})\left(\frac{\text{m}}{\text{s}^2}\right)} (\text{m}) = \text{N} \cdot \text{m} = \text{joule} = \text{J}$$

The U.S. Customary unit for kinetic energy is the pound force-foot (lbf · ft). As shown next, note that you must first divide the value of the mass by a correction factor of $32.2\frac{\text{lbm} \cdot \text{ft}}{\text{lbf} \cdot \text{s}^2}$. As we mentioned in Chapter 2, this correction factor is needed, because in U.S. Customary units the relationship between mass and weight is not defined using Newton's law. Consequently, a correction factor must be introduced to make the distinction between mass (pound mass) and weight (pound force).

$$\text{kinetic energy} = \left(\frac{1}{2}\right)\left(\overbrace{\frac{\text{lbm}}{32.2\frac{\text{lbm} \cdot \text{ft}}{\text{lbf} \cdot \text{s}^2}}}^{\text{mass}}\right)(\text{ft/s})^2 = \left(\frac{1}{2}\right)\left(\overbrace{\frac{\cancel{\text{lbm}}}{32.2\frac{\cancel{\text{lbm}} \cdot \text{ft}}{\text{lbf} \cdot \cancel{\text{s}^2}}}}^{\text{mass}}\right)\left(\frac{\text{ft}^2}{\cancel{\text{s}^2}}\right)$$

$$= \underbrace{\left(\frac{1}{2}\right)}_{\text{no units}}\left(\frac{\text{lbf}}{\cancel{\text{ft}}}\right)(\text{ft}^{\cancel{2}}) = \text{lbf} \cdot \text{ft}$$

---

**EXAMPLE 3.1**

Pete Saloutos / Shutterstock.com

robert_s / Shutterstock.com

Senohrabek / Shutterstock.com

To develop a feel for what the magnitudes of kinetic energy represent, consider the following situations.

(a) A sprinter with a mass of 80 kilograms (kg) running at a peak speed of 10 meters/second (m/s):

$$\begin{matrix}\text{kinetic} \\ \text{energy}\end{matrix} = \left(\frac{1}{2}\right)(\text{mass})(\text{speed})^2 = \left(\frac{1}{2}\right)(80 \text{ kg})(10 \text{ m/s})^2 = 4{,}000 \text{ J}$$

(b) A car with a mass of 1,000 kilograms (kg) moving at a speed of 100 kilometers/hour (km/hr) or 27.8 meters/second (m/s):

$$\begin{matrix}\text{kinetic} \\ \text{energy}\end{matrix} = \left(\frac{1}{2}\right)(\text{mass})(\text{speed})^2 = \left(\frac{1}{2}\right)(1{,}000 \text{ kg})(27.8 \text{ m/s})^2$$

$$= 386{,}420 \text{ J}$$

When comparing the kinetic energy values for the sprinter and the car, note that the car has a kinetic energy that is almost 100 times greater than that of the sprinter.

(c) A commercial plane with a mass of 180,000 kilograms (kg) flying at a speed of 900 kilometers/hour (km/h) or 250 meters/second (m/s):

$$\begin{matrix}\text{kinetic} \\ \text{energy}\end{matrix} = \left(\frac{1}{2}\right)(\text{mass})(\text{speed})^2 = \left(\frac{1}{2}\right)(180{,}000 \text{ kg})(250 \text{ m/s})^2$$

$$= 5{,}625{,}000{,}000 \text{ J}$$

Compared to the car, the plane has a kinetic energy that is almost 15,000 times greater!

When we do work on an object, we change its kinetic energy.

Let's now think more carefully about how *kinetic energy* is related to *work*. When you apply a force by pulling or pushing on an object and moving through a distance, you perform work; and when *you do work* on an object, you change its speed and *kinetic energy*.

For example, when you push on a lawn mower that is initially at rest, you apply a force, and as you move it, you do work on the lawn mower. Consequently, you change its speed and kinetic energy from a zero value to some nonzero value. So, it is important to understand that when we do work on an object, we *change* the kinetic energy of the object. The relationship between work (force times distance) and the change in kinetic energy is given by

$$\text{work} = \text{change in kinetic energy of the object} =$$
$$\begin{matrix}\text{(force)}\\\text{(distance)}\end{matrix} = \left[\left(\frac{1}{2}\right)(\text{mass})(\text{speed})^2\right]_{\text{final}} - \left[\left(\frac{1}{2}\right)(\text{mass})(\text{speed})^2\right]_{\text{initial}}$$

3.3

Note the work on the left-hand side of Equation (3.3) and the change in kinetic energy on the right-hand side; both have the same units (J or lbf·ft). The next time you are in a grocery store, think about the relationship between work and the change in kinetic energy shown in Equation (3.3) as you push on a shopping cart. The harder you push on a shopping cart, the faster it will move and the higher its kinetic energy becomes.

Initial position                    Final position

As you know, the harder you push, the faster the cart will move.

Next, we look at two examples to see how you can use Equation (3.3).

EXAMPLE 3.2

Consider an object with a mass of one kilogram initially at rest on a smooth surface. What is the speed of the object if you were to push on it with a force of one newton after it has moved a distance of one meter?

Speed = 0                    Speed = ?
1 N                          1 N
1 kg                         1 kg
|← 1 m →|

$$\text{Work} = (\text{force})(\text{distance})$$
$$= \left[\left(\frac{1}{2}\right)(\text{mass})(\text{speed})^2\right]_{\text{final position}} - \left[\left(\frac{1}{2}\right)(\text{mass})(\text{speed})^2\right]_{\text{initial position}}$$

$$(1\text{ N})(1\text{ m}) = \left[\left(\frac{1}{2}\right)(1\text{ kg})(\text{speed})^2\right]_{\text{final position}} - \overset{0}{\overbrace{\left[\left(\frac{1}{2}\right)(1\text{kg})(0)^2\right]}}_{\text{initial position}}$$

And solving for speed at the final position, we get

recall from Chapter 2 that $1\ N = (1\ kg)(1\frac{m}{s^2})$

$$(\text{speed})^2 = \frac{2(1\ N)(1\ m)}{1\ kg} = \frac{2 \quad \overbrace{(1\ N)} \quad (1\ m)}{1\ kg}$$

$$= \frac{(2)(1)(1)kg\frac{m^2}{s^2}}{1\ kg} = 2\frac{m^2}{s^2}$$

$$\text{speed} = \sqrt[2]{2\frac{m^2}{s^2}} = 1.4\frac{m}{s}$$

Think about the following now.

(a) How much work was done? $(1\ N)(1\ m) = 1\ N\cdot m$ or 1 joule;

(b) What is the initial kinetic energy of the object? 0 joule;

(c) What is the kinetic energy of the object after it has moved 1 m? 1 joule;

How are these three values related? Work is equal to change in kinetic energy, that is, $1\ J = 1\ J - 0\ J$.

**EXAMPLE 3.3**

Tyler Olson / Shutterstock.com

Consider the following situation. You have been pushing a shopping cart filled with your groceries. You stop for a moment to add another item to your cart. How hard do you need to push the cart to get it moving again at the speed of, say, 0.3 meter per second (m/s) or one foot per second (ft/s) over a distance of one meter or 3.28 feet? Your groceries and the cart have a combined mass of 40 kilograms or 88 pound mass.

Let us now apply Equation (3.3) to this example.

$$\text{work} = \text{change in kinetic energy}$$

$$\begin{matrix}(\text{force})\\(\text{distance})\end{matrix} = \left[\left(\frac{1}{2}\right)(\text{mass})(\text{speed})^2\right]_{\text{final}} - \left[\left(\frac{1}{2}\right)(\text{mass})(\text{speed})^2\right]_{\text{initial}}$$

$$(\text{force})(1\ m) = \left[\left(\frac{1}{2}\right)(40\ kg)(0.3)^2\right]_{\text{final}} - \overbrace{\left[\left(\frac{1}{2}\right)(40\ kg)(0)^2\right]_{\text{initial}}}^{\text{zero}}$$

$$(\text{force})(1\ m) = 1.8 - 0\ (\text{N·m})$$

Solve for force to get

$$\text{force} = 1.8\ N\ (\text{newton})$$

From our discussion in Chapter 2, recall that 1 lbf (pound force) = 4.448 N (newton)

$$\text{force} = 1.8 \text{ N}\left(\frac{1 \text{ lbf}}{4.448 \text{ N}}\right) = 0.4 \text{ lbf(pound force)}$$

As you can see, it does not take much effort to move the cart at a speed of 1 foot/second. Of course, we have neglected the rolling friction of the wheels. Where does the energy that does the work come from? In this case, the food that you ate provides the fuel (energy content) to move the cart.

hxdbzxy / Shutterstock.com

Next, we consider potential energy, which is the amount of energy needed to lift an object.

## Potential Energy—What It Takes to Lift Things

How do we quantify how much energy we need to lift things? The energy required to lift an object over a vertical distance is called *potential energy*. It is the work that must be performed to overcome the gravitational pull of the Earth on the object. The change in the potential energy of the object when its elevation is changed could be quantified provided that we know its weight and the change in the elevation. The change in the potential energy is given by

$$\begin{aligned}\text{change in potential energy} &= (\text{weight of the object})(\text{change in elevation})\\[1em] &= \overbrace{(\text{mass of the object}) \times}^{\text{weight of the object}}\ (\text{change in elevation})\\ &\phantom{=}\ (\text{acceleration due to gravity})\end{aligned}$$

3 . 4

The SI unit for potential energy is also the joule, and it is obtained by substituting newton for weight or kilogram (kg) for the units of mass, meter per second squared (m/s²) for the units of acceleration due to gravity, and meter (m) for the elevation change:

$$\text{change in potential energy} = \overbrace{(\text{kg})\left(\frac{\text{m}}{\text{s}^2}\right)}^{\text{weight of the object in newtons}} (\text{m}) = \text{N} \cdot \text{m} = \text{joule} = \text{J}$$

The U.S. Customary units for potential energy is also the lbf·ft as

$$\begin{aligned}\text{potential energy} &= (\text{weight in pound force})(\text{ft})\\ &= \overbrace{(\text{pound force})}^{\text{lbf}}(\text{ft}) = \text{lbf} \cdot \text{ft}\end{aligned}$$

As in the case with kinetic energy, keep in mind that it is the *change* in the potential energy that is of importance. For example, the energy required to lift an elevator from the first floor to the second floor is the same as lifting the elevator from the third floor to the fourth floor, provided that the distance between each floor is the same. This point is demonstrated in Example 3.5.

---

**EXAMPLE 3.4**

This example demonstrates what one joule of energy represents. Consider an average-size apple with a mass of 100 grams (g) or 0.22 pounds, which is almost a quarter of a pound. How much energy does it take to raise the apple by a vertical distance of one meter (3.3 feet)?

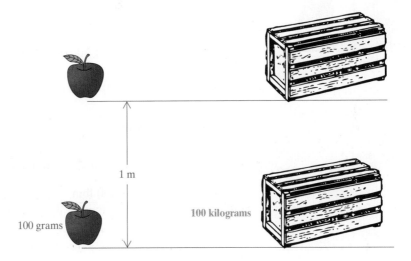

1 m

100 grams

100 kilograms

$$\begin{array}{l}\text{change in}\\ \text{potential}\\ \text{energy}\end{array} = (\text{mass})(\text{acceleration due to gravity})(\text{change in elevation})$$

$$= (100 \text{ grams})\left(\overbrace{\frac{1 \text{ kg}}{1{,}000 \text{ grams}}}^{0.98 \text{ N}}\right)\left(9.8\frac{\text{m}}{\text{s}^2}\right)(1 \text{ m}) = 0.98 \text{ joule} \approx 1 \text{ J}$$

So it takes approximately one joule of energy to raise an average-size apple by a vertical distance of one meter. Now you have a good idea of what one joule represents. So in a fruit warehouse, how much energy does it take to raise a large crate of apples with a mass of 100 kilograms by one meter? 1,000 joules!

$$\begin{array}{l}\text{change in}\\ \text{potential}\\ \text{energy}\end{array} = (\text{mass})(\text{acceleration due to gravity})(\text{change in elevation})$$

$$= (100 \text{ kg})\left(9.8\frac{\text{m}}{\text{s}^2}\right)(1 \text{ m}) = 980 \text{ joules} \approx 1{,}000 \text{ J}$$

**EXAMPLE 3.5**

In this example, we calculate the energy required to lift an elevator and its occupant with a combined mass of 2,000 kilograms or 4,400 pound mass for the following situations:

(a) between the first and the second floors

(b) between the third and the fourth floors

(c) between the first and the fourth floors

The vertical distance between each floor is 4.5 m or 14.8 feet.
We can use Equation (3.4) to analyze this problem; the energy required to lift the elevator is equal to the change in its potential energy.

(a)

$$\text{change in potential energy} = (2{,}000 \text{ kg})\overbrace{\left(9.81\,\frac{\text{m}}{\text{s}^2}\right)}^{19{,}620 \text{ newton}}(4.5 \text{ m}) = 88{,}290 \text{ N}\cdot\text{m}$$

$$= 88{,}290 \text{ J}$$

or in U.S. Customary units,

$$\text{change in potential energy} = \overbrace{\left(\frac{4{,}400 \text{ lbm}}{32.2\,\dfrac{\text{lbm}\cdot\text{ft}}{\text{lbf}\cdot\text{s}^2}}\right)}^{4{,}400 \text{ lbf}}\left(32.2\,\frac{\text{ft}}{\text{s}^2}\right)(14.8 \text{ ft}) = 65{,}120 \text{ lbf}\cdot\text{ft}$$

(b)

$$\text{change in potential energy} = (2{,}000 \text{ kg})\overbrace{\left(9.81\,\frac{\text{m}}{\text{s}^2}\right)}^{19{,}620 \text{ newton}}(4.5 \text{ m}) = 88{,}290 \text{ N}\cdot\text{m}$$

$$= 88{,}290 \text{ J}$$

or in U.S. Customary units,

$$\text{change in potential energy} = \overbrace{\left(\frac{4{,}400 \text{ lbm}}{32.2\,\dfrac{\text{lbm}\cdot\text{ft}}{\text{lbf}\cdot\text{s}^2}}\right)}^{4{,}400 \text{ lbf}}\left(32.2\,\frac{\text{ft}}{\text{s}^2}\right)(14.8 \text{ ft}) = 65{,}120 \text{ lbf}\cdot\text{ft}$$

(c)

$$\text{change in potential energy} = (2{,}000 \text{ kg})\overbrace{\left(9.81\,\frac{\text{m}}{\text{s}^2}\right)}^{19{,}620 \text{ newton}}(13.5 \text{ m}) = 264{,}870 \text{ N}\cdot\text{m}$$

$$= 264{,}870 \text{ J}$$

or in U.S. Customary units,

$$\text{change in potential energy} = \overbrace{\left(\dfrac{4{,}400 \text{ lbm}}{32.2 \dfrac{\text{lbm} \cdot \text{ft}}{\text{lbf} \cdot \text{s}^2}}\right)}^{4{,}400 \text{ lbf}}\left(32.2 \dfrac{\text{ft}}{\text{s}^2}\right)(44.4 \text{ ft}) = 195{,}360 \text{ lbf} \cdot \text{ft}$$

Note that the amount of energy required to lift the elevator from the first to the second floor and from the third to the fourth floor is the same. Also realize that we have neglected any frictional effect in our analysis. The actual energy requirement would be greater in the presence of friction.

Besides moving or lifting something, we need energy to perform many other tasks. Consequently, we define other forms of energy. Next, we discuss thermal energy as a means by which we quantify the amount of energy that we would need to heat (raise temperature of) or cool (lower temperature of) something.

## Thermal Energy—What It Takes to Heat or Cool Things

We also need energy to heat or cool our homes, to cook, and to heat water to shower or bathe. But what is "heat?" Before we explain what thermal energy means, you need to understand the concept of internal energy. ***Internal energy*** is a measure of the molecular activity of a substance and is related to the temperature of the substance. The higher the temperature of an object, the higher its molecular activity (the more excited the molecules are) and thus the higher the internal energy of the object. Moreover, ***thermal energy*** transfer occurs whenever a temperature difference exists within an object or between an object and its surroundings. So when we heat water, we transfer energy from a heating element that has a higher internal energy to water that has a lower internal energy. This form of energy transfer is called *heat*. Another important fact that you should remember is that *heat always flows from a high-temperature region to a low-temperature region*. We discuss heat transfer in greater detail in Chapter 5.

> Thermal energy quantifies the amount of energy required to heat or cool something.

There are three units commonly used to quantify thermal energy: *the British thermal unit or Btu, the calorie,* and *the joule*.

1. One British thermal unit (***Btu***) represents the amount of thermal energy needed to raise the temperature of one pound mass (lbm) of water by one degree Fahrenheit (°F).

Sklep Spozywczy / Shutterstock.com

2. The *calorie* represents the amount of heat required to raise the temperature of one gram (g) of water by one degree Celsius (°C). Note, however, that the energy content of food is typically expressed in *Calories* (with an uppercase C), which is equal to 1,000 calories.

3. In SI units, no distinction is made between the units of thermal, kinetic, and potential energy; for all forms of energy, the unit of *joule* is used. The relationship among different energy units are: 1 Btu = 1,055 joules; 1 Btu = 252 calories; and 1 calorie = 4.186 joules.

Let us now focus on how we would use the thermal energy concepts that you have learned to estimate your energy consumption resulting from a common daily activity.

---

**EXAMPLE 3.6**

### How much energy does it take to heat twenty gallons of water?

When you take a long shower, you can use up to 20 gallons of hot water. Let us look at how much energy it takes to heat 20 gallons of water from room temperature, say at 70°F to hot water at 120°F —an increase of 50°F.

Di Studio / Shutterstock.com

Each gallon of water has a mass of 8.34 pounds, so 20 gallons of water will have a total mass of 166.8 (20 × 8.34) pounds. Recall that one British thermal unit (Btu) represents the amount of thermal energy needed to raise the temperature of one pound mass (lbm) of water by one degree Fahrenheit (°F). In this example, we need to raise the temperature of 166.8 pound mass of water by 50°F, which is equal to (166.8)(50) = 8,340 Btu. The steps to arrive at the final result are shown next. Pay close attention to the way units cancel out for the final result, and you are left with only British thermal unit (Btu).

$$(20 \text{ \sout{gallons of water}})\left(\frac{8.34 \text{ \sout{lbm}}}{1 \text{ \sout{gallon of water}}}\right)\left(\frac{1 \text{ Btu}}{(1 \text{ \sout{lbm}})(1°F)}\right)\left(\overset{50}{\overline{(120 - 70) °F}}\right) = (8,340 \text{ Btu})$$

We can also express the results in joules by noting that 1 British thermal unit is equal to 1,055 joules in the following manner:

$$(8,340 \text{ Btu})\left(\frac{1,055 \text{ joules}}{1 \text{ Btu}}\right) = (8,340 \text{ \sout{Btu}})\left(\frac{1,055 \text{ joules}}{1 \text{ \sout{Btu}}}\right)$$

$$= 8,798,700 \text{ joules}$$

$$\approx 8.8 \text{ MJ (Mega joules)}$$

In reality, you would need to spend more energy, because your hot water heating system is not 100% efficient, and you would have heat losses!

# Before You Go On

Answer the following questions to test your understanding of the preceding section:

1. What does kinetic energy quantify?

2. What does potential energy quantify?

3. What does thermal energy quantify?

4. What are the SI and U.S. Customary units for kinetic energy, potential energy, and thermal energy?

*Vocabulary—State the meaning of the following words:*

Kinetic Energy

Potential Energy

Thermal Energy

Btu

Joule

calorie

Calorie

## LO³ 3.3 Difference Between Energy and Power

> Power represents the amount of work done or energy expended per unit of time.

People commonly confuse energy with power. *Power* shows how fast you are expending energy. The value of power required to do the work (perform a task) represents how fast you want the work (task) done. If you want the work done in a shorter period of time, then you need to spend more power. You should understand clearly that in order to perform the same task in a shorter period of time, more power is required. For example, to lift an elevator, say, with 10 people from the first floor to the 50th floor, you need to spend a certain amount of energy. Now, to determine the power requirement for this task, you need to ask yourself how fast you want to get the people to the 50th floor. The shorter the time period, the bigger the power requirement would be.

$$\text{power} = \frac{\text{work}}{\text{time}} = \frac{\text{energy}}{\text{time}}$$

Here is another example. Do you require more energy to walk up a flight of stairs or to run up the stairs? Which requires more power? To walk up a flight of stairs or to run up the stairs requires the same amount of energy, because your weight and

Tom Wang / Shutterstock.com

the height associated with the flight of stairs remain constant. However, you need more power to run up the stairs, since you want to do the work (or change your potential energy) in a shorter time period.

For the sake of demonstrating the difference between energy and power, imagine that in order to perform a task, 3,600 joules of energy is required. The next question then becomes, how fast do we want this work done? If we want the task done in 1 second, $\frac{3,600 \text{ joules}}{1 \text{ second}} = 3,600 \frac{\text{joules}}{\text{second}}$ of power is required; if we want the work done in 1 minute or 60 seconds, then $\frac{3,600 \text{ joules}}{60 \text{ second}} = 60 \frac{\text{joules}}{\text{second}}$ of power is needed; and if we want the task done in 1 hour or 3,600 seconds, then the required power is $\frac{3,600 \text{ joules}}{3,600 \text{ second}} = 1 \frac{\text{joules}}{\text{second}}$. From this simple example, you should see clearly that in order to perform the same task in a shorter period of time, more power is required. Again, *more power means more energy expenditure per second.*

Many managers understand the concept of power well, for they understand the benefit of teamwork. In order to finish a project in a shorter period of time, instead of assigning a task to an individual, the task is divided among several team members. More useful energy expenditure per day is expected from a team than from a single person, thus the project or the task can be done in less time.

## Units of Power: Watts, Kilowatts, and Horsepower

**SI Units**  The SI unit for power is defined in the following manner:

$$\text{power} = \frac{\text{work}}{\text{time}} = \frac{(\text{force})(\text{distance})}{\text{time}} = \frac{(\text{netwon})(\text{meter})}{\text{second}} = \frac{\text{N} \cdot \text{m}}{\text{s}} = \frac{\text{J}}{\text{s}} = \text{Watt} = \text{W}$$

Note that $1 \text{ N} \cdot \text{m}$ is called 1 joule (J), and 1 joule/second (J/s) is called 1 *watt* (W). The electric power consumption of various devices is also expressed in watts; however, note that for electrical/electronic devices, the power formula $P = (V)(I)$ is used. In this relationship, $P$ is power in watts, $V$ is the voltage, and $I$ is the current in amps. In the United States, most electronic devices and small appliances are connected to a 120-volt source. By using this power relationship between voltage and current, you can determine their power consumption, provided that you know how much electric current the device draws. For example, a hair dryer that draws 10 amps consumes 1,200 watts of power. We discuss electricity in greater detail in Chapter 4.

It is also important to mention here that a unit which is often confused for the unit of power is the kilowatt-hour. The *kilowatt-hour* (kWh) is a unit of *energy*—not power, and it represents the amount of energy consumed during 1 hour by a device that uses 1,000 watts or one kilowatt (kW). We explain this concept along with energy consumption for common products from everyday life in more detail in Chapter 4.

**U.S. Customary Units**  In U.S. Customary units, the units of power are expressed in pound force-ft per second $\left( \frac{\text{lbf} \cdot \text{ft}}{\text{s}} \right)$ and *horsepower* (hp) in the following manner:

$$\text{power} = \frac{\text{work}}{\text{time}} = \frac{(\text{force})(\text{distance})}{\text{time}} = \frac{(\text{pound force})(\text{ft})}{\text{second}} = \frac{\text{lbf} \cdot \text{ft}}{\text{s}}$$

and

$$1 \text{ horsepower} = 1 \text{ hp} = 550 \frac{\text{lbf} \cdot \text{ft}}{\text{s}}$$

The U.S. Customary units of power are related to the SI unit watt (W) through:

$$1 \frac{\text{lbf} \cdot \text{ft}}{\text{s}} = 1.36 \text{ Watts}$$

and

$$1 \text{ hp} = 746 \text{ W} = 0.746 \text{ kW}$$

In the above relationships, note that $1 \left( \frac{\text{lbf} \cdot \text{ft}}{\text{s}} \right)$ is slightly greater in magnitude than 1 watt. Also keep in mind that 1 horsepower is smaller than 1,000 watts or 1 kilowatt (kW).

In the United States, for heating, ventilating, and air-conditioning (HVAC) applications, Btu per hour (Btu/h) is used to represent the heat loss from a building during cold months and the heat gained by the building during summer months. We discuss heat loss and gain in buildings in detail in Chapter 5.

The definitions and relationships among various units of energy and power are summarized in Table 3.1.

| TABLE 3.1 | The Units of Energy and Power |
|---|---|
| **SI Units of Energy** | **U.S. Customary Units of Energy** |
| The unit for kinetic energy and potential energy is the joule, which is equal to 1 newton-meter, that is<br><br>$1 \text{ J} = (1 \text{ N})(1 \text{ m})$ | The unit for kinetic energy and potential energy is the pound force-foot (lbf · ft) |
| The unit of thermal energy also is the joule. However, the calorie, which represents the amount of heat required to raise the temperature of one gram (g) of water by one degree Celsius (°C) is also used. | The unit for thermal energy is British thermal unit (Btu). One Btu represents the amount of thermal energy needed to raise the temperature of one pound mass (lbm) of water by one degree Fahrenheit (°F). |
| **SI Units of Power** | **U.S. Customary Units of Power** |
| The unit of power is the watt, which is equal to 1 joule per second, that is<br><br>$1 \text{ W} = 1 \text{ J/s};$<br><br>and<br><br>$1 \text{ kilowatt} = 1 \text{ kW} = 1,000 \text{ W}$ | The units of power are expressed in $\frac{\text{lbf} \cdot \text{ft}}{\text{s}}$ and horsepower (hp), where $1 \text{ hp} = 550 \frac{\text{lbf} \cdot \text{ft}}{\text{s}}$<br><br>In thermal applications such as the heating and cooling of buildings, Btu per hour $\left( \frac{\text{Btu}}{\text{h}} \right)$ is used. |
| **The Relationship Among SI and U.S. Customary Units** | |
| Energy = 1 joule (J) ≈ 0.74 lbf · ft; 1 lbf · ft ≈ 1.4 joules (J) | |
| 1 Btu = 778 lbf · ft; 1 cal = 4.186 J; 1 Btu = 1,055 J; 1 Btu = 252 cal | |
| Power : 1 cal/s = 4.186 W; 1 W = 3.4123 Btu/h; hp = 746 W | |

**EXAMPLE 3.7**

This example demonstrates what one watt and one kilowatt of power represent. Consider the apple in Example 3.4. How much power does it take to raise the apple by a vertical distance of one meter in one second?

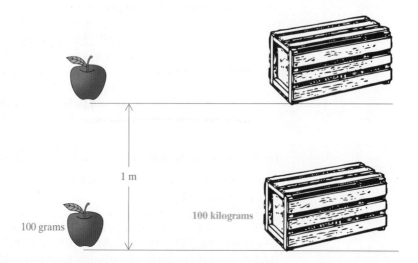

1 m

100 grams    100 kilograms

In Example 3.4, we calculated that it takes approximately one joule of energy to raise an average-size apple by one meter. Then to raise the apple by one meter in one second, it would require a power of one watt as shown.

$$\text{power} = \frac{\text{work or energy}}{\text{time}} = \frac{1 \text{ joule}}{1 \text{ second}} = 1 \text{ watt}$$

Now can you estimate how much power it would take to raise the large crate of apples in Example 3.4?

$$\text{power} = \frac{\text{work or energy}}{\text{time}} = \frac{1,000 \text{ joules}}{1 \text{ second}} = 1000 \text{ watts} = 1 \text{ kW}$$

**EXAMPLE 3.8**

**How powerful are you? Horsepower versus kilowatt**

Horsepower and kilowatts represent units of power. One horsepower, which is equal to 550 pound force-feet per second (lbf · ft/s) can be interpreted as lifting 220 pounds a distance of 2.5 feet every second, so

$$\left[ \frac{(220 \text{ lbf})(2.5\text{ft})}{1 \text{ second}} = 550 \frac{\text{lbf} \cdot \text{ft}}{\text{s}} \right]$$

Can you do it? If so, how long can you keep it up? On the other hand, one kilowatt is equal to 1,000 N · m/s and could be understood as lifting 1,000 newtons a distance of one meter every second. One kilowatt is

larger than one horsepower (1 kilowatt = 1.34 horsepower) and when expressed in its equivalent U.S. Customary units, one kilowatt could be explained as lifting 295 pounds a distance of 2.5 feet every second. This way, you see that one kilowatt represents more power than one horsepower.

You have a good feel for how much effort you have to exert to lift a gallon of water. If a gallon of water weighs 8.34 pounds, how many of these one-gallon water-filled containers do you need to lift simultaneously at a distance of 2.5 feet every second to produce enough power that is equal to one horsepower or one kilowatt? About 26 of them to produce one horsepower and about 35 of them to produce one kilowatt! By now you should be able to verify these results on your own.

**EXAMPLE 3.9**

**A garage door opener**

When an automatic garage door opener is activated, you notice that it takes 12 seconds to completely open the garage door. The garage door weighs approximately 300 pound force (lbf), and when it is fully open its mass center is raised by a vertical distance of 4.5 feet. Estimate the size of the motor for the garage door opener.

$$\text{power} = \frac{\text{work}}{\text{time}} = \frac{(\text{force})(\text{distance})}{\text{time}} = \frac{\text{change in potential energy}}{\text{second}}$$

$$= \frac{(300 \text{ lbf})(4.5 \text{ ft})}{12 \text{ s}}$$

$$= 112.5 \left( \frac{\text{lbf} \cdot \text{ft}}{\text{s}} \right)$$

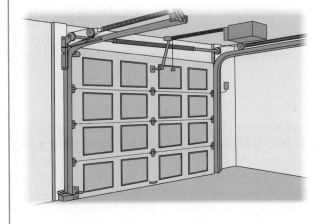

You can convert this value to horsepower (hp) in the following manner.

$$\text{power} = 112.5 \left( \frac{\text{lbf} \cdot \text{ft}}{\text{s}} \right) \left( \frac{1 \text{ hp}}{550 \frac{\text{lbf} \cdot \text{ft}}{\text{s}}} \right)$$

$$= 0.2 \text{ hp}$$

This value makes sense, because most garage door opener motors are rated at one-quarter horsepower (0.25 hp). What is the size of the garage door opener motor in your or a friend's home?

## *Before You Go On*

Answer the following questions to test your understanding of the preceding section:

1. What is the difference between energy and power?

2. What are the SI and U.S. Customary units for power?

3. Which represents more power, kilowatt or horsepower?

**Vocabulary—State the meaning of the following words:**

Power

Kilowatt

Horsepower

## LO⁴  3.4  **Energy Content (Heating Values) of Fuels**

To generate energy, we use fuels such as coal and natural gas in power plants. To move our cars, we burn gasoline or diesel fuels. To heat our homes, we may burn natural gas in a furnace or wood in a fireplace. When a fuel is burned, thermal energy is released. The ***heating value of a fuel*** quantifies the amount of energy that is released when a unit mass (kilogram or pound) or a unit volume (cubic meter, cubic foot, or gallon) of a fuel is burned. The energy contents of common fuels are given in Table 3.2.

Let us now focus on how we use the information given in Table 3.2 to estimate fuel consumption resulting from some activities such as taking a shower or burning fuel to keep a home warm during a cold winter day. It will soon become evident from the results of these examples that we need to burn a large amount of fuel to address our daily needs. Remember, here we focus only on two examples; think about all of the other activities, such as driving, cooking, clothes drying, and so on that also require fuel expenditure.

### Efficiency

In the previous sections, we discussed how to estimate the energy and power requirements for some of our daily activities. It also is important to understand that there are always some losses associated with systems that move us (e.g., our cars and elevators), heat or cool water and our buildings, or make consumer products. Not all of the energy that we generate by burning fuels or other means can be used completely. Some of it will be lost in a given system due to friction, heat loss, and other factors. When we wish to show how well

**Efficiency is a measure of how much input is required to have a desired output.**

**EXAMPLE 3.10**

**How much natural gas do you need to burn to heat twenty gallons of water?**

In Example 3.6, we showed that you need to expend 8,340 Btu to heat up 20 gallons of water from room temperature at 70°F to 120°F to produce hot water to take a shower. Let's now look at how much natural gas, with a heating value of 1,000 British thermal units (Btu) per cubic foot (ft³) (see Table 3.2), we need to burn to generate this amount of thermal energy.

$$8,340 \text{ Btu} = (\text{Amount of natural gas in cubic feet})\left(1,000\,\frac{\text{Btu}}{\text{cubic foot}}\right)$$

$$8,340 \;\cancel{\text{Btu}} = (\text{Amount of natural gas in } \cancel{\text{cubic feet}})\left(1,000\,\frac{\cancel{\text{Btu}}}{\cancel{\text{cubic foot}}}\right)$$

Recall from Section 2.3 that we emphasized that *all formulas used in any analysis must be homogeneous in dimensions and units*. Note that the unit of Btu on the left-hand side of the equation cancels out with the unit of Btu on the right-hand side of the equation, and the units of cubic feet in the numerator and denominator cancel out, resulting in:

$$(\text{Amount of natural gas}) = 8.34 \text{ ft}^3$$

Jo Ann Snover / Shutterstock.com

**EXAMPLE 3.11**

**How much fuel do you need to burn to keep a home warm during a cold winter day?**

In the Midwest part of the United States, to keep a single family home warm and cozy during a cold winter day, you may need to generate as much as 50,000 British thermal units (Btu) of thermal energy per hour. Then the total of amount of energy that needs to be generated for a cold spell during a 24-hour period is equal to

$$\text{energy needed} = \left(50,000\,\frac{\text{Btu}}{\text{hour}}\right)\left(24\,\frac{\text{hours}}{\text{day}}\right)$$

$$= \left(50,000\,\frac{\text{Btu}}{\cancel{\text{hour}}}\right)\left(24\,\frac{\cancel{\text{hours}}}{\text{day}}\right)$$

$$= 1,200,000\,\frac{\text{Btu}}{\text{day}}$$

Note the unit of hour and hours in the denominator and numerator cancel out, and you are left with the units of Btu/day.

Now we can determine how much fuel we need to burn to generate 1,200,000 Btu in one day.

Maria Dryfhout / Shutterstock.com

**Natural gas:**

$$1,200,000\ \frac{\text{Btu}}{\text{day}} = \left(\text{amount of natural gas in }\frac{\text{cubic feet}}{\text{day}}\right)\left(1,000\ \frac{\text{Btu}}{\text{cubic foot}}\right)$$

$$(\text{amount of natural gas}) = 1,200 \text{ cubic feet/day}$$

**Fuel oil:**

$$1,200,000\ \frac{\text{Btu}}{\text{day}} = \left(\text{amount of fuel oil in }\frac{\text{gallons}}{\text{day}}\right)\left(139,000\ \frac{\text{Btu}}{\text{gallon}}\right)$$

$$(\text{Amount of fuel oil}) = 8.6 \text{ gallons/day}$$

**Coal:**

$$1,200,000\ \frac{\text{Btu}}{\text{day}} = \left(\text{amount of coal in }\frac{\text{pounds}}{\text{day}}\right)\left(10,000\ \frac{\text{Btu}}{\text{pound}}\right)$$

$$(\text{amount of coal}) = 120 \text{ pounds/day}$$

**Wood:**

$$1,200,000\ \frac{\text{Btu}}{\text{day}} = \left(\text{amount of wood in }\frac{\text{cord}}{\text{day}}\right)\left(20,000,000\ \frac{\text{Btu}}{\text{cord}}\right)$$

$$(\text{amount of wood}) = 0.06\ \frac{\text{cord}}{\text{day}} = \left(0.06\ \frac{\text{cord}}{\text{day}}\right)\left(\frac{288 \text{ ft}^3}{1 \text{ cord}}\right)$$

$$\approx 8\ \frac{\text{ft}^3}{\text{day}}$$

a system is functioning, we express its efficiency. *Efficiency is a measure of how much input is required to have a desired output.* In general, the overall efficiency of a system is defined as

$$\text{efficiency} = \frac{\text{desired output}}{\text{required input}} \qquad \text{3.5}$$

or in other words as

$$\text{efficiency} = \frac{\text{what you want to get out of a system}}{\text{what you need to put into the system in terms of energy by burning fuel}}$$

| TABLE 3.2 | The Energy Content of Common Fuels | |
|---|---|---|
| **Fuel** | **Quantity** | **Average Energy Content** |
| **Coal** Siberia - Video and Photo / Shutterstock.com | One pound | 10,000 Btu (10.5 $\times$ 10$^6$ J) |
| **Diesel** Melinda Fawver / Shutterstock.com | One gallon | 139,000 Btu (146.6 $\times$ 10$^6$ J) |
| **Gasoline** | One gallon | 124,000 Btu (130.8 $\times$ 10$^6$ J) |
| **Fuel Oil** (Home Heating Oil) | One gallon | 139,000 Btu (146.6 $\times$ 10$^6$ J) |
| **Natural Gas** pixelsnap / Shutterstock.com | One cubic foot | 1,000 Btu (1.05 $\times$ 10$^6$ J) |
| **Wood** Michael Dechev / Shutterstock.com | One cord (128 ft$^3$) (4 $\times$ 4 $\times$ 8 feet pile of wood stacked neatly) | 20,000,000 Btu (21.1 $\times$ 10$^9$ J) |

All systems require more input than what they put out. For example, for a power plant that generates electricity, the overall efficiency is defined as

$$\text{efficiency} = \frac{\text{net energy generated by the power plant}}{\text{energy input from burning the fuel in the plant}}$$

The efficiency of today's power plants where a fossil fuel (oil, gas, coal) is burned in the boiler is near 40%. In the next chapter, we discuss electricity in greater detail.

**Internal Combustion Engines** The thermal efficiency of a typical gasoline engine is approximately 25 to 30 percent and for a diesel engine is 35 to 40 percent. The thermal efficiency of an internal combustion engine is defined as

$$\text{efficiency} = \frac{\text{energy output of the car}}{\text{heat energy input to the car as fuel is burned}}$$

Keep in mind that when expressing the overall efficiency of a car, one must also account for the mechanical losses due to friction as well.

**Refrigeration and Cooling Systems** Refrigeration and air-conditioning systems also play significant roles in our daily lives. Their main purpose is to remove heat from inside a refrigerated space (such as a refrigerator) or a building and to transfer that heat to the surroundings (air in the kitchen or outside air). Most of today's air-conditioning and refrigeration systems are designed according to a vapor–compression cycle. We explain how a refrigerator and an air-conditioning unit work in Chapter 4. For now, our focus is on the efficiency of these appliances. In the United States, it is customary to express the energy efficiency ratio (EER) or the *seasonal energy efficiency ratio (SEER)* of refrigeration or air-conditioning systems using mixed SI and U.S. Customary units in the following manner:

$$\text{EER} = \frac{\text{heat removal from the unit (Btu)}}{\text{energy input to the unit (Watt-hour)}} \qquad \boxed{3.5}$$

The reason for using the units of watt-hour (Wh) for energy input is that refrigeration and air conditioning units are powered by electricity, and electricity consumption is measured (even in the United States) in kilowatt-hour (kWh). Many of today's air-conditioning units have SEER values that range from approximately 10 to 17.

**Furnaces** The sizes of home gas and oil furnaces in the United States are expressed in units of British thermal unit per hour (Btu/h), and in other countries the kilowatt (kW) is used. In 1992, the United States government established a minimum *annual fuel utilization efficiency (AFUE)* rating of 78% for furnaces installed in new homes, so manufacturers must design their gas furnaces to adhere to this standard. Today, most high-efficiency furnaces offer AFUE ratings in the range of 80 to 96 percent.

Next, we will look at an example to show you how to use an efficiency relationship to determine the amount of fuel that must be burned to produce a certain amount of energy.

**EXAMPLE 3.12**

Let us determine the power required to move 30 people, with an average mass of 61 kilograms (kg) or 135 pound mass (lbm) per person, between two floors of a building at a vertical distance of 5 meters (m) or 16 feet (ft) in 2 seconds (s).

The required energy is equal to:

$$\text{change in potential energy} = (30 \text{ persons})\left(61\frac{\text{kg}}{\text{person}}\right)\left(9.8\frac{\text{m}}{\text{s}^2}\right)(5 \text{ m})$$

$$= 90,000 \text{ joules}$$

The next time you feel lazy and are thinking about taking the elevator to go up only one floor, reconsider and think about the total amount of energy that could be saved if people would take the stairs instead of taking the elevator to go up one floor. For example, if one million people decided to take the stairs on a daily basis, the minimum amount of energy saved during a year, based on an estimate of 220 working days in a year, would be

$$\text{energy savings} = \left(\frac{90,000 \text{ joules}}{30 \text{ persons}}\right)\left(\frac{1}{\text{day}}\right)(1,000,000 \text{ persons})(220 \text{ days})$$

$$= 660 \times 10^9 \text{ joules} = 660 \text{ GJ (gigajoules)}$$

Let us now estimate the amount of fuel, such as coal, that can be saved in a power plant. Let's assume a 36 percent overall efficiency for the power plant, a 6 percent loss in the power transmission lines, and a energy content (heating value) of approximately 7.5 megajoules/kilogram (MJ/kg) for coal.

$$\text{efficiency} = \frac{\text{net energy generated by the power plant}}{\text{energy input from burning the fuel in the boiler}}$$

$$0.30 = \frac{660 \text{ GJ}}{\text{energy input from burning the fuel in the plant}}$$

$$\text{energy input from fuel} = \frac{660 \text{ GJ}}{0.3} = 2.2 \times 10^{12} \text{ J} = 2,200 \text{ GJ}$$

$$= 2.2 \text{ TJ (terajoules)}$$

$$\text{amount of coal required} = \frac{2.2 \times 10^{12} \text{ joules}}{7.5 \times 10^6 \frac{\text{joules}}{\text{kilogram}}} \approx 293,000 \text{ kg (646,000 lbm)}$$

As you can see, the amount of coal that could be saved is quite large! Before you get on an elevator next time, think about the amount of fuel that can be saved if people just walk up a floor! Also, consider the amount of pollution that can be prevented by reducing the amount of fuel burned.

## Before You Go On

Answer the following questions to test your understanding of the preceding section:

1. What does the heating value or energy content of a fuel represent?

2. What are typical heating values for a pound of coal, a gallon of gasoline, and a cubic foot of natural gas?

3. What do we mean by efficiency and why is it important to know the efficiency of products that we use in our daily lives?

## SUMMARY

### LO¹ Energy and Work

We need *energy* to create goods, to build shelter, to cultivate and process food, and to maintain our living places at comfortable temperatures. Energy can have different forms, and to better *explain quantitatively* the requirements to move objects such as our cars, to lift things like an elevator, or to heat or cool our homes, energy is defined and classified into different categories such as *kinetic energy, potential energy, and thermal energy*. In a physics context, work is performed when *a force moves an object through a distance [work = (force)(distance)]*, and when we do work on an object we change its *energy*. Another way of thinking about the relationship between work and energy is that we need to expend *energy* to do *work*.

### LO² Forms of Energy

In order to quantify the requirements to move objects such as our cars, to lift things like an elevator, or to heat or cool our homes, energy is defined and classified into different categories such as kinetic energy, potential energy, and thermal energy.

#### Kinetic Energy

Kinetic energy is the way we quantify how much energy or work is required to move something. The amount of kinetic energy associated with an object having a known mass and moving with a known speed is given by

$$\text{kinetic energy} = \left(\frac{1}{2}\right)(\text{mass})(\text{speed})^2$$

The SI and U.S. Customary units for kinetic energy are the joule (J) and pound force-foot (lbf · ft), respectively. Moreover, when we do work on or against an object, we change the kinetic energy of the object according to

$$\text{work} = (\text{force})(\text{distance})$$
$$= \text{change in kinetic energy}$$
$$= \left[\left(\frac{1}{2}\right)(\text{mass})(\text{speed})^2\right]_{\text{final}}$$
$$- \left[\left(\frac{1}{2}\right)(\text{mass})(\text{speed})^2\right]_{\text{initial}}$$

#### Potential Energy

The work that must be done or the energy required to lift an object over a vertical distance is called potential energy. The change in the potential energy of the object when its elevation is changed is given by

change in potential energy

$$= \overbrace{\left(\begin{array}{c}\text{mass of}\\\text{the object}\end{array}\right)\left(\begin{array}{c}\text{acceleration}\\\text{due to gravity}\end{array}\right)}^{\text{weight of the object}}\left(\begin{array}{c}\text{change in}\\\text{elevation}\end{array}\right)$$

The SI and U.S. Customary units for potential energy are the joule (J) and pound force-foot (lbf·ft), respectively.

## Thermal Energy

Thermal energy or heat transfer occurs whenever a temperature difference exists within an object or between a body and its surroundings. Heat always flows from a high-temperature region to a low-temperature region. The three units that are commonly used to quantify thermal energy are the British thermal unit (Btu), the calorie, and the joule (J). One British thermal unit represents the amount of thermal energy needed to raise the temperature of one pound mass (lbm) of water by one degree Fahrenheit (°F). The *calorie* represents the amount of heat required to raise the temperature of one gram (g) of water by one degree Celsius (°C).

## LO³ Difference between Energy and Power

You should clearly understand the definition of power, its common units, and how it is related to work and energy. *Power* is the time rate of doing work or how fast you are expending energy. The value of power required to do the work (perform a task) represents how fast you want the work (task) done. If you want the work done in a shorter period of time, then you need to spend more power.

$$\text{power} = \frac{\text{work}}{\text{time}} = \frac{\text{energy}}{\text{time}}$$

The SI and U.S. Customary units of power are the watt (W) and pound force-foot per second (lbf·ft/s), respectively, and

$$1 \text{ horsepower} = 1 \text{ hp} = 550\frac{\text{lbf} \cdot \text{ft}}{\text{s}}$$

## LO⁴ Energy Contents (Heating Values) of Fuels

The heating value of a fuel quantifies the amount of energy that is released when a unit mass (kilogram or pound) or a unit volume (cubic meter, cubic foot, or gallon) of a fuel is burned. For example, when you burn one gallon of gasoline, 124,000 Btu of energy is released. Or when you burn one pound of coal, 10,000 Btu of energy is released.

### Efficiency

You should know the basic definition of efficiency and be familiar with its various forms, including the definitions of thermal efficiency, SEER, and AFUE, which are commonly used to express the efficiencies of systems including heating, ventilating, and air-conditioning (HVAC) equipment and household appliances. In general, the overall efficiency of a system is defined as

$$\text{efficiency} = \frac{\text{what you want to get out of a system}}{\text{what you need to put into the system}}$$

All machines and systems require more input than what they put out. For example, the thermal efficiency of a typical gasoline engine is approximately 25 to 30 percent. The thermal efficiency of an internal combustion engine is defined as

$$\text{efficiency} = \frac{\text{energy output of the car}}{\text{heat energy input as fuel is burned}}$$

## KEY TERMS

AFUE 90
calorie 80
Calories 80
Efficiency 88
Energy 70
Heating Value of a Fuel 86
Horsepower 82
Joule 71
Kilowatt 82
Kilowatt-hour 82
Kinetic Energy 72
Potential Energy 76
Power 81
SEER 90
Thermal Energy 79
Watt 82
Work 71

# Apply What You Have Learned

Identify ways that you can save energy; for example, walking up a floor instead of taking the elevator, or walking or riding your bike an hour a day instead of taking the car. Estimate the amount of energy that you could save every year with your proposal. Also, estimate the amount of fuel that can be saved at the same time. State your assumptions, and present your analysis in a brief report.

Christian Mueller / Shutterstock.com

# PROBLEMS

*Problems that promote life-long learning are denoted by* 🔑

**3.1** If you were to push a lawn mower with a constant horizontal force of 10 N, how much work would you do while pushing the lawn mower a total distance of 5 m on level ground?

**3.2** Which of the following requires more work: to change the speed of a car from 0 to 40 mph (0 to 64 km/h) or from 40 to 65 mph (64 to 105 km/h)?

**3.3** Look up the manufacturer's horsepower ratings for the most recent year of the following cars:

a. Toyota Camry
b. Honda Accord
c. BMW 750 Li
d. A car of your choice

**3.4** An elevator has a rated capacity of 2,000 pound mass (907 kg). It can transport people at the rated capacity between the first and the fifth floors in 7 seconds with a vertical distance of 15 ft (4.6 m) between each floor. Estimate the power requirement for such an elevator.

**3.5** Determine the gross force needed to bring a car that is traveling at 110 km/h to a full stop in a distance of 100 m. The mass of the car is 2,100 kg. What happens to the initial kinetic energy? Where does it go or to what form of energy does the kinetic energy convert?

**3.6** A power plant has an overall efficiency of 30%. The plant generates 20 MW of electricity and uses coal as fuel. Determine how much coal must be burned to sustain the generation of 20 MW of electricity.

**3.7** 🔑 Estimate the amount of gasoline that could be saved if all of the passenger cars in the United States were driven 1,000 miles (1,609 km) less each year. State your assumptions and write a brief report discussing your findings.

**3.8** 🔑 Investigate the typical power consumption range of the following products:

a. home refrigerator
b. 46-inch television set
c. clothes washer
d. electric clothes dryer
e. vacuum cleaner
f. hair dryer

Discuss your findings in a brief report.

**3.9** 🔑 Investigate the typical power consumption range of the following products:

a. personal computer with a 19-inch monitor
b. laser printer
c. smart phone

Discuss your findings in a brief report.

**3.10** Look up both the furnace size and the air-conditioning unit size in your own or a friend's home or apartment. Investigate the SEER and the AFUE of the units.

**3.11** Investigate the size of a gas furnace used in a typical single-family dwelling in Colorado, and compare that size to the furnaces used in Minnesota and in Kansas. If you live outside of the U.S., investigate furnace sizes in three towns in different climatic zones in your country.

**3.12** An air-conditioning unit has a cooling capacity of 18,000 Btu/h. If the unit has a rated energy efficiency ratio (EER) of 11, how much electrical energy is consumed by the unit in one hour? If a power company charges 12 cents per kWh for usage, how much would it cost to run the air-conditioning unit for a month (31 days), assuming the unit runs eight hours a day?

**3.13** Visit a store that sells air-conditioning units. Obtain information on their rated cooling capacities and EER values. Contact your local power company and determine the cost of electricity in your area. Estimate how much it will cost for you to run the air-conditioning unit during the summer. Write a brief report to your instructor discussing your findings and assumptions.

**3.14** In Example 3.10, we calculated the amount of natural gas that you would burn to heat 20 gallons of water from room temperature at 70 °F to 120 °F to take a shower. Determine the additional amount of gas that must be burned if the water heater has an efficiency of (a) 80 percent, (b) 85 percent, and (c) 90 percent.

**3.15** For Example 3.12, how much fuel would be saved if the efficiency of the power plant is increased from 36 to 40 percent?

**3.16** Estimate the amount of fuel that could be saved if the efficiency of 10 million cars is increased by five percent. Assume an annual driving distance of 12,000 miles (19,300 km) and an average gasoline consumption of 25 miles per gallon (10.6 km/liter).

**3.17** Estimate the amount of natural gas that a person may consume annually for the exclusive purpose of showering? State all your assumptions.

**3.18** Estimate the annual gasoline and energy consumption rates for your car. State all your assumptions. If you don't have a car, perform the analysis for your dream car.

**3.19** For Example 3.11, estimate the amount of natural gas that would be consumed by a furnace with an efficiency of 90% during a cold spell lasting one week.

**3.20** Calculate the power requirement in horsepower for an escalator in a mall that transports 50 people with an average weight of 170 lbf (756 N) a vertical distance of 30 feet (9.1 m) in 25 seconds.

*"My grandfather once told me that there are two kinds of people: those who do the work and those who take the credit. He told me to try to be in the first group; there was much less competition there."*

—INDIRA GANDHI (1917–1984)

# CHAPTER 4

# Electricity

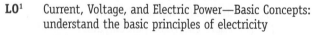

## LEARNING OBJECTIVES

**LO¹** Current, Voltage, and Electric Power—Basic Concepts: understand the basic principles of electricity

**LO²** Residential Power Distribution and Consumption: be familiar with a typical residential power distribution system and consumption

**LO³** Lighting Systems: be familiar with different lighting systems and their power consumption rates

**LO⁴** Electric Power Generation, Transmission, and Distribution: be familiar with how electricity is generated, transmitted, and distributed

# *Discussion Starter*

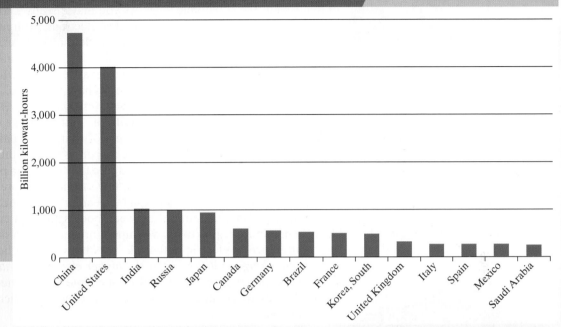

*Source:* U.S. Department of Energy

We have come a long way from 1879 when electricity was first sold in San Francisco to power only 21 electric lights. Can you imagine your life today without electricity? Most of us own at least one TV set, a computer, a printer, a hair dryer, a cell phone, a microwave, a refrigerator, and many other electronic devices and home appliances. We turn on the lights when it gets dark and turn on the air conditioning when it gets hot outside. Despite its vital role in our lives, most of us take the flow of electricity to our homes for granted. We don't pause to think about how electricity is generated, what the primary sources of electricity generation are, and how not to waste it, when electricity is delivered to our homes. As we explained in Chapter 1, the world population is expected to reach 9.3 billion by the year 2050. The most recent available data ranks China ahead of the United States with nearly 4,750 billion kilowatt-hours of total electricity generation, as shown here. By now you should know what one kilowatt-hour represents.

**To the Students:** Think about your activities during the past week and name at least five that you think consumed large quantities of electricity. How much electricity in kilowatt-hours do you think you consume each day, week, month, and year?

## LO[1]  4.1  Current, Voltage, and Electric Power—Basic Concepts

We use electricity in our homes for lighting, to power our electronic devices, and to run appliances. As a good global citizen, in order to better recognize how much electricity you consume every day, you must first understand some of the basic concepts associated with electricity, such as ampere, voltage, and power.

The *ampere* is a base unit for electric current. To understand what the ampere represents, we need to take a closer look at the behavior of material at the subatomic level. As you may already know, an atom has three major subatomic particles: electrons, protons, and neutrons. Neutrons and protons form the nucleus of an atom. How material conducts electricity is influenced by the number and the arrangement of *electrons*. Electrons have negative charge, whereas protons have a positive charge and neutrons have no charge.

Simply stated, the basic law of electric charges states that *unlike charges attract each other while like charges repel*. In SI units, the unit of charge is the *coulomb* (C). One coulomb is defined as the amount of charge that passes a point in a wire in one second when a current of one ampere is flowing through the wire. You may already know that in order for water to flow through a pipe, a pressure difference must exist. Moreover, the water flows from a high-pressure region to a lower-pressure region. As we also explained earlier, whenever there is a temperature difference in a medium or between bodies, thermal energy flows from a high-temperature region to a low-temperature region. In a similar way, whenever a difference in electric potential exists between two bodies, *electric charge will flow from the higher electric potential to the lower potential region*. This flow of charge occurs when the two bodies are connected by an electrical conductor, such as a copper wire.

The flow of electric charge is called *electric current* or simply, *current*. The electric current, or the flow of charge, is measured in *amperes* (A). One ampere or "amp" is defined as the flow of one unit of charge per second. For example, a toaster that draws six amps has six units of charge flowing through the heating element each second. The amount of *current* that flows through an electrical element depends on the *electrical potential*, or voltage, available across the element and the resistance the element offers to the flow of charge. *Voltage* represents the amount of work required to move charge between two points, and the amount of charge that is moving between the two points per unit time is called *current*. The voltage induces current to flow in a circuit.

Voronina Svetlana / Shutterstock.com

## Direct Current and Alternating Current

*Direct current* (dc) is the flow of electric charge that occurs in one direction, as shown in Figure 4.1a. Direct current is typically produced by batteries or direct current generators. In the late 19th century, given the limited understanding of the fundamentals and technology of electricity and for economic reasons, direct current could not be transmitted over long distances. Therefore, it was succeeded by alternating current (ac). Direct current was not economically feasible to transform because of the high voltages needed for long-distance transmission. However, developments in the 1960s led to techniques that now allow the transmission of direct current over long distances.

*Alternating current* (ac) is the flow of electric charge that periodically reverses. As shown in Figure 4.1b, the magnitude of the current starts from zero, increases to a maximum value, and

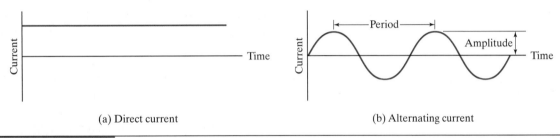

(a) Direct current                                            (b) Alternating current

| **FIGURE 4.1** | Direct and alternating currents. |

then decreases to zero; the flow of electric charge reverses direction, reaches a maximum value, and returns to zero again. This flow pattern is repeated in a cyclic manner. The time interval between the peak values of the current on two successive cycles is called a *period*, and the number of cycles per second is called the *frequency*. The peak (maximum) value of the alternating current in either direction is called the *amplitude*. Alternating current is created by generators at power plants (we explain how a power plant generates electricity in Section 4.4). The current drawn by various electrical devices at your home is alternating current. The alternating current in domestic and commercial power use is 60 cycles per second or hertz (Hz) in the United States.

## Electrical Circuits and Components

Bruno Ferrari / Shutterstock.com

An **electrical circuit** refers to the combination of various electrical components that are connected together. Examples of electrical components include wires, switches, outlets, resistors, and lights. First, let us take a closer look at electrical wires. In a wire, the **resistance** to electrical current depends on the material that the wire is made from, its length, diameter, and temperature. Different materials show varying amounts of resistance to the flow of electric current. *Resistivity* is a measure of resistance of a piece of material to electric current.

## Ohm's Law

Ohm's law describes the relationship among voltage, $V$, resistance, $R$, and current, $I$, according to

| voltage = (resistance)(current) | **4.1** |

Winai Tepsuttinun / Shutterstock.com

or using variables $V$, $R$, and $I$ as

$$V = (R)(I)$$

Note from Ohm's law in Equation (4.1) that current is directly proportional to voltage and inversely proportional to resistance. As electric potential is increased, so is the current; if the resistance is increased, the current will decrease. The electric resistance is measured in units of **ohms** ($\Omega$). An element with one ohm of resistance allows a current flow of one amp when there exists a potential of one volt across the element. Stated another way, when an electrical potential of one volt across a

DK.samco / Shutterstock.com

conductor (such as a wire) with a resistance of one ohm exists, then one ampere of electric current flows through the conductor. Let us now demonstrate how you would use Ohm's law.

---

**EXAMPLE 4.1**

Assume the electric resistance of a device is 60 ohms ($\Omega$). Determine the value of current flowing through the device when it is connected to a 120-volt source.

Using Ohm's law given as Equation (4.1), we have

$$V = RI$$

$$I = \frac{V}{R} = \frac{120 \text{ Volts}}{60 \text{ Ohms}} = 2 \text{ Amperes}$$

---

## Electric Power

The ***electric power*** consumption of various electrical components can be determined using the following power formula:

$$P = (V)(I)$$

4.2

In Equation (4.2), $P$ is the power in watts, $V$ is the voltage, and $I$ is the current in amps. In the United States, most electronic devices and small appliances

---

**EXAMPLE 4.2**

gmstockstudio / Shutterstock.com

ER_09 / Shutterstock.com

A unit that is often confused for the unit of power is the kilowatt-hour (kWh). The kilowatt-hour is used to measure the consumption of electricity by home appliances, electronic devices, lighting systems, and so on. First, you should remember that the kilowatt-hour is a unit of energy—*not* power. By now you should know the difference between energy and power. Second, one kilowatt-hour represents the amount of energy consumed during one hour by a device that uses 1,000 watts (W) or one kilowatt (kW). Let us now determine the energy consumption of some items used at home.

Here are some examples:

**46-inch LCD TV** This type of TV can consume about 250 watts. So if you watch this TV for four hours, then the TV will consume:

(250 watts)(4 hours) = 1,000 watt-hour = 1 kilowatt-hour of energy.

**Clothes Dryers** Depending on their sizes, dryers can consume between 2,000 to 5,000 watts; so if you run a clothes dryer with a power rating of 5,000 watts for two hours, it will consume:

(5,000 watts)(2 hours) = 10,000 watt-hour = 10 kilowatt-hour of energy.

**A 100-watt Incandescent Light Bulb** If left on for 10 hours, an incandescent bulb will consume:

(100 watts)(10 hours) = 1,000 watt-hour = 1 kilowatt-hour of energy.

Africa Studio / Shutterstock.com

are connected to a 120-volt source. Using Equation (4.2), you can then determine their power consumption—provided that you know how much electric current they draw. For example, a hair dryer that draws 10 amps consumes 1,200 watts of power.

---

**EXAMPLE 4.3**

Assume that your electric power company is charging you 10 cents for each kilowatt-hour (kWh) of usage. Estimate the energy cost of leaving five 100-watt light bulbs on from 6 p.m. until 11 p.m. every night for 30 nights.

$$(5 \text{ light bulbs})\left(100 \ \frac{\text{W}}{\text{light bulb}}\right)\left(\frac{1 \text{ kW}}{1,000 \text{ W}}\right)\left(5 \ \frac{\text{hours}}{\text{night}}\right)(30 \text{ nights})\left(10 \ \frac{\text{cents}}{\text{kWh}}\right)$$

$$= (5 \ \cancel{\text{light bulbs}})\left(100 \ \frac{\text{W}}{\cancel{\text{light bulb}}}\right)\left(\frac{1 \ \cancel{\text{kW}}}{1,000 \ \cancel{\text{W}}}\right)\left(5 \ \frac{\cancel{\text{hours}}}{\cancel{\text{night}}}\right)(30 \text{ nights})\left(10 \ \frac{\text{cents}}{\cancel{\text{kWh}}}\right)$$

$$= 750 \text{ cents} = \$7.50$$

---

**EXAMPLE 4.4**

In Example 3.9, we estimated the amount of power required to lift a garage door. Assume a garage door opener has a rated one-quarter hp motor, operates near rated power, and takes 12 seconds to lift the door. Estimate the amount of electrical energy consumed during a year for a case when the garage door is lifted four times daily.

Recall

$$\text{power} = \frac{\text{work or energy}}{\text{time}}$$

then

energy required to lift the garage door = (power)(time)

$$= \left(\frac{4}{\cancel{\text{day}}}\right)(0.25 \ \cancel{\text{hp}})(12 \ \cancel{\text{s}})\left(\frac{0.746 \text{ kW}}{1 \ \cancel{\text{hp}}}\right)\left(\frac{1 \text{ h}}{3,600 \ \cancel{\text{s}}}\right)\left(\frac{365 \ \cancel{\text{days}}}{1 \text{ year}}\right)$$

$$= 0.91 \ \frac{\text{kWh}}{\text{year}}$$

Note the conversion factors used in this example: 1 hp = 0.746 kW; 1h = 3,600 s; and 1 year = 365 days.

This is a good place to mention that the annual electrical energy consumption of a typical home in the United States is approximately between 4,000 and 10,000 kWh.

## Before You Go On

Answer the following questions to test your understanding of the preceding section:

1. In what unit is the electric current measured?

2. What do we mean by voltage?

3. What do we mean by an electric circuit?

4. In your own words, explain Ohm's law.

*Vocabulary—State the meaning of the following words:*

Ampere

Direct current (dc)

Alternating current (ac)

Voltage

Electric power

## LO² 4.2 Residential Power Distribution and Consumption

As a good citizen, it is also important that you become familiar with your home's power distribution system and consumption. An example of a typical residential power distribution system is shown in Figure 4.2, which gives examples of amperage requirements for outlets, lights, kitchen appliances, and central air conditioning. Today, *a typical home has a total 200 amperage rating.* Moreover, an electrical plan for the building is developed first in order to wire a building. In the plan, the location and the types of switches and outlets, including outlets for the range and dryer, must be specified. Examples of electrical symbols used in a house plan are shown in Figure 4.3. An example of an electrical plan for a residential building is shown in Figure 4.4. It is worth noting here that, as an intelligent citizen, you only need to be familiar with these facts; we don't expect you to become an expert or an electrician—unless of course, it is your desire to become one.

Today, a typical home has a total 200 amperage rating.

Various types of wires are used for general wiring, ranging from American Wire Gauge Numbers of 00 to 14.

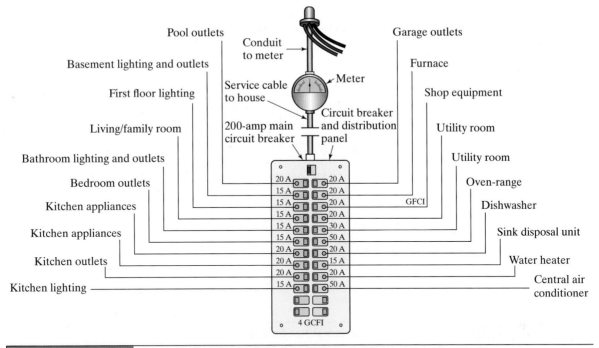

Pool outlets
Conduit to meter
Garage outlets
Basement lighting and outlets
Furnace
First floor lighting
Service cable to house
Meter
Shop equipment
Living/family room
200-amp main circuit breaker
Circuit breaker and distribution panel
Utility room
Bathroom lighting and outlets
Utility room
Bedroom outlets
Oven-range
Kitchen appliances
GFCI
Dishwasher
Kitchen appliances
Sink disposal unit
Kitchen outlets
Water heater
Kitchen lighting
Central air conditioner

20 A, 15 A, 15 A, 15 A, 15 A, 15 A, 20 A, 20 A, 20 A, 15 A
20 A, 20 A, 20 A, 20 A, 30 A, 50 A, 20 A, 15 A, 20 A, 50 A
4 GCFI

**FIGURE 4.2**   An example of an electrical distribution system for a residential building
*Source:* Based on Electrical Wiring, Second Edition, 1981, p. 18, AAVIM.

When examining Figure 4.2, note the electric current ratings (in amperes) for lighting, outlets, kitchen appliances, oven-range, dishwasher, water heater, air conditioning, and furnace.

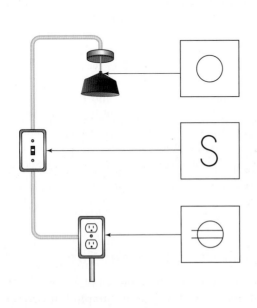

| Common Electrical Symbols | | | |
|---|---|---|---|
| ○ | Ceiling outlet | ▨ | Service entrance panel |
| ─○ | Wall outlet | S | Single-pole switch |
| ○$_{PS}$ | Ceiling outlet with pull switch | S$_2$ | Double-pole switch |
| ─○$_{PS}$ | Wall outlet with pull switch | S$_3$ | 3-way switch |
| ⊖ | Duplex convenience outlet | S$_4$ | 4-way switch |
| ⊖$_{WP}$ | Weatherproof convenience outlet | S$_P$ | Switch with pilot light |
| ⊖$_{1,3}$ | Convenience outlet  1 = single   3 = triple | ▣ | Push button |
| ⊜$_R$ | Range outlet | CH◖ | Bell or chimes |
| ⊖$_S$ | Convenience outlet with switch | ◀ | Telephone |
| ⊜$_D$ | Dryer outlet | TV | Television outlet |
| ⊖ | Split-wired duplex outlet | S⁻⁻ | Switch wiring |
| ◓ | Special-purpose outlet | ⊐⊏ | Fluorescent ceiling fixture |
| D | Electric door opener | ⚬⌣⚬ | Fluorescent wall fixture |

**FIGURE 4.3**   Examples of electrical symbols in a house plan
*Source:* Based on Electrical Wiring, Second Edition, 1981, p. 10, AAVIM.

**General-Purpose Circuits**

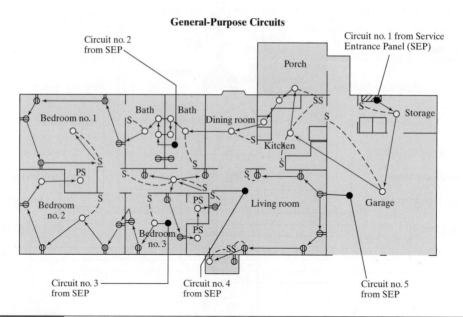

**FIGURE 4.4**    An example of an electrical plan for a house
*Source:* Teia / Shutterstock.com, with added data based on Electrical Wiring, Second Edition, 1981, AAVIM.

Mariusz Szczygiel / Shutterstock.com

## The American Wire Gage (AWG)

At the heart of every home electrical distribution system is wire. Electrical wires are typically made of copper or aluminum. The actual size of the wires is commonly expressed in terms of gage number. The *American Wire Gage (AWG)* is based on successive gage numbers having a constant ratio of approximately 1.12 between their diameters. For example, the ratio of the diameter of No. 1 AWG wire to No. 2 is 1.12 (289 mils / 258 mils = 1.12; note 1 mil = 1/1000 inch). Table 4.1 shows the gage number, the diameter, and their typical residential applications. When examining Table 4.1, note that the smaller the gage number, the bigger the wire diameter. Also, from the data given in Table 4.1, note that 00 AWG wire is used for the service cable to the house.

The National Electrical Code, published by the Fire Protection Association, contains specific information on the type of wires used for general wiring. The code describes the wire types, maximum operating temperatures, insulating materials, outer cover sheaths, type of usage, and the specific location where a wire should be used.

## Home Appliances and Electronic Power Consumption

Let us now turn our attention to the power consumption of typical home appliances and electronic devices. The ranges of power consumption for common appliances and electronics such as clock radios, coffee makers, clothes washers and dryers, fans, hair dryers, televisions, toasters and toaster ovens, and vacuum cleaners are shown in Table 4.2.

| TABLE 4.1 | Examples of American Wire Gage (AWG) for Solid Copper Wire | | |

| American Wire gage (AWG) Number | Diameter (mils) | Current | Common Use |
|---|---|---|---|
| 00 | 365.0 | 200 A | Service entrance |
| 0 | | 150 A | |
| 1 | 289.0 | | |
| 2 | 258.0 | 100 A | Service panels |
| 5 | 182.0 | | |
| 6 | 162.0 | 60 A | Electric furnaces |
| 7 | 144.0 | 40 A | Kitchen appliances, receptacles, and light fixtures |
| 10 | 91.0 | 30 A | |
| 12 | 81.0 | 20 A | Residential wiring |
| 14 | 64.0 | 15 A | Lamps and light fixtures |
| 16 | 51.0 | | |
| 18 | 40.0 | | |
| 20 | 32.0 | | |

| TABLE 4.2 | Examples of Home Appliances and Electronic Devices and Their Range of Power Consumption |

| Item | Range of Power Consumption (watts) |
|---|---|
| Aquarium | 50–1,210 |
| Clock radio | 10 |
| Coffee maker | 900–1,200 |
| Clothes washer | 350–500 |
| Clothes dryer | 1,800–5,000 |
| Dishwasher | 1,200–2,400* |
| Dehumidifier | 785 |
| Electric blanket—single/double | 60 / 100 |
| Fans | |
|   Ceiling | 65–175 |
|   Window | 55–250 |
|   Furnace | 750 |
|   Whole house | 240–750 |
| Hair dryer | 1,200–1,875 |
| Heater (portable) | 750–1,500 |
| Clothes iron | 1,000–1,800 |
| Microwave oven | 750–1,100 |

*(continues)*

| TABLE 4.2 | Examples of Home Appliances and Electronic Devices and Their Range of Power Consumption *(continued)* |
|---|---|
| **Item** | **Range of Power Consumption (watts)** |
| Personal computer | |
|    CPU—awake /asleep | 120 / 30 or less |
|    Monitor—awake /asleep | 150 / 30 or less |
|    Laptop | 50 |
| Radio (stereo) | 70–400 |
| Refrigerator (frost-free, 16 cubic feet) | 725 |
| Televisions | 65 – 250 |
| Toaster | 800–1,400 |
| Toaster oven | 1,225 |
| VCR/DVD | 17–21 / 20–25 |
| Vacuum cleaner | 1,000–1,440 |
| Water heater (40 gallon) | 4,500–5,500 |
| Water pump (deep well) | 250–1,100 |
| Water bed (with heater, no cover) | 120–380 |

*\*Note:* Using the drying feature greatly increases energy consumption.

***Source of data:*** *U.S. Department of Energy.*

As you can see, hot water heaters, dishwashers, and clothes dryers are energy hogs! Moreover, you will find electric motors running all types of appliances and electronic devices in homes. These electric motors also consume lots of energy. Here are a few examples of household appliances with motors.

- Refrigerator: compressor motor, fan motor
- Garbage disposer
- Microwave with a turning tray
- Stove hood with a fan
- Exhaust fan in the bathroom
- Room ceiling fan
- Hand-held power screwdriver or hand-held drill
- Heating, ventilating, or cooling system fan
- Vacuum cleaner
- Hair dryer
- Electric shaver
- Computer: cooling fan, hard drive

arka38 / Shutterstock.com

sue yassin / Shutterstock.com

Tatiana Popova / Shutterstock.com

## Refrigerators and Air Conditioners

Refrigeration and air-conditioning systems also play significant roles in our everyday lives. Therefore, as a well-educated citizen, you need to know how they function. Most of today's air-conditioning and refrigeration systems are designed according to *a vapor–compression cycle*. A schematic diagram of a simple vapor–compression cycle is shown in Figure 4.5. As you can see, the main components of refrigeration or air-conditioning systems include a *condenser*, an *evaporator*, a *compressor*, and a *throttling device*, such as an expansion valve or a capillary tube.

*Refrigerant* is the fluid that transports thermal energy from the *evaporator*, where thermal energy or heat is absorbed, to the **condenser**, where the thermal energy is ejected to the surroundings. After leaving the evaporator, the refrigerant enters the *compressor*, where the temperature and the pressure of the refrigerant are raised. The discharge side of the compressor is connected to the inlet side of the condenser, where the refrigerant enters the condenser in gaseous phase at a high temperature and pressure. Because the refrigerant in the condenser has a higher temperature than the surrounding air, heat transfer to the surrounding air occurs, and consequently, thermal energy is ejected to the surroundings. Both the evaporator and the condenser are made of a series of tubes with good thermal conductivity. After leaving the condenser, the liquid refrigerant flows through an expansion valve or a long capillary tube that makes the refrigerant expand. This expansion is followed by a drop in the refrigerant's temperature and pressure. The refrigerant leaves the expansion valve or the capillary tube and flows into the evaporator to complete the cycle.

The efficiency of a refrigeration system or an air-conditioning unit is given by the coefficient of performance (COP), which is defined as

$$\text{COP} = \frac{\text{heat removal from the evaporator}}{\text{energy input to the compressor}}$$

The COP of most vapor–compression units is 2.9 to 4.9. As mentioned in the previous chapter, it is customary in the United States to express the coefficient

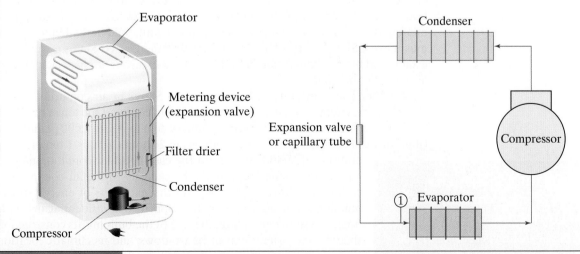

**FIGURE 4.5**   Vapor–compression cycle
*Source:* Designua / Shutterstock.com

**AIR CONDITIONING**

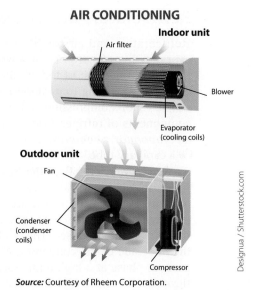

*Source:* Courtesy of Rheem Corporation.

of performance of a refrigeration or an air-conditioning system using mixed SI and U.S. Customary units. Quite often, the coefficient of performance is called the energy efficiency ratio (EER) or the seasonal energy efficiency ratio (SEER). In such cases, the heat removal is expressed in British thermal units (Btu), and the energy input to the compressor is expressed in watt-hours (Wh). Because 1 Wh = 3.412 Btu, EER or SEER values of greater than 10 are obtained for the coefficient of performance.

$$\text{EER} = \frac{\text{heat removal from the evaporator (Btu)}}{\text{energy input to the compressor (Wh)}}$$

4 . 3

As you recall, the reason for using the units of watt-hour (Wh) for energy input to the compressor is that compressors are powered by electricity, and electricity consumption is measured (even in the United States) in kilowatt-hours. Today's air-conditioning units have SEER values that range from approximately 10 to 17. In fact, all new air-conditioning units sold in the United States must have a SEER value of at least 10. In 1992, the United States government established the minimum standard efficiencies for various appliances, including air-conditioning units.

Another common unit used in the United States in air-conditioning and refrigeration systems is *ton of refrigeration or ton of cooling*. One ton of refrigeration represents the capacity of a refrigeration system to freeze 2,000 pound mass (lbm) or 1 ton of liquid water at 32°F into ice in 24 hours. That is, 1 ton of refrigeration = 12,000 Btu/hour. In the case of an air-conditioning unit, one ton of cooling represents the capacity of the air-conditioning system to remove 12,000 British thermal units of energy from a building in one hour. Clearly, the capacity of a residential air-conditioning system depends on the size of the building, its construction, shading, the orientation of its windows, and its climatic location. Residential air-conditioning units generally have a one-to five-ton capacity.

A SEER rating represents the seasonal energy efficiency ratio of an air-conditioning unit.

A ton of cooling is equal to 12,000 Btu of energy expended per hour.

**EXAMPLE 4.5**

An air-conditioning unit has a cooling capacity of 24,000 Btu/hour (two tons of cooling). If the unit has a rated energy efficiency ratio (EER) of 10, how much electrical energy is consumed by the unit in one hour? If a power company charges 12 cents per kilowatt-hour (kWh) of usage, how much would it cost to run the air-conditioning unit for a month (30 days), assuming the unit runs 10 hours a day?

First, note that a kilowatt-hour is a unit of energy—not power—and represents the amount of energy consumed during one hour by a device that uses 1,000 watts or one kilowatt.

We can now compute the energy consumption of the given air-conditioning unit using Equation (4.3).

$$\text{EER} = \frac{\text{heat removal from the unit (Btu)}}{\text{energy input to the unit (Wh)}}$$

$$10 = \frac{24{,}000 \text{ Btu}}{\text{energy input to the unit (Wh)}}$$

$$\text{energy input to the unit (Wh)} = 2{,}400 \text{ Wh}$$

$$= 2.4 \text{ kWh}$$

The cost to run the unit for 10 hours a day over a period of 30 days is calculated in the following manner:

$$\text{cost to operate the unit} = \left(\frac{2.4 \text{ kWh}}{1 \text{ h}}\right)\left(\frac{10 \text{ h}}{\text{day}}\right)\left(\frac{\$0.12}{\text{kWh}}\right)(30 \text{ days}) = \$86.40$$

Pay close attention to how the units of kilowatt-hour (KWh), hour (h), day, and days in the numerator and denominator cancel out, so you are left with a dollar ($) value.

$$\text{cost to operate the unit} = \left(\frac{2.4 \ \cancel{\text{kWh}}}{\cancel{1\text{h}}}\right)\left(\frac{10 \ \cancel{\text{h}}}{\cancel{\text{day}}}\right)\left(\frac{\$0.12}{\cancel{\text{kWh}}}\right)(30 \ \cancel{\text{days}})$$

$$= \$86.40$$

## Standards and Codes

Today's existing standards and codes ensure that we have safe appliances, lighting systems, and electronic devices. They also ensure that we have safe structures, safe transportation systems, safe drinking water, safe indoor/outdoor air quality, safe products, and reliable services. Standards also encourage uniformity in the size of parts and components that are made by various manufacturers around the world. It is also important to note that there are many standardization organizations in the world. Among the more well-known and internationally recognized organizations are the National Fire Protection Association (NFPA) and the Underwriters Laboratories (UL).

**National Fire Protection Association (NFPA)** Losses from fires total billions of dollars per year. Fire, formally defined as a process during which rapid oxidization of a material occurs, gives off radiant energy that can be not only felt but also seen. Fires can be caused by malfunctioning electrical systems, hot surfaces, and overheated materials. The ***National Fire Protection Association (NFPA)*** is a not-for-profit organization established in 1896 to provide codes and standards to reduce the burden of fire. The NFPA publishes the *National Electrical Code®*, the *Life Safety Code®*, the *Fire Prevention Code™*, the *National Fuel Gas Code®*, and the *National Fire Alarm Code®*. It also provides training and education.

*Source:* Underwriters Laboratories Inc.

**Underwriters Laboratories (UL)** The ***Underwriters Laboratories Inc. (UL)*** is a nonprofit organization that performs product safety tests and certifications. Founded in 1894, today Underwriters Laboratories has laboratories in the United States, England, Denmark, Hong Kong, India, Italy, Japan, Singapore, and Taiwan. Its certification mark is one of the most recognizable marks on products.

## Before You Go On

Answer the following questions to test your understanding of the preceding section:

1.  What is a typical amperage rating for a residential building?

2.  What are electric wires typically made of, and how are wire sizes expressed?

3.  Give examples of power ratings for home appliances and electronics.

4.  What is the mission of the NFPA?

*Vocabulary—State the meaning of the following terms:*

AWG

COP

SEER

Ton of cooling

## LO³  4.3  Lighting Systems

In this section, we provide a brief introduction to lighting systems. Lighting systems account for a major portion of electricity use in buildings, and have received a great deal of attention recently due to energy and sustainability concerns. As one would expect, energy is saved by reducing illumination levels, increasing lighting efficiency, or by taking advantage of daylighting. *Daylighting* refers to using windows and skylights to bring natural light into a

S_E / Shutterstock.com

building to reduce the need for artificial lighting. As is the case with any new areas you explore, lighting has its own terminology. Therefore, make sure you spend a little time to familiarize yourself with the terms so you can follow the example problems later.

**Illumination** Let us begin by defining illumination. *Illumination* refers to the distribution of light on a horizontal surface, and the *amount of light emitted by a lamp* is expressed in **lumens**. As a reference, a 100-watt incandescent bulb may emit 1,700 lumens. Another important lighting characteristic is the intensity of illumination. The *intensity of illumination* is a measure of how light is distributed over an area. A common unit of illumination intensity is called **footcandle** and is equal to one lumen distributed over an area of one square foot ($ft^2$). To give you an idea of what a footcandle represents, you need between 5 to 20 footcandles to find your way around at night. As another example, 30 to 50 footcandles are needed for office work. If you have to do detailed work, such as fixing electronic equipment or a spring-driven watch, you need around 200 footcandles of illumination intensity.

**Efficacy** Another term used by lighting engineers, *efficacy* is the ratio of how much light is produced by a lamp (in lumens) to how much energy is consumed by the lamp (in watts).

$$\text{efficacy} = \frac{\text{light produced (lumens)}}{\text{energy consumed by the lamp (watts)}} \qquad 4.4$$

Efficacy is used by lighting engineers when designing an optimal lighting system for a building or by someone who is doing an energy audit of a building to determine if the lighting system is energy efficient. When engineers design a lighting system for a building, they consider many factors such as activity, safety, and task. Sometimes the lighting system is designed to draw attention to a feature or to something special in a building; this is called accent lighting, and you often see it in places like department stores.

**Lighting Systems** As you know, there are many types of light bulbs and fixtures. According to the U.S. Department of Energy, in 2009, *incandescent* lights account for 85 percent of lights used in homes. Unfortunately, incandescent lights have very low efficacy values (10 to 17 lumens per watt). They also have a short service life (750 to 2,500 hours). Because of the advent of new technologies, we are now using different types of light bulbs that are much more efficient.

Another important factor in choosing a lighting system for an application is its source color. As shown in Figure 4.6, in an incandescent lamp, the electric current runs through the lead wires and heats up the filament (a tiny coil of tungsten wire), which makes the tungsten glow or produce light. The light produced in this manner is a yellowish color. In general, the colors of light sources are classified into warm or cool categories. The yellow to red range of colors are considered warm, whereas the blue to green range of colors are considered cool. For a light source, it is common to define a color temperature in Kelvin. The higher color Kelvin temperatures (3,600 to 5,500 K) are considered cool, while lower color temperatures (2,700 to 3,000 K) are considered warm. Warm light sources are preferred for general indoor tasks. Be careful with the counterintuitive way the warm and cool light sources are defined (high temperatures are cool, whereas low temperatures are warm)! How true the colors of an object appear when illuminated by a light source is more important than the color temperature of the light source. For this reason, a variable called the ***color rendition index*** (CRI) is defined. The CRI provides a measure of how well a light source renders the true colors of an object when compared with direct sunlight. The color rendition index has a scale of 1 to 100, with a 100-W incandescent light bulb having a CRI value of approximately 100. For most indoor applications, light sources with a CRI of 80 or higher are preferred.

> The color rendition index (CRI) provides a measure of how well a light source renders the true colors of an object as compared with direct sunlight. The color rendition index has a scale of 1 to 100.

There are different types of incandescent light bulbs. The standard incandescent light—referred to as a screw-in, A-type—is becoming obsolete. There are also tungsten, halogen, and R-type incandescent light bulbs. The tungsten and halogen lamps have higher efficiencies than A-types because they have inner coatings that reflect heat; consequently, they require less energy to keep the filament hot at a certain temperature. The reflective R-type incandescent

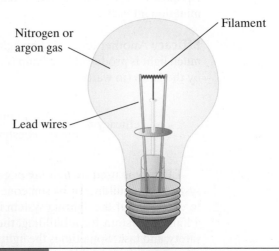

| FIGURE 4.6 | A schematic of an incandescent bulb |
|---|---|

*Source:* DOE's Office of Energy Efficiency and Renewable Energy.

| TABLE 4.3 | Comparison of Incandescent Lights | | | |
|---|---|---|---|---|
| Incandecent Lighting Type | Efficacy (lumens/W) | Life (hours) | Color Rendition Index (CRI) | Color Temperature (K) |
| Standard A | 10–17 | 750–2,500 | 98–100 | 2,700–2,800 (warm) |
| Tungsten halogen | 12–22 | 2,000–4,000 | 98–100 | 2,900–3,200 (warm to neutral) |
| Reflector | 12–19 | 2,000–3,000 | 98–100 | 2,800 (warm) |

*Source:* U.S. Department of Energy

lights also spread and direct light over a specific area. They are commonly used as floodlights or spotlights. The comparisons of performance of incandescent lights are shown in Table 4.3.

The second most common type of lighting systems are *fluorescent* lamps, which use 25 to 35 percent of energy compared to incandescent lamps and produce the same amount of illumination. The efficacy of fluorescent lamps is somewhere between 30 to 110 lumens/watts. When compared to incandescent lamps, they also have longer service life, in the range of 7,000 to 24,000 hours. In a fluorescent tube, electric current is conducted through mercury and inert gases to produce light. The fluorescent lights of the past had a poor color rendition, but because of improvements in technology, they now have high CRI values. The 40-W, 4-foot (1.2-meter) and 75-W, 8-foot (2.4-meter) lamps are the two most common fluorescent lamps. These lamps require special fixtures, but the new generation of compact fluorescent lamps (CFLs) fit into the incandescent fixtures (Figure 4.7). Although CFLs are more expensive than incandescent

> Fluorescent lamps use 25 to 35% of energy compared to incandescent lamps and produce the same amount of illumination.

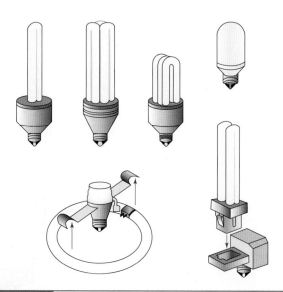

| FIGURE 4.7 | Examples of compact fluorescent lights that fit into screw-in, A-type fixtures |
|---|---|

*Source:* DOE's Office of Energy Efficiency and Renewable Energy.

| TABLE 4.4 | | Comparison of Fluorescent Lights | | |
|---|---|---|---|---|
| **Fluorescent Lighting Type** | **Efficacy (lumens/watt)** | **Lifetime (hours)** | **Color Rendition Index (CRI)** | **Color Temperature (K)** |
| Straight tube | 30–100 | 7,000–24,000 | 50–90 (fair to good) | 2,700–6,500 (warm to cold) |
| Compact flourescent lamp (CFL) | 50–70 | 10,000 | 65–88 (good) | 2,700–6,500 (warm to cold) |

light bulbs (3 to 10 times), because of their long service lives (6,000 to 15,000 hours) and high efficacy values, their use results in net savings. The comparison among different types of fluorescent lights is shown in Table 4.4.

Another common type of lighting system is the *high-intensity discharge* (HID) lamps (Figure 4.8). They have the highest efficacy values and the longest service life of any lighting systems. HID lamps are commonly used in indoor arenas and outdoor stadiums. As you know from your experience, they have low color rendition index, and when you turn them on, it takes a few minutes before they produce light. The comparison among different types of high-intensity discharge lights is shown in Table 4.5.

The newest type of lighting system is light-emitting diode (LED) lights. LEDs have become a popular alternative to incandescent lights, because they last longer than conventional incandescent lights with a service life of approximately 20,000 hours. They also use much less power and operate at cooler temperatures, so they reduce fire hazards—particularly for Christmas lighting during the holiday season. Increasingly, LEDs are becoming popular alternatives in other applications such as traffic lights, street lights, indoor lights, large display screens, and TV screens. The U.S. Department of Energy estimates that the widespread use of LED lights by 2027 could result in energy savings of $350 \times 10^9$ kWh.

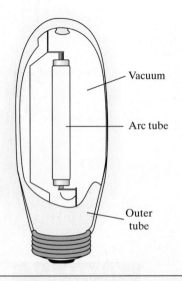

Vacuum

Arc tube

Outer tube

| FIGURE 4.8 | A schematic of a high-intensity discharge lamp |
|---|---|

*Source:* DOE's Office of Energy Efficiency and Renewable Energy

| TABLE 4.5 | Comparison Between High-Intensity Discharging Lights | | | |
|---|---|---|---|---|
| High-Intensity Discharging Lighting Type | Efficacy (lumens/W) | Life (hours) | Color Rendetion Index (CRI) | Color Temperature (K) |
| Mercury Vapor | 25–60 | 16,000–24,000 | 50 (poor to fair) | 3,200–7,000 (warm to cold) |
| Metal Halide | 70–115 | 5,000–20,000 | 70 (fair) | 3,700 (cold) |
| High Pressure Sodium | 50–140 | 16,000–24,000 | 25 (poor) | 2,100 (warm) |

**EXAMPLE 4.6**

According to Sylvania, a light bulb manufacturer, its 75-W CFL floodlight consumes 23 W and produces 1,250 lumens. What is the efficacy of the floodlight?

$$\text{efficacy} = \frac{\text{light produced (lumens)}}{\text{energy consumed (watts)}} = \frac{1,250}{23} = 54$$

**EXAMPLE 4.7**

A 100-W CFL light manufactured by Buyer's Choice consumes 23 watts, has an illumination rating of 1,600 lumens, a service life of 8,000 hours, and costs $1.81. As an alternative, a generic 100-W incandescent light bulb costs $0.38, produces 1,500 lumens, and has a service life of 750 hours. Compare the performance of each light bulb by calculating the efficacy for each light and also estimating how much it would cost to run each light for eight hours per day for 220 days in a year. Assume electricity costs nine cents per kWh.

For the Buyer's Choice 100-W CFL light,

$$\text{efficacy} = \frac{1,600}{23} = 70$$

$$\text{cost} = \left(\frac{8 \text{ hours}}{\text{day}}\right) (220 \text{ days})(23 \text{ W}) \left(\frac{1 \text{ kW}}{1,000 \text{ W}}\right) \left(\frac{\$0.09}{\text{kWh}}\right) = \$3.64$$

For the generic incandescent 100-W light bulb,

$$\text{efficacy} = \frac{1,500}{100} = 15$$

$$\text{cost} = \left(\frac{8 \text{ hours}}{\text{day}}\right) (220 \text{ days})(100 \text{ W}) \left(\frac{1 \text{ kW}}{1,000 \text{ W}}\right) \left(\frac{\$0.09}{\text{kWh}}\right) = \$15.84$$

It should be obvious that the CFL light is more efficient and economical to operate than the generic incandescent light bulb.

## Lighting System Audit

As we said at the beginning of this section, lighting systems account for a major portion of electricity use in buildings and have received much attention recently due to energy and sustainability concerns. A lighting energy audit starts with space classification. That is, what is the space used for? Is it used as an office, a warehouse, or a manufacturing plant? Next, an energy auditor determines the space characteristics (length, width, height), light fixtures (lamp types, their number, and lamp wattage), and their controls. The auditor then talks to the users about the lighting level, their tasks, and the occupancy profile, and using a light meter, the auditor measures the light level in the space. A comparison between the measurements and the Illuminating Engineering Society (IES) recommendation values for a given task is then made. The auditor also calculates power consumption of the lighting system per unit area (watts/ft$^2$) and compares it to design guidelines. Finally, the energy auditor prepares a report discussing his or her findings, including an estimate of annual lighting energy cost and ways by which the energy consumption due to the lighting system may be reduced (for example, by reducing illumination levels, taking advantage of daylighting, or increasing the efficiency of lighting systems).

## Before You Go On

Answer the following questions to test your understanding of the preceding section:

1. What do we mean by illumination, and how is it expressed?

2. What are the common types of lighting systems?

3. Which type of lighting system has the highest CRI?

4. Which type of lighting system has the longest life span?

5. Which type of lighting system has the highest efficacy?

*Vocabulary–State the meaning of the following terms:*

Footcandle

Efficacy

CRI

CFL

LED

## LO⁴  4.4  Electric Power Generation, Transmission, and Distribution

The most common means by which we produce electricity are magnetism (e.g. in power plants), chemical reaction (batteries), light (photovoltaic cells), and converting wind and water energy.

We use electricity in our homes for lighting, space heating and cooling, running our appliances, and powering our electronic devices. Manufacturers use electricity to make consumer goods and provide services that allow us to enjoy a high standard of living. By now, you should have a good understanding of the basic principles of electricity. As a well-educated global citizen, it is also important that you understand how electricity is generated. The most common means by which we produce electricity are *magnetism* (power plants), *chemical reaction* (batteries), and *light* (photovoltaic).

### Power Plants

Everyone should have a basic understanding of how electricity is generated in a conventional power plant, because we rely on it for just about everything we do. It is also important to understand what is required to produce electricity so that we won't waste it.

Water is used in all conventional steam power-generating plants to produce electricity. A simple schematic of a power plant is shown in Figure 4.9. Fuel such as coal or natural gas is burned in a **boiler** to generate heat, which in turn is used to to change the phase of the water that runs through tubes in a heat exchanger from a liquid to a high temperature and pressure steam; this steam then passes through turbine blades, turning the blades that in effect run the generator connected to the turbine, creating electricity. The low-pressure steam coming out of the turbine is cooled down back to liquid water in a condenser by drawing cold water into the condenser from nearby rivers or lakes. The liquefied water is pumped through the boiler again, completing a cycle, as shown in Figure 4.9.

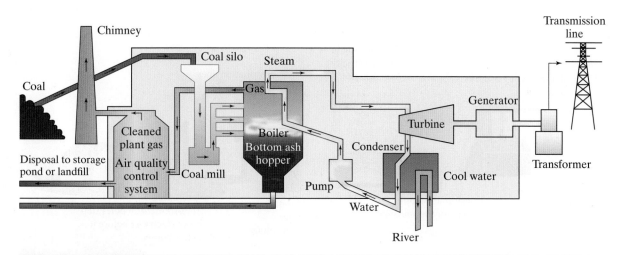

**FIGURE 4.9**  A schematic of a steam power plant
*Source:* Based on Xcel Energy

As we also mentioned in Chapter 3, there is always some loss associated with all systems, including power plants. All systems require more input than what they put out. Recall that when we wish to show how well a system is functioning, we express its efficiency. The overall ***efficiency of a steam power plant*** is defined as:

$$\text{power plant efficacy} = \frac{\text{net energy generated}}{\text{energy input from fuel}}$$

The efficiency of today's power plants where a fossil fuel (natural gas, coal) is burned in the boiler is near 40 percent. What does this number mean? It means that for every 100 pounds of fuel burned, only 40 pounds of fuel (i.e., its energy content) goes to producing electricity; the remaining 60 pounds is lost in the process.

The electricity that is generated in a power plant is transmitted to homes, factories, and other facilities through *transmission lines*. As much as six additional percent of the power generated is lost during transmission in power

lines. A schematic of electric power generation, transmission, and distribution is shown in Figure 4.10. The electricity that is generated at power plants is stepped up in voltage and is transmitted through a network of approximately 160,000 miles of high-voltage lines. As shown in Figure 4.10, the transmission lines then deliver the electricity to neighborhood transformer stations, where the voltage is stepped down to be carried by the local distribution lines that finally carry the electricity to our homes. The process of transporting electricity is further explained in the box, "The Process of Transporting Electricity."

hans engbers / Shutterstock.com

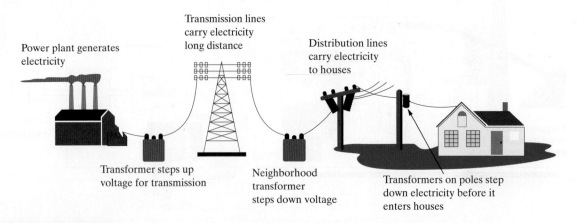

| FIGURE 4.10 | A schematic of power generation, transmission, and distribution |

*Source:* Based on National Energy Education Development Project

# The Process of Transporting Electricity

Getting electricity from power-generating stations to our homes and workplace is quite a challenging process. Electricity must be produced at the same time as it is used because large quantities of electricity cannot be stored effectively. High-voltage transmission lines (those lines between tall metal towers that you often see along the highway) are used to carry electricity from power-generating stations to the places where it is needed. However, when electricity flows over these lines, some of it is lost. One of the properties of the high voltage lines is that the higher the voltage, the more efficient they are at transmitting electricity, that is, the lower the losses are. Using transformers, high-voltage electricity is "stepped down" several times to a lower voltage before arriving over the distribution system of utility poles and wires to homes and workplaces so that it can be used safely.

There is no "national" power grid. There are actually three power grids operating in the 48 contiguous states:(1) the Eastern Interconnected System (for states east of the Rocky Mountains), (2) the Western Interconected System (from the Pacific Ocean to the Rocky Mountain states), and (3) the Texas Interconnected System. These systems generally operate independently of each other, although there are limited links between them. Major areas in Canada are totally interconnected with our Western and Eastern power grids, while parts of Mexico have limited connection to the Texas and Western power grids.

## The "Smart Grid"

The "Smart Grid" consists of devices connected to transmission and distribution lines that allow utilities and customers to receive digital information from and communicate with the grid. These devices allow a utility to find out where an outage or other problem is on the line and sometimes even fix the problem by sending digital instructions. Smart devices in the home, office, or factory inform consumers of times when an appliance is using relatively high-cost energy and allow consumers to remotely adjust its settings. Smart devices make a Smart Grid, as they help utilities reduce line losses, detect and fix problems faster, and help consumers conserve energy, especially at times when demand reaches significantly high levels or an energy demand reduction is needed to support system reliability.

U.S. Energy Information Administration

Shutter_M / Shutterstock.com

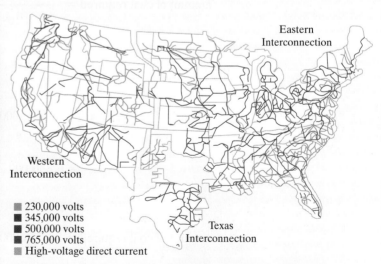

- 230,000 volts
- 345,000 volts
- 500,000 volts
- 765,000 volts
- High-voltage direct current

"The National Power Grid," Microsoft® Encarta® Encyclopedia. http://encarta.msn.com ©1993-2004 Microsoft Corporation. All rights reserved. Anita Potter/Shuttershock.com, data from OffGridWorld, What is the Electric Power Grid, http://www.offgridworld.com/what-is-the-electric-power-grid-u-s-grid-map/, EIA U.S. Energy Information Administration

**EXAMPLE  4.8**

A 100-megawatt (MW) coal-fired power plant has an efficiency of 40 percent. Estimate the daily consumption of coal burned in this plant to generate electricity. Assume a heating value of approximately 7.5 megajoules per kilogram (MJ/kg) for coal.

The efficiency of a power plant is given by:

$$\text{power plant efficiency} = \frac{\text{net energy generated}}{\text{energy input from fuel}}$$

$$0.4 = \frac{\overbrace{100 \times 10^6 \left(\dfrac{J}{s}\right)}^{100\ \text{MW}} \left(\dfrac{3{,}600\ s}{1\ h}\right)\left(\dfrac{24\ h}{\text{day}}\right)}{\text{energy input from fuel}}$$

And solving for energy input from fuel, we get

$$\text{energy input from fuel} = \frac{\overbrace{100 \times 10^6 \left(\dfrac{J}{\cancel{s}}\right)}^{100\ \text{MW}} \left(\dfrac{3{,}600\ \cancel{s}}{1\ \cancel{h}}\right)\left(\dfrac{24\ \cancel{h}}{\text{day}}\right)}{0.4}$$

$$= 2.16 \times 10^{13}\ \frac{J}{\text{day}}$$

$$\text{amount of coal required} = \frac{2.16 \times 10^{13}\ \dfrac{\cancel{J}}{\text{day}}}{7.5 \times 10^6\ \dfrac{\cancel{J}}{\text{kg}}} = 2{,}880{,}000\ \frac{\text{kg}}{\text{day}}$$

or in terms of pound mass and ton as

$$\text{amount of coal required} = \left(2{,}880{,}000\ \frac{\text{kg}}{\text{day}}\right)\left(\frac{2.2\ \text{lbm}}{1\ \text{kg}}\right)$$

$$= 6{,}336{,}000\ \frac{\text{lbm}}{\text{day}}$$

$$\text{amount of coal required} = \left(6{,}336{,}000\ \frac{\text{lbm}}{\text{day}}\right)\left(\frac{1\ \text{ton}}{2{,}000\ \text{lbm}}\right) = 3{,}168\ \frac{\text{ton}}{\text{day}}$$

Take a moment and think about these values!

## Nuclear Energy

Nuclear power plants are similar in many ways to conventional fossil-fuel-fired power plants that produce electricity. The main difference is that, instead of fuel such as coal or natural gas, nuclear fuel is used to generate steam. Again, as is the case with any new concepts, the energy sector has its own terminology, and as good citizens you need to become aware of them. There are two processes by which *nuclear energy* is harnessed: nuclear fission and nuclear fusion.

hxdyl / Shutterstock.com

Nuclear power plants use nuclear fission to produce electricity. To release energy in *nuclear fission*, atoms of uranium are bombarded by small particles called neutrons. This process splits the atoms of uranium and releases more neutrons and energy in the form of heat and radiation. The additional neutrons go on to bombard other uranium atoms, and the process keeps repeating itself, leading to a chain reaction. This process is depicted in Figure 4.11.

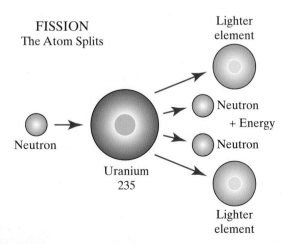

| **FIGURE 4.11** | The nuclear fission process |

The fuel most widely used by nuclear power plants is uranium 235 or simply U-235. The major components (such as the turbine and generator) of a nuclear power plant are similar to those of conventional plants. Also note that for nuclear power plants the overall efficiency is nearly 34 percent.

The energy in the nucleus or core of atoms also can be released by nuclear fusion. In ***nuclear fusion***, energy is released when atoms are combined or fused together to form a larger atom. This is the process by which the sun's energy is produced.

## Hydro Energy

Electricity is also generated by using water stored behind dams. The water is guided into turbines located in hydroelectric power plants housed within the dam to generate electricity. The potential energy due to the height of water stored behind the dam is converted to kinetic energy (moving energy) as the water flows through and consequently spins the turbine, which turns the generator. Hydropower accounts for approximately 6 percent of the total United States electricity generation.

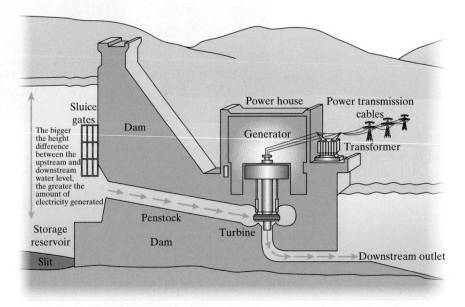

Based on Environment Canada

## Chemical Reaction—Batteries

All of you have used batteries for different purposes (Figure 4.12) at one time or another. In all batteries, electricity is produced by the chemical reaction that takes place within the battery. When a device that uses batteries is on, its circuits create paths for the electrons to flow through. When the device is turned off, there is no path for the electrons to flow through, thus the chemical reaction stops.

A battery cell consists of chemical compounds, internal conductors, positive and negative connections, and the casing. Examples of cells include sizes N, AA, AAA, C, and D. A cell that cannot be recharged is called a *primary cell*. An alkaline battery is an example of a primary cell. On the other hand, a *secondary cell* is a cell that can be recharged. Recharging is accomplished by reversing the current flow from the positive to the negative areas. Lead acid

**FIGURE 4.12**     Different types of batteries

cells in your car battery, nickel-cadmium (NiCd) cells, and nickel-metal hydride (NiMH) cells are examples of secondary cells. The NiCd batteries are some of the most common rechargeable batteries used in cordless phones, toys, and some cellular phones. The NiMH batteries, which are smaller, are used in many smaller cellular phones because of their size and capacity.

To increase the voltage output, batteries are often placed in a series arrangement. If we connect batteries in a series arrangement, the batteries produce a net voltage, which is the sum of the individual batteries. For example, if we were to connect four 1.5-volt batteries in series, the resulting potential would be 6 volts, as shown in Figure 4.13. Batteries connected in a parallel arrangement, as shown in Figure 4.14, produce the same voltage but more current.

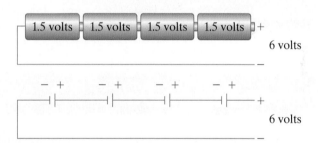

**FIGURE 4.13**     Batteries connected in a series arrangement

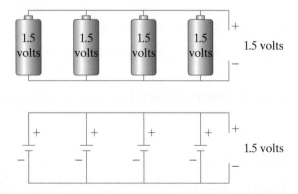

**FIGURE 4.14**     Batteries connected in a parallel arrangement

## Light—Photoemission

*Photoemisssion* is another principle used to generate electricity. When light strikes a surface that has certain properties, electrons can be freed; thus electric power is generated. You have seen examples of photovoltaic devices such as light meters used in photography, photovoltaic cells in hand-held calculators, and solar cells used to generate electricity.

## Photovoltaic Systems

A *photovoltaic system* converts light energy directly into electricity. It consists of a photovoltaic array, batteries, a charge controller, and an inverter (a device that converts direct current into alternating current). The backbone of any

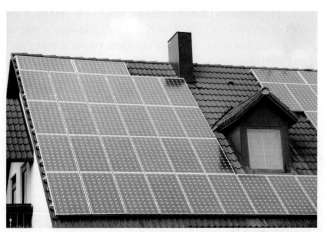

photovoltaic system are the cells. The photovoltaic cells are combined to form a module, and modules are combined to form an array. Photovoltaic systems come in all sizes and shapes and are generally classified into stand-alone systems, hybrid systems, or grid-tied systems. The systems that are not connected to a utility grid are called stand-alone. Hybrid systems are those which use a combination of photovoltaic arrays and some other form of energy, such as diesel generation or wind. As the name implies, grid-tied systems are connected to a utility grid. We discuss photovoltaic systems and wind energy in more detail in Chapter 7.

manfredxy / Shutterstock.com

# Before You Go On

Answer the following questions to test your understanding of the preceding section:

1.  In your own words, explain how electricity is produced in a conventional power plant.

2.  In your own words, explain how nuclear energy is harnessed.

3.  How does a photovoltaic system work?

4.  In your own words, explain how the power of water is harnessed to generate electricity.

*Vocabulary–State the meaning of the following words:*

Nuclear energy

Photovoltaic system

Power plant efficiency

# SUMMARY

## LO¹ Current, Voltage, and Electric Power—Basic Concepts

You should be familiar with basic principles of electricity. The flow of electric charge is called *electric current* or simply *current. The electric current, or the flow of charge, is measured in amperes* (A). One ampere or "amp" is defined as the flow of 1 unit of charge per second. *Voltage* represents the amount of work required to move charge between two points, and the amount of charge that is moving between the two points per unit time is called *current.* Moreover, *direct current (dc)* is the flow of electric charge that occurs in one direction. Batteries and photovoltaic systems create direct current. *Alternating current (ac)* is the flow of electric charge that periodically reverses. Alternating current is created by generators at power plants. The current drawn by various electrical devices at your home is alternating current. The alternating current in domestic and commercial power use is 60 cycles per second (hertz) in the United States.

*Resistivity* is a measure of the resistance of a piece of material to electric current. *Ohm's law* describes the relationship among voltage (*V*), resistance (*R*), and current (*I*), according to

$$\text{voltage} = (\text{resistance})(\text{current})$$

The electric resistance is measured in units of ohms (Ω). An element with a 1-ohm resistance allows a current flow of 1 amp when a potential of 1 volt across the element exists. The electric power consumption of various electrical components can be determined using the power formula:

$$\text{power} = (\text{voltage})(\text{current})$$

## LO² Residential Power Distribution and Consumption

As a good citizen, it is important that you become familiar with your home's power distribution system and consumption. You should know that, in the United States, a typical home has a total 200 amperage rating and be aware that, in order to wire a building, an electrical plan for the building is first developed. Moreover, at the heart of every home

electrical-distribution system is wire. Electrical wires are typically made of copper or aluminum. The actual size of the wires is commonly expressed in terms of the gage number as denoted by the American Wire Gage (AWG). The smaller the gage number, the bigger the wire diameter. The National Electrical Code, which is published by the Fire Protection Association, contains specific information on the types of wires used for general wiring. You should also be familiar with the power consumption of typical home appliances and electronics such as refrigerators; electric stoves; microwaves; dishwashing machines; clothes washing machines; clothes dryers; heating, cooling, and ventilating units; TVs; computers; and radios.

## LO³ Lighting Systems

You should be familiar with basic lighting terminology and be able to calculate power consumption rates for lighting systems. Illumination refers to the distribution of light on a horizontal surface, and the *amount of light emitted by a lamp* is expressed in lumens. As a reference, a 100-watt incandescent lamp may emit 1,700 lumens. A common unit of *illumination intensity* is called a *footcandle* and is equal to one lumen distributed over an area of 1 square foot. For example, to find your way around at night, you need between 5 to 20 footcandles. Efficacy is another term used in lighting vocabulary. *Efficacy* is the ratio of how much light is produced by a lamp (in lumens) to how much energy is consumed by the lamp (in watts).

$$\text{efficacy} = \frac{\text{light produced (lumens)}}{\text{energy consumed by the lamp (watts)}}$$

How true the colors of an object appear when illuminated by a light source is represented by the *color rendition index* (CRI). The color rendition index has a scale of 1 to 100 with a 100-W incandescent light bulb having a CRI value of approximately 100. There are many different types of lighting systems, including incandescent light bulbs, fluorescent lamps, compact fluorescent lamps, high-intensity discharge (HID) lamps, and LED (light emitting diode) lights.

## LO⁴  Electric Power Generation, Transmission, and Distribution

You should know how electricity is generated. The most common means of producing electricity are *magnetism*, *chemical reaction*, and *light*. Water is used in all conventional steam power-generating plants to produce electricity. Fuel is burned in a boiler to generate heat, which in turn is added to liquid water to change the phase of water to high temperature and pressure steam; steam then passes through turbine blades, turning the blades, which in effect runs the generator (magnetism) connected to the turbine, creating electricity. You should also know that the efficiency of today's power plants where a fossil fuel (fuel oil, gas, coal) is burned in the boiler is nearly 40 percent.

Nuclear power plants use nuclear fission to produce electricity. To release energy by nuclear fission, atoms of uranium are bombarded by small particles called neutrons. This process splits the atoms of uranium and releases more neutrons and energy in the form of heat and radiation. The additional neutrons go on to bombard other uranium atoms, and the process keeps repeating itself, leading to a chain reaction. The fuel most widely used by nuclear power plants is Uranium 235 or simply U-235.

Electricity is also generated by liquid water stored behind dams. The water is guided into water turbines (that are connected to generators) located in hydroelectric power plants housed within the dam to generate electricity.

In all batteries, electricity is produced by the chemical reaction that takes place within the battery. When a device that uses batteries is on, its circuits create paths for the electrons to flow through. When the device is turned off, there is no path for the electrons to flow through, and the chemical reaction stops. Photoemisssion is another principle used to generate electricity. A Photovoltaic system converts light energy directly into electricity. This system consists of a photovoltaic array, batteries, a charge controller, and an inverter.

## KEY TERMS

Alternating Current  98
Ampere  98
AWG  104
Boiler  117
Color Rendition Index  112
Condenser  107
Direct Current  98
Efficiency of a Steam Power
   Plant  118

Efficacy  111
Electric Charge  98
Electric Circuit  99
Electric Current  98
Electric Power  100
Electric Resistance  99
Footcandle  111
Hydro Energy  122
Lumens  111

NFPA  110
Nuclear Energy  121
Nuclear Fission  121
Ohm  99
Photoemission  124
Photovoltaic System  124
UL  110
Voltage  98

## *Apply What You Have Learned*

This is a class project. You are to perform a lighting energy audit for an indoor gym or sport arena. Collect information about the arena or gym size and an occupancy profile— that is, the number of people that use the facility every 15 or 30 minutes. Also obtain information about the lighting systems (fluorescent, incandescent, mercury vapor, sodium, metal halide, etc.) and controlling devices used in the space. Calculate the watts per square foot ($W/ft^2$) for the space as a function of time. Write a brief report to your instructor discussing your findings. Suggest ways the lighting energy consumption may be reduced.

Pavel L Photo and Video / Shutterstock.com

# PROBLEMS

*Problems that promote life-long learning are denoted by* 🔑

**4.1** If a 1500-W hair dryer is connected to a 120-V line, what is the maximum current drawn?

**4.2** A toaster connected to a 120-V line draws 7 amps. What is the power consumption of the toaster?

**4.3** A 20-MW coal-fired power plant has an efficiency of 40%. Estimate the annual consumption of coal burned in this plant. Assume a heating value of approximately 7.5 MJ/kg (megajoules per kilogram) for coal.

**4.4** A 2,000-W dishwasher is run for approximately one hour for 122 days during a year. If the electric power company charges 12 cents per kWh, calculate the annual electric energy consumption of the dishwasher and the associated cost.

**4.5** A 500-W clothes washing machine is run for approximately two hours a week for 52 weeks during a year. If the electric power company charges 12 cents per kWh, calculate the annual electric energy consumption of the clothes washer and the associated cost.

**4.6** A 1,200-W hair dryer is run for approximately 5 minutes every day. If the electric power company charges 12 cents per kWh, calculate the annual electric energy consumption of the hair dryer and the associated cost.

**4.7** A TV set consumes 150 W and is left on for approximately 5 hours every night. If the electric power company charges 12 cents per kWh, calculate the annual electric energy consumption of the TV and the associated cost.

**4.8** A relatively small house in the United States has an annual electrical energy consumption of 4,000 kWh. Estimate the annual consumption of coal burned in a power plant to generate the required electrical energy. Assume a heating value of approximately 7.5 MJ/kg (megajoules per kilogram) for coal, 40% efficiency at the power plant, and 6% loss for the power transmission lines.

**4.9** Rework Problem 4.8 for a larger house with an annual electrical energy consumption of 10,000 kWh.

**4.10** 🔑 The National Electrical Code (NEC) covers the safe and proper installation of wiring, electrical devices, and equipment in private and public buildings. The NEC is published by the National Fire Protection Association (NFPA) every three years. As an example of an NEC provision, the receptacle outlets in a room in a dwelling should be placed such that no point on the wall space is more than 6 ft away from the outlet in order to minimize the use of extension cords. After performing a Web search or obtaining a copy of the NEC handbook, give at least three other examples of National Electric Codes for a family dwelling.

**4.11** Imagine that you are given three items: a battery, a light bulb, and a piece of wire. How would you make a flashlight using these items?

**4.12** According to Sylvania (a light bulb manufacturer), its 40-W CFL light consumes 9 W and produces 495 lumens. What is the efficacy for this light?

**4.13** The Sylvania Super Saver 75-W light uses 20 W, produces 1,280 lumens, and costs $4.49. As an alternative, a generic 75-W incandescent light bulb costs $0.40 and produces 1,200.

a. Compare the performance of each light by calculating the efficacy for each light.

b. Estimate how much it would cost to run each light for 4 h per day for 300 days in a year.

Assume electricity costs 9 cents per kWh.

**4.14** 🔑 Visit the lighting section of a hardware store and look up the following information for comparable CFL and LED lights. Read the manufacturer's ratings on the packaging and record lumens, light source color temperature, and power consumption in terms of watts.

**4.15** Estimate the annual cost of electricity for the lighting system at your home or in your dorm room. State your assumptions.

**4.16**   Estimate the annual electricity usage and cost for your computer and for any electronic entertainment devices you own. State your assumptions.

**4.17**   Estimate the annual electricity usage and cost for your TV. State your assumptions.

**4.18**   Investigate the amount of electric power generated by power plants in your state. Discuss the types of plants and the energy sources (fuel types) used to generate electricity.

**4.19**   Estimate the amount of electrical energy consumed during a year for a case when a garage door is lifted four times each day. Assume the garage door opener has a ½-hp-rated motor and it takes 14 s to lift the garage door.

**4.20**   Estimate the annual electrical energy consumption of a laptop computer that consumes approximately 65 W. Assume the laptop is used on average for 8 hours per day and for 220 days during a year.

yurchello108 / Shutterstock.com

*"By failing to prepare you are preparing to fail."* — BENJAMIN FRANKLIN (1706–1790)

# CHAPTER

# 5

# Thermal Energy: Heat Loss and Gain in Buildings

## LEARNING OBJECTIVES

**LO¹** Temperature Difference and Heat Transfer—Basic Concepts: understand how heat transfer occurs

**LO²** Modes of Heat Transfer: explain various modes of heat transfer—including conduction, convection, and radiation—and how to quantify them

**LO³** Daylighting: understand the basic daylighting concepts and window ratings

**LO⁴** Degree-Days and Energy Estimation: explain the meaning of degree-days and how this is used to estimate heating energy consumption

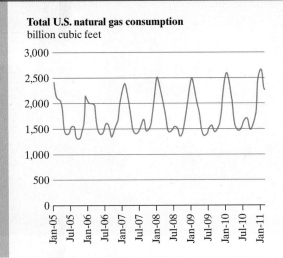

**Total U.S. natural gas consumption**
billion cubic feet

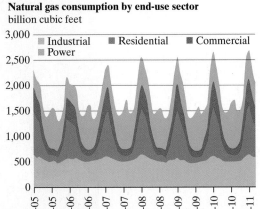

**Natural gas consumption by end-use sector**
billion cubic feet

## Natural Gas Consumption Has Two Peaks Each Year

Consumption of natural gas is seasonal, with consumption patterns among end-use sectors highly driven by the weather. Total natural gas consumption peaks during the winter when cold weather increases demand for natural gas heating. A second, smaller peak occurs during the summer, when electricity generation using natural gas increases to serve the air-conditioning load.

The residential and commercial demand for heating accounts for over 50 percent of the natural gas delivered for end-users in the United States during the winter. During the summer, total consumption of natural gas is, on average, about 30 percent lower than in the winter, with about half the gas used to generate electricity for air-conditioning.

In contrast to these seasonal patterns, natural gas demand in the industrial sector is more evenly distributed throughout the year, although it has varied from about 20 to 40 percent of total consumption over the past six years.

*Source:* U.S. Energy Information Administration, July, 2011

**To the Students:** Why do buildings need to be heated or cooled? What are the mechanisms that drive the need for heating or cooling? What do you think is your campus' or home's annual heating requirement in Btu per hour or kW? On average, how much fuel or electricity do you think you consume to keep your residence comfortable?

## LO¹ 5.1 Temperature Difference and Heat Transfer—Basic Concepts

We spend a great deal of time indoors. Think about all of the human activities that take place inside buildings. We spend time at home, in schools, offices, malls, supermarkets, restaurants, sport arenas, airports, hospitals, and so on. The air temperature inside all of these buildings is usually kept comfortable—regardless of whether it is cold or hot outside.

Songquan Deng / Shutterstock.com

In the United States, space heating and air-cooling accounts for nearly 40 and 8 percent, respectively, of energy use in homes. As a good global citizen, it is therefore important to understand how we spend energy to heat or cool a building so we won't waste it. In order to better understand the heating and cooling of buildings, we must first understand how heat transfer occurs.

Thermal energy transfer occurs whenever a temperature difference inside something or between two things exists, such as the warm air inside a room and the cold air outside a building during the winter months. This form of energy transfer that occurs between bodies of different temperatures is called ***heat transfer***. Another important point to remember is that heat always flows from a high-temperature region to a low-temperature region. This statement can be confirmed easily from your daily observations. When hot coffee in a cup is left in a room with a lower temperature, the coffee cools down. Thermal energy transfer takes place from the hot coffee through the cup to the surrounding room air. This occurs as long as there is a temperature difference between the coffee and its surrounding air. As another example, during the winter when it is cold outside, thermal energy transfer occurs from the warmer air inside a building to the colder air outside through the walls, roof, windows, and doors. The opposite is true in the summer; heat transfer occurs from the hotter air outside to the cooler air inside a building.

Make sure you understand what we mean by *temperature* and *heat*. Heat is a form of energy that is transferred from one place to another place as a result of a temperature difference, whereas temperature represents the level of microscopic molecular movements in something by a single number, say, 77°F (22°C). In winter, the warm air inside a house has higher molecular activity than the colder air outside.

> Heat transfer is a form of energy transfer that occurs as the result of temperature difference.

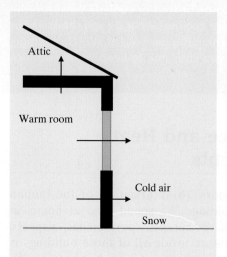

As we discussed in Chapter 3, there are three units that are commonly used to quantify thermal energy: (1) the *British thermal unit* (Btu), (2) the calorie, and (3) the joule (J). Recall that the British thermal unit (Btu) is defined as the amount of heat required to raise the temperature of one pound mass (1 lbm) of water by one degree Fahrenheit (°F). The *calorie* is defined as the amount of heat required to raise the temperature of 1 gram of water by 1 degree Celsius (°C), and 1 joule = 0.239 calorie (or 1 cal = 4.186 J). These units allow us to keep track of how much energy is consumed to heat or cool something.

Now that you know heat transfer occurs as a result of temperature differences in an object or between things, let us look at different *modes* of heat transfer. These are three different mechanisms by which energy is transferred from a high-temperature region to a low-temperature region.

## LO² 5.2  Modes of Heat Transfer

Nebojsa Markovic / Shutterstock.com

The three modes of heat transfer are *conduction*, *convection*, and *radiation*. **Conduction** refers to that mode of heat transfer that occurs when a temperature difference exists in a medium. The energy is transported within the object from the region with more energetic molecules to the region with less energetic molecules. Of course, it is the interaction of the molecules with their neighbors that makes the transfer of energy possible. To better demonstrate the idea of molecular interactions, consider the following example of heat transfer by conduction. All of you have experienced what happens when you heat up some food in an aluminum or steel pot on a stove. Why do the handles or the lid of the pot get hot, even though the handles and the lid are not in direct contact with the heating element? Well, let us examine what is happening. Because of the energy transfer from the heating element, the molecules of the pot in the region near the heating element are more energetic than those molecules further away. The more energetic molecules transfer some of their energy to their neighboring regions, and those neighboring regions do the same thing, until the energy transfer eventually reaches the handles and the lid of the pot. The energy is transported from the high-temperature region (bottom of the pot near the heating element) to the low-temperature region in a medium (the handles of the pot) by molecular activity. The rate of heat transfer by conduction is given by *Fourier's law*, which states that *the rate of heat transfer through a material is proportional to the temperature difference, area through which heat transfer occurs, and the type of material involved*. Fourier's law also states that the heat transfer rate is inversely proportional to the material thickness over which the temperature difference exists. For example, referring to the window shown in Figure 5.1, heat transfer occurs from the warmer glass surface inside in contact with the warm room air to the colder glass surface facing the colder air outside.

> Thermal conductivity is a property of materials that shows how good the material is in transferring thermal energy (heat) from a high-temperature region to a low-temperature region within the material.

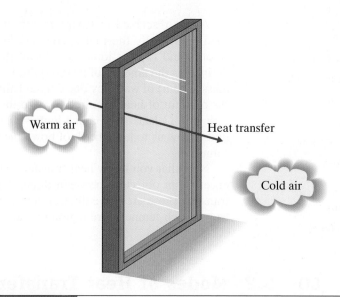

A window under winter conditions.

Note that our medium is the glass, and the heat transfer is occurring through the glass. We can write Fourier's law for a single-pane glass window as

$$q = kA\left(\frac{T_1 - T_2}{L}\right)$$

5.1

where

$q$ = heat transfer rate (W or Btu/h)

$k$ = thermal conductivity $\left(\dfrac{\text{W}}{\text{m} \cdot {}^{\circ}\text{C}} \text{ or } \dfrac{\text{Btu}}{\text{h} \cdot \text{ft} \cdot {}^{\circ}\text{F}}\right)$

$A$ = cross-sectional area normal to heat flow (m$^2$ or ft$^2$)

$T_1 - T_2$ = temperature difference across the material of $L$ thickness (${}^{\circ}$C or ${}^{\circ}$F)

$L$ = material thickness (m or ft)

As we mentioned in Chapter 2, laws are based on observations, and Fourier's law is no exception. Also, notice that Equation (5.1) is homogenous in units; that is, if we substitute the appropriate units for each variable, the left-hand-side and the right-hand-side of Equation (5.1) will have the unit of watts or Btu/hour.

Again, remind yourself that a temperature difference must exist in order for heat transfer to occur. *Thermal conductivity* is a property of materials that shows how good the material is in transferring thermal energy (heat) within the material. In general, solids have a higher thermal conductivity than liquids, and liquids have a higher thermal conductivity than gases. The thermal conductivity of some materials is given in Table 5.1. We discuss some of these common materials in Chapter 10.

Next, we look at an example problem.

| TABLE 5.1 | Thermal Conductivity of Some Common Materials |
|---|---|
| **Material** | **Thermal Conductivity (W/m · °C)** |
| Air (at atmospheric pressure) | 0.0263 |
| Aluminum (pure) | 237 |
| Aluminum alloy (4.5% copper, 1.5% magnesium, 0.6% manganese) | 177 |
| Asphalt | 0.062 |
| Bronze (90% copper, 10% aluminum) | 52 |
| Brass (70% copper, 30% zinc) | 110 |
| Brick | 1.0 |
| Concrete | 1.4 |
| Copper (pure) | 401 |
| Glass | 1.4 |
| Gold | 317 |
| Human fat layer | 0.2 |
| Human muscle | 0.41 |
| Human skin | 0.37 |
| Iron (pure) | 80.2 |
| Stainless steels | 13.4 to 15.1 |
| Lead | 35.3 |
| Paper | 0.18 |
| Platinum (pure) | 71.6 |
| Sand | 0.27 |
| Silicon | 148 |
| Silver | 429 |
| Zinc | 116 |
| Water (liquid) | 0.61 |

**EXAMPLE 5.1**

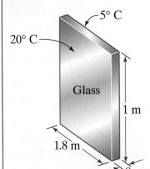

Calculate the heat transfer rate from a single-pane glass window with an inside surface temperature of approximately 20°C and an outside surface temperature of 5°C. The glass is 1 m tall, 1.8 m wide, and 8 mm thick, as shown in Figure 5.2. The thermal conductivity of the glass is approximately $k = 1.4$ W/m · °C.

We start solving the problem by converting the unit of thickness from millimeter to meter.

$$L = (8 \text{ mm})\left(\frac{1 \text{ m}}{1{,}000 \text{ mm}}\right) = 0.008 \text{ m}$$

Next, we calculate the area and the temperature difference.

$$A = (1 \text{ m})(1.8 \text{ m}) = 1.8 \text{ m}^2$$
$$T_1 - T_2 = 20°C - 5°C = 15°C$$

**FIGURE 5.2**

Single-pane window.

Finally, we substitute for the values of $k, A, (T_1 - T_2)$, and $L$ in Equation (5.1) to get

$$q = kA\frac{T_1 - T_2}{L} = \left(1.4\,\frac{W}{m\cdot°C}\right)(1.8\text{ m}^2)\left(\frac{15°C}{0.008\text{ m}}\right)$$

$$= \left(1.4\,\frac{W}{\cancel{m\cdot°C}}\right)(1.8\text{ }\cancel{m^{-2}})\left(\frac{15°\cancel{C}}{0.008\text{ }\cancel{m}}\right)$$

$$= 4,725\text{ W}$$

Note how the units in the numerator and denominator cancel out, and the result is expressed in watts only.

## Thermal Resistance

In this section, we explain what the R-values of insulating materials mean. ***Thermal resistance*** or R-value provides a measure of resistance to heat flow. Most of you understand the importance of having a well-insulated house, because the better insulated it is, the cheaper it will be to heat or cool the house. For example, you may have heard that in order to reduce heat loss through the attic, some people add enough insulation to their attic so that the R-value of insulation is 40 or higher. But what does the R-value of 40 mean? Let us start with Equation (5.1):

$$q = kA\left(\frac{T_1 - T_2}{L}\right)$$

Rearranging it, we have

$$q = \frac{T_1 - T_2}{\dfrac{L}{kA}} = \frac{\text{temperature difference}}{\text{thermal resistance}} \qquad \textbf{5.2}$$

In Equation (5.2), the term $\dfrac{L}{kA}$ is called thermal resistance. Figure 5.3 depicts the idea of thermal resistance and how it is related to the material's thickness ($L$), area ($A$), and thermal conductivity ($k$). When examining Equation (5.2), you should note the following:

> Thermal resistance or R-value provides a measure of resistance to heat flow. The higher the R-value, the lower the value of heat flow would be.

1. The heat transfer (flow) rate is directly proportional to the temperature difference—the larger the temperature difference, the larger the heat transfer rate.

2. The heat flow rate is inversely proportional to the thermal resistance—the higher the value of thermal resistance, the lower the heat transfer rate.

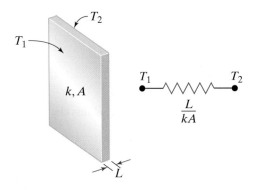

**FIGURE 5.3**  A slab of material and its thermal resistance.

When expressing Fourier's law in the form of Equation (5.2), we are making an analogy between the flow of heat and the flow of electricity in a wire. Recall Ohm's law from Chapter 4, which relates the voltage $(V)$ to current $(I)$ and the electrical resistance $(R)$, according to

$$V = RI$$

or

$$I = \frac{V}{R} \quad \text{or} \quad \text{electric current} = \frac{\text{voltage}}{\text{electrical resistance}}$$

$$q = \frac{T_1 - T_2}{\dfrac{L}{kA}} \quad \text{or} \quad \text{heat flow} = \frac{\text{temperature difference}}{\text{thermal resistance}}$$

Comparing these equations, note that the heat flow $(q)$ is analogous to the electric current $(I)$, the temperature difference $(T_1 - T_2)$ is analogous to the voltage $(V)$, and the thermal resistance $(L/kA)$ is analogous to the electrical resistance $(R)$.

Now turning our attention back to thermal resistance and Equation (5.2), we realize that the thermal resistance for a piece of material is defined as

$$R_{\text{thermal}} = \frac{L}{kA} \qquad\qquad \text{5.3}$$

with the units of:

$$R_{\text{thermal}} = \frac{L}{kA} = \frac{\cancel{m}}{\left(\dfrac{W}{\cancel{m} \cdot {}^\circ C}\right)\left(\cancel{m}^{2}\right)} = \frac{{}^\circ C}{W}$$

Therefore, $R_{\text{thermal}}$ has the SI units of °C/W or the U.S. Customary units of °F·h/Btu. When Equation (5.3) is expressed per unit area of the material, it is referred to as the R-value or the R-factor.

$$R = \frac{L}{k}$$

In this case, $R$ has the units of

$$R = \frac{L}{k} = \frac{m}{\dfrac{W}{m \cdot °C}} = \frac{m^2 \cdot °C}{W} \text{ or } \frac{ft^2 \cdot °F}{\dfrac{Btu}{h}}$$

Also, note that some manufacturers of insulating materials express the R-value of their products per unit thickness. Notwithstanding, remind yourself that neither $R_{thermal}$ nor $R$ is dimensionless, and sometimes the R-values are expressed per unit thickness. The R-value or R-factor of a material provides a measure of resistance to heat flow: The higher the value, the more resistance to heat flow the material offers. Finally, when the materials used for insulation purposes consist of various components, the total R-value of the composite material is the sum of the resistance offered by the various components.

---

**EXAMPLE 5.2**

Determine the thermal resistance ($R_{thermal}$) and the R-value for the glass window of Example 5.1.

The thermal resistance $R_{thermal}$ and the R-value of the window can be determined from Equations (5.3) and (5.4), respectively.

$$R_{thermal} = \frac{L}{kA} = \frac{0.008 \text{ m}}{1.4 \left(\dfrac{W}{m \cdot °C}\right)(1.8 \text{ m}^2)} = 0.00317 \frac{°C}{W}$$

So the R-value or the R-factor for the given glass pane is

$$R = \frac{L}{k} = \frac{0.008 \text{ m}}{1.4 \left(\dfrac{W}{m \cdot °C}\right)} = 0.00571 \frac{m^2 \cdot °C}{W}$$

---

To reduce heat loss from windows, the manufacturers often use two panes of glass that are separated by an air gap, because air has a relatively low thermal conductivity. The next example demonstrates this concept.

As evident from the results of Examples 5.2 and 5.3, ordinary glass windows do not offer much resistance to heat flow. To further increase the R-value of windows, some manufacturers make windows that use triple glass panes and fill the spacing between the glass panes with argon gas, which has a relatively small thermal conductivity value. We discuss the thermal resistance of walls and ceilings later in this section, after we explain the convective heat transfer mode. See Example 5.5 for a sample calculation for total thermal resistance of a typical exterior frame wall of a house consisting of siding, sheathing, insulation material, and gypsum wallboard (drywall).

---

**EXAMPLE 5.3**

A double-pane glass window consists of two pieces of glass, each having a thickness of 8 mm, with a thermal conductivity of $k = 1.4$ W/m·°C. The two glass panes are separated by an air gap of 10 mm, as shown in Figure 5.4. Assuming the thermal conductivity of air to be $k = 0.025$ W/m·°C, determine the total R-value for this window.

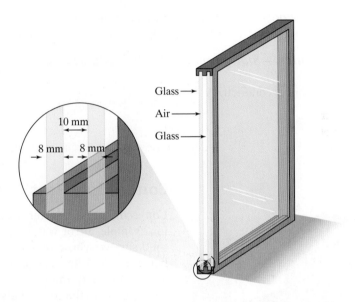

**FIGURE 5.4**   Double-pane window.

The total thermal resistance of the window is obtained by adding the resistance offered by each pane of glass and the air gap in the following manner:

$$R = R_{glass} + R_{air} + R_{glass} = \frac{L_{glass}}{k_{glass}} + \frac{L_{air}}{k_{air}} + \frac{L_{glass}}{k_{glass}}$$

substituting for $L_{glass}, k_{glass}, L_{air}, k_{air}$, we have

$$R = \frac{0.008 \text{ m}}{1.4 \left( \dfrac{W}{m \cdot °C} \right)} + \frac{0.01 \text{ m}}{0.025 \left( \dfrac{W}{m \cdot °C} \right)} + \frac{0.008 \text{ m}}{1.4 \left( \dfrac{W}{m \cdot °C} \right)} = 0.4 \frac{m^2 \cdot °C}{W}$$

As you can see, by adding an extra pane of glass and an air gap, we can increase the thermal resistance of a window significantly.

## Before You Go On

Answer the following questions to test your understanding of the preceding section:

1.  When does heat transfer occur?

2.  What are the modes of heat transfer?

3.  What do we mean by conduction heat transfer?

4.  What does the R-value of a material represent?

*Vocabulary—State the meaning of the following terms:*

Thermal conductivity

R-value

Rebbeck Images / Shutterstock.com

## Convection Heat Transfer

*Convection* heat transfer occurs when a fluid (a gas or a liquid, for example, air or water) in motion comes into contact with a solid surface whose temperature differs from the moving fluid. For example, when you sit in front of a fan to cool down on a hot summer day, the heat transfer that occurs from your warm (solid) body to the cooler moving air (fluid) is by convection. Or, when you are cooling hot food, such as freshly baked cookies, by blowing air on them, you are using the principles of convection, heat transfer.

There are two broad areas of convection heat transfer: *forced convection* and *free (natural) convection*. Forced convection refers to situations where the flow of fluid is caused or forced by a fan or a pump. The cooling of computer chips by blowing air across them with a fan is an example of forced convection. Free convection, on the other hand, refers to situations where the flow of fluid occurs naturally. When you leave a hot pie to cool on the kitchen counter, the heat transfer is by natural convection. As another example, the heat loss from the exterior surfaces of a hot oven is also by natural convection.

For both forced and free convection situations, the overall heat transfer rate between the fluid and the surface is governed by Newton's law of cooling, which is given by

$$q = hA(T_s - T_f)$$

5 . 5

In Equation (5.5), $h$ is the heat transfer coefficient in $W/m^2 \cdot °C$ (or $Btu/h \cdot ft^2 \cdot °F$), $A$ is the area of the exposed surface in meter squared ($m^2$) or foot squared ($ft^2$), $T_s$ is the surface temperature in degrees Celsius (°C) or Fahrenheit (°F), and $T_f$ represents the temperature of moving fluid in degrees °C or °F.

The value of the *heat transfer coefficient* ($h$) for a particular situation is determined from experiments; these values are available in many books about heat transfer. Also, it is important to know that the value of $h$ is higher for forced convection than it is for free convection. Of course, you already know this! When you are trying to cool down rapidly, do you sit in front of a fan or do you sit in an area of the room where the air is still? Moreover, the heat transfer coefficient is higher for liquids than it is for gases. Have you noticed that you can walk around comfortably in a T-shirt when the outdoor air temperature is 70°F, but if you went into a swimming pool whose water temperature was 70°F you would feel cold? That is because the liquid water has a higher heat transfer coefficient than air; therefore, according to Newton's law of cooling, Equation (5.5), water removes more heat from your body. The typical range of heat transfer coefficient values is given in Table 5.2.

Similar to the R-value in conduction, it is also common to define a resistance term for the convection process. The thermal convection resistance $R_{thermal}$ is defined as

$$q = \frac{T_1 - T_2}{\frac{1}{hA}} = \frac{\text{temperature difference}}{\text{thermal resistance}}$$

$$R_{thermal} = \frac{1}{hA}$$

**5.6**

Again, $R_{thermal}$ has units of °C/W or °F·h/Btu. Equation (5.6) is often expressed per unit area of solid surface exposure and is called *film resistance*.

$$R = \frac{1}{h}$$

**5.7**

Then the R-value for film resistance has the units of

$$\left( \frac{m^2 \cdot {}^\circ C}{W} \text{ or } \frac{ft^2 \cdot {}^\circ F}{\frac{Btu}{h}} \right)$$

Once again, it is important to realize that neither $R_{thermal}$ nor $R$ is dimensionless, and they provide a measure of resistance to heat flow; the higher the values of $R$, the more resistance to heat flow to or from the surrounding fluid.

### TABLE 5.2   Typical Values of Heat Transfer Coefficients

| Convection Type | Heat Transfer Coefficient, $h$ (W/m² · °C) | Heat Transfer Coefficient, $h$ (Btu/h · ft² · °F) |
|---|---|---|
| *Free Convection* | | |
| Gases | 2 to 25 | 0.35 to 4.4 |
| Liquids | 50 to 1,000 | 8.8 to 175 |
| *Forced Convection* | | |
| Gases | 25 to 250 | 4.4 to 44 |
| Liquids | 100 to 20,000 | 17.6 to 3,500 |

**EXAMPLE 5.4**

Calculate the R-factor (film resistance) for the following situations:

(a)  Wind blowing over a wall where $h = 5.88$ Btu/h·°F·ft²

(b)  Still air inside a room near a wall where $h = 1.47$ Btu/h·°F·ft²

(a)  For the situation where wind is blowing over a wall,

$$R = \frac{1}{h} = \frac{1}{5.88 \frac{\text{Btu}}{\text{h}\cdot\text{ft}^2\cdot°\text{F}}} = 0.17 \frac{\text{h}\cdot\text{ft}^2\cdot°\text{F}}{\text{Btu}}$$

(b)  For still air inside a room near a wall,

$$R = \frac{1}{h} = \frac{1}{1.47 \frac{\text{Btu}}{\text{h}\cdot\text{ft}^2\cdot°\text{F}}} = 0.68 \frac{\text{h}\cdot\text{ft}^2\cdot°\text{F}}{\text{Btu}}$$

As you can see from the results, the heat transfer situation with the wind blowing has a lower thermal resistance. Therefore, more heat transfer (loss) occurs from a building on a windy day compared to a calm winter day.

**EXAMPLE 5.5**

A typical exterior frame wall (made up of 2 × 4 studs) of a house contains the materials shown in Table 5.3 and in Figure 5.5. For most residential buildings, the inside room temperature is kept around 70°F. Assuming an outside temperature of 20°F and an exposed area of 150 ft², we are interested in determining the heat loss through the wall.

In general, the heat loss through the walls, windows, doors, or roof of a building occurs due to conduction heat losses through the building materials—including siding, insulation material, gypsum wallboard (drywall), glass, and so on—and convection losses through the wall surfaces exposed to the indoor warm air and the outdoor cold air. The total resistance to heat flow is the sum of resistances offered by each component in the path of heat flow. For a composite wall, we can write

$$q = \left[\frac{T_{\text{inside}} - T_{\text{outside}}}{\text{sum of resistances}}\right](\text{area})$$

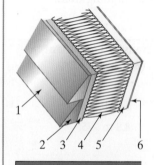

**FIGURE 5.5**

Wall layers.

| TABLE 5.3 | Thermal Resistance of Wall Materials |
|---|---|
| **Items** | **Thermal Resistance (h·ft²·°F/Btu)** |
| 1. Outside film resistance (winter, 15 mph wind) | 0.17 |
| 2. Siding, wood (1/2 × 8 lapped) | 0.81 |
| 3. Sheathing (1/2 in. regular) | 1.32 |
| 4. Insulation batt (3 − 3½ in.) | 11.0 |
| 5. Gypsum wallboard (1/2 in.) | 0.45 |
| 6. Inside film resistance (winter) | 0.68 |

The total resistance to heat flow is given by

$$\text{sum of resistances} = 0.17 + 0.81 + 1.32 + 11.0 + 0.45 + 0.68$$

$$= 14.43 \frac{\text{h} \cdot \text{ft}^2 \cdot {}^\circ\text{F}}{\text{Btu}}$$

$$q = \left[\frac{T_{\text{inside}} - T_{\text{outside}}}{\text{sum of resistances}}\right](\text{area}) = \left(\frac{(70-20)\,\cancel{{}^\circ\text{F}}}{14.43 \dfrac{\text{h} \cdot \cancel{\text{ft}^2} \cdot \cancel{{}^\circ\text{F}}}{\text{Btu}}}\right)(150 \ \cancel{\text{ft}^2})$$

$$= 520 \ \frac{\text{Btu}}{\text{h}}$$

Note how similar units were cancelled out and the final result is expressed in Btu per hour. The equivalent thermal resistance circuit for this problem is shown in Figure 5.6.

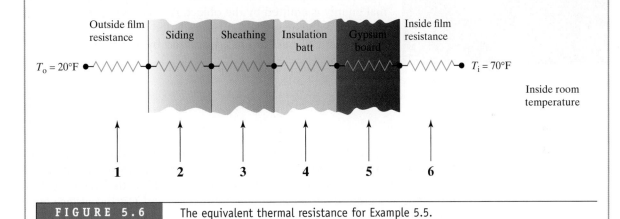

**FIGURE 5.6**   The equivalent thermal resistance for Example 5.5.

When performing a heating load analysis to select a furnace to heat a building, instead of

$$q = \left[\frac{T_{\text{inside}} - T_{\text{outside}}}{\text{sum of resitances}}\right](\text{area})$$

it is common to calculate the heat loss through the walls, roofs, windows, and doors of a building from

$$q = U(T_{\text{inside}} - T_{\text{outside}})(\text{area}) \qquad \boxed{5.8}$$

In Equation (5.8), $U$ represents the ***overall heat transfer coefficient***, or simply the U-factor for a wall, roof, window, or a door. As you can see from

Equation (5.8), the **U-factor** is the reciprocal of the total thermal resistance and has units of $\frac{\text{Btu}}{\text{h} \cdot \text{ft}^2 \cdot °\text{F}}$. For Example 5.5, the U-factor for the wall is equal to

> The U-factor or the overall heat transfer coefficient represents the combined effect of all thermal resistances in a wall, door, or a window.

$$U = \frac{1}{\text{sum of resistances}} = \frac{1}{14.43} = 0.0693 \ \frac{\text{Btu}}{\text{h} \cdot \text{ft}^2 \cdot °\text{F}}$$

Using this U-value, the heat loss through the wall is then calculated from

$$q = (0.0693 \frac{\text{Btu}}{\text{h} \cdot \text{ft}^2 \cdot °\text{F}})(70 - 20)°\text{F} \, (150 \text{ ft}^2) = 520 \frac{\text{Btu}}{\text{h}}.$$

Skylines / Shutterstock.com

## Radiation Heat Transfer

All matter emits thermal **radiation**. This rule is true as long as the body in question is at a nonzero absolute temperature. The higher the temperature of the surface of the object, the more thermal energy is emitted by the object. A good example of thermal radiation is the heat you can feel radiated by a fire in a fireplace. The amount of radiant energy emitted by a surface is given by the equation

$$q = \epsilon \sigma A T_s^4 \qquad \textbf{5.9}$$

In Equation (5.9), $q$ represents the rate of thermal energy per unit time that is emitted by the surface in watts; $\epsilon$ is the emissivity of the surface, $0 < \epsilon < 1$, and $\sigma$ is the Stefan–Boltzmann constant ($\sigma = 5.67 \times 10^{-8}$ W/m$^2 \cdot$K$^4$); $A$ represents the area of the surface in meter squared, and $T_s$ is the surface temperature of the object expressed in Kelvin. **Emissivity** ($\epsilon$) is a property of the surface of the object, and its value indicates how well the object emits thermal radiation compared to a black body (an ideal perfect emitter).

It is important to note here that, unlike the conduction and convection modes, heat transfer by radiation can occur in a vacuum. A daily example of this is the radiation of the sun reaching the Earth's atmosphere as it travels through a vacuum in space. Because all objects emit thermal radiation, it is the net energy exchange among the bodies that is of interest to us. Because of this fact, thermal radiation calculations are generally complicated in nature and require an in-depth understanding of the underlying concepts and geometry of the problem. As a result, we will not discuss this in greater detail in this book. However, it is important for you, as a good global citizen, to understand the basics, that: All matter emits thermal radiation, and the higher the temperature of the surface of an object, the more radiant energy is emitted by the object.

As you can see from the following example, the higher the temperature of the surface, the more radiant energy is emitted by the surface, resulting in a faster cooling rate.

**EXAMPLE 5.6**

On a hot summer day, the temperature of the flat roof of a tall building reaches 50°C. The area of the roof is 400 m². Estimate the heat radiated from this roof to the sky in the evening when the temperature of the surrounding air or sky is at 20°C. The temperature of the roof decreases as it cools down. Estimate the rate of energy radiated from the roof, assuming roof temperatures of 50, 40, 30, and 25°C. Assume $\epsilon = 0.9$ for the roof.

We can determine the amount of thermal energy radiated by the surface from Equation (5.9). For a roof temperature of 50°C, we get

$$q = \epsilon \sigma A T_s^4 = (0.9)\left(5.67 \times 10^{-8} \frac{W}{m^2 \cdot K^4}\right)(400 \text{ m}^2)(323 \text{ K})^4 = 222,000 \text{ W}$$

The rest of the solution is shown in Table 5.4.

**TABLE 5.4**   **The Results of Example 5.6**

| Surface Temperature °C | Surface Temperature, $T(K) = T(°C) + 273$ | Energy Emitted by the Surface (W) |
|---|---|---|
| 50 | 323 | $(0.9)(5.67 \times 10^{-8})(400)(323)^4 = 222,000$ W |
| 40 | 313 | $(0.9)(5.67 \times 10^{-8})(400)(313)^4 = 196,000$ W |
| 30 | 303 | $(0.9)(5.67 \times 10^{-8})(400)(303)^4 = 172,000$ W |
| 25 | 298 | $(0.9)(5.67 \times 10^{-8})(400)(298)^4 = 161,000$ W |

# Before You Go On

Answer the following questions to test your understanding of the preceding section:

1. What do we mean by convection heat transfer?

2. What do we mean by heat transfer coefficient?

3. What does the overall heat transfer coefficient value represent?

4. When does radiation heat transfer occur?

*Vocabulary—State the meaning of the following terms:*

Film resistance

U-factor or U-value

Emissivity

## LO³  5.3  Daylighting

Up to this point, we have discussed heat transfer from a building and how to quantify it. As you would expect, building characteristics and construction affect energy usage. A building is characterized by its "envelope," which consists of the foundation, walls, windows, doors, roof, and insulation along with their age and condition. Building orientation and the amount and orientation of glazing (windows) also affects the energy consumption rates of a building. Moreover, energy can be saved by taking advantage of daylighting. Let's examine daylighting in more detail now.

*Daylighting* refers to the use of windows and skylights to bring natural light into a building. A building that makes use of daylighting minimizes the use of artificial lighting, such as ceiling lights or lamps, to brighten the inside of a house during daylight hours.

Properly selected and positioned windows and skylights also can provide buildings with heat and ventilation. However, if not properly selected and positioned, they can negatively impact the home's energy use.

Courtesy of The National Fenestration Rating Council (NFRC).

### Window Basics

The National Fenestration Rating Council (NFRC) defines the ratings for window energy performance. Factors that you must consider when selecting windows and skylights include:

- U-factor
- Solar heat gain coefficient (SHGC)
- Air leakage

**Window Technologies**

Energy-efficient window technologies are available to produce windows with the U-factor, SHGC, and VT properties needed for any application.

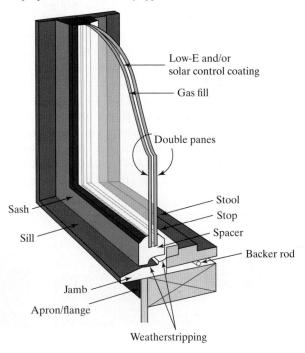

- Visible transmittance (VT)
- Types of window frames
  - Aluminum or metal
  - Composite
  - Fiberglass
  - Vinyl
  - Wood

As you know by now, the U-factor (the overall heat transfer coefficient) provides an indication of how much heat loss you could expect through a window, door, or skylight. In the United States, it usually is expressed in units of Btu/h·ft²·°F, and the lower the U-factor, the more energy-efficient the window, door, or skylight is. Typical values are between 0.2 and 1.2 Btu/h·ft²·°F.

The **solar heat gain coefficient** (SHGC) represents how well a window blocks unwanted heat yet allows sunlight to go through. The SHGC is the fraction of the heat from the sun that enters through a window. It is expressed as a number between 0 and 1. The lower the value, the better the window is at blocking unwanted heat. The best SHGC rating for a window depends on the climate where the window is installed. It is important to consider the SHGC value when selecting a window for hot climates that predominantly require cooling (see Figure 5.7). The **visible transmittance** (VT) index measures how much light comes through a window. VT is expressed as a number between 0 and 1, and the higher the VT value, the clearer the glass. This means that more visible light is transmitted through windows with higher VT values.

The *air leakage* (AL) rating is expressed as the equivalent cubic feet of air passing through a square foot of window area ($ft^3/ft^2$). The lower the AL, the less the window leaks. Typical values range between 0.1 and 0.3.

The ***condensation resistance*** (CR) measures how well a window resists the formation of moisture (condensation) on its inside glass surface. The higher the CR rating, the better the window resists condensation formation; it is expressed as a number between 1 and 100. The higher value represents more resistance to the formation of condensation.

As well-educated citizens, you need to be aware of these definitions and how these factors affect heating and cooling loads in our homes. Finally, the ENERGY STAR® qualification criteria for residential windows, doors, and skylights in the United States are shown in Figure 5.7.

**Windows**

| Climate Zone | U-Factor[1] | SHGC[2] | |
|---|---|---|---|
| Northern | ≤0.30 | Any | Prescriptive |
| | =0.31 | ≥0.35 | Equivalent energy performance |
| | =0.32 | ≥0.40 | |
| North-central | ≤0.32 | ≤0.40 | |
| South-central | ≤0.35 | ≤0.30 | |
| Southern | ≤0.60 | ≤0.27 | |

**Doors**

| Glazing level | U-Factor[1] | SHGC[2] |
|---|---|---|
| Opaque | ≤0.21 | No rating |
| ≤ ½-Lite | ≤0.27 | ≤0.30 |
| > ½-Lite | ≤0.32 | ≤0.30 |

**Skylights**

| Climate Zone | U-Factor[1] | SHGC[2] |
|---|---|---|
| Northern | ≤0.55 | Any |
| North-central | ≤0.55 | ≤0.40 |
| South-central | ≤0.57 | ≤0.30 |
| Southern | ≤0.70 | ≤0.30 |

[1] $Btu/h \cdot ft^2 \cdot °F$
[2] Fraction of incident solar radiation

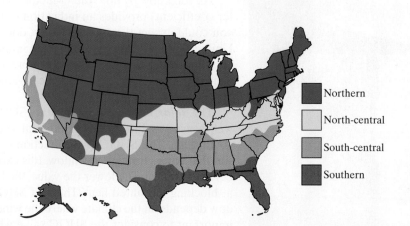

Northern
North-central
South-central
Southern

**FIGURE 5.7**    The ENERGY STAR® qualification criteria for residential windows, doors, and skylights.
*Source:* ENERGY STAR Program Requirements for Windows, Doors, and Skylights: Version 5.0 (April 7, 2009).

## LO⁴   5.4   Degree Days and Energy Estimation

How do you estimate how much energy is needed on a monthly and an annual basis to heat a building? One way is to use degree days. *A degree day (DD) is the difference between 65°F (typically) and the average temperature of the outside air during a 24-hour period.* For example, for a city in Minnesota, on November 10, 2016, the low temperature was 28°F and the high temperature was 38°F. Then the degree day for November 10, 2016 for that city was $DD = 65°F - [(38°F + 28°F / 2)] = 32°F.$ Now, if we were to add up the degree days for each day in a month, we would get the total degree days for that month. Similarly, if we were to add up the degree days for each month, we would then obtain the annual degree days. In practice, historical degree-day values (based on average values over many years, see Figure 5.8) are used to estimate monthly and annual energy consumption to heat buildings from the following relationships.

> A degree day (DD) is the difference between 65°F (typically) and the average temperature of the outside air during a 24-hour period.

$$Q_{DD} = \frac{\text{building heat loss} \left( \dfrac{\text{Btu}}{\text{h}} \right) \times 24 \text{ h}}{\text{design temperature difference (°F)}}$$

**5.10**

$$Q_{monthly} = (Q_{DD})(\text{monthly degree days})$$

**5.11**

$$Q_{yearly} = (Q_{DD})(\text{yearly degree days})$$

**5.12**

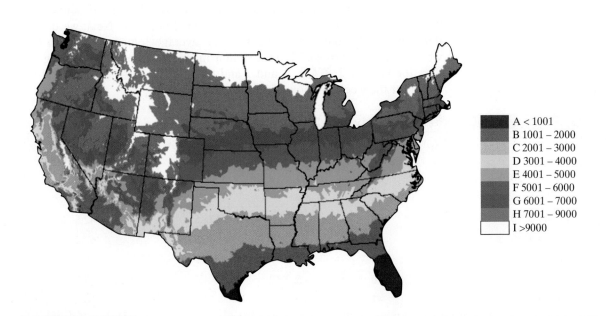

| | |
|---|---|
| A | < 1001 |
| B | 1001 – 2000 |
| C | 2001 – 3000 |
| D | 3001 – 4000 |
| E | 4001 – 5000 |
| F | 5001 – 6000 |
| G | 6001 – 7000 |
| H | 7001 – 9000 |
| I | >9000 |

**FIGURE 5.8**   The annual degree days for the United States.
*Source:* NOAA.

> An AFUE rating represents the annual fuel utilization efficiency of a furnace.

Most homes in cold climates within the United States are kept warm by furnaces. In a furnace, natural gas or fuel oil is burned, and the hot combustion products go through the inside of a heat exchanger where thermal energy is transported to cold indoor air that is passing over the heat exchanger. The warm air is then distributed to various parts of the house through conduits.

The sizes of home gas furnaces in the United States are expressed in units of Btu/h. The size of a typical single-family-home gas furnace used in the Midwest is about 60,000 Btu/hour.

**EXAMPLE 5.7**

The temperature data for the first seven days of January 2015 for a city is given in Table 5.5. Calculate the degree days for each day and for the entire week.

| **TABLE 5.5** | **The Temperature Data in Degrees Fahrenheit for a Given City During January 1–7, 2015** | |
|---|---|---|
| **Date** | **Hi** | **Lo** |
| January 1, 2015 | 30 | 11 |
| January 2, 2015 | 28 | 6 |
| January 3, 2015 | 33 | 12 |
| January 4, 2015 | 12 | –7 |
| January 5, 2015 | 1 | –11 |
| January 6, 2015 | 8 | –5 |
| January 7, 2015 | –1 | –9 |

First, we need to calculate the daily average. For example, for January 1, 2015, the average temperature was $(30 + 11)/2 = 20.5°F$. Next, we determine the degree days by subtracting the average temperature from 65, which is $65 - 20.5 = 44.5°F$. In a similar manner, we calculate the degree days for January 2, 3, 4, 5, 6, and 7, and then add them to get the total of 401°F for the week, as shown in Table 5.6.

| **TABLE 5.6** | **The Degree-Days for January 1–7, 2015 for the Given City** | | | |
|---|---|---|---|---|
| **Date** | **Hi** | **Lo** | **Daily Average (Hi + Lo)/2** | **Degree Days = 65-Daily Average** |
| January 1, 2015 | 30 | 11 | 20.5 | 44.5 |
| January 2, 2015 | 28 | 6 | 17 | 48 |
| January 3, 2015 | 33 | 12 | 22.5 | 42.5 |
| January 4, 2015 | 12 | –7 | 2.5 | 62.5 |
| January 5, 2015 | 1 | –11 | –5 | 70 |
| January 6, 2015 | 8 | –5 | 1.5 | 63.5 |
| January 7, 2015 | –1 | –9 | –5 | 70 |
| | | | | Total = 401°F |

**EXAMPLE 5.8**

Calculate the annual degree days for Minneapolis, Minnesota using the monthly values given in Table 5.7.

**TABLE 5.7** The Monthly Degree-Day Values for Minneapolis, Minnesota

| Month | Degree Days |
|---|---|
| January | 1,631 |
| February | 1,380 |
| March | 1,166 |
| April | 621 |
| May | 288 |
| June | 81 |
| July | 22 |
| August | 31 |
| September | 189 |
| October | 505 |
| November | 1,014 |
| December | 1,454 |

The annual degree day is determined by simply adding the monthly values:
1,631 + 1,380 + 1,166 + 621 + 288 + 81 + 22 + 31 + 189 + 505 + 1,014 + 1,454 = 8,382.

**EXAMPLE 5.9**

**Can you estimate your home's annual natural gas consumption?**

We may burn natural gas to heat water, to cook, and to heat our homes during cold winter months. Most of the natural gas consumption in a home is due to heating the house during winter. For a building located in Minneapolis, Minnesota with annual heating degree-days of 8,382, a heating load (heat loss) of 62,000 Btu/h, and a design temperature difference of 82°F (68°F indoor and −14°F outdoor), estimate the annual energy consumption. If the building is heated with a furnace with an efficiency of 94%, how much gas is burned to keep the home at 68°F?

We solve this problem using Equations (5.10) and (5.12).

$$Q_{\text{DD}} = \frac{62,000\left(\dfrac{\text{Btu}}{\text{h}}\right) \times 24\ \text{h}}{\left[68 - (-14)\right]\ (\text{°F})} = \frac{62,000\left(\dfrac{\text{Btu}}{\text{h}}\right) \times 24\ \text{h}}{82\ (\text{°F})} = 18,146\ \text{Btu/DD}$$

$$Q_{\text{yearly}} = 18,146\left(\frac{\text{Btu}}{\text{DD}}\right)(8,382\ \text{DD}) = 152 \times 10^6\ \text{Btu/year}$$

Assuming the gas used in Minnesota has a heating value of 1,000 Btu/ft³, the amount of gas burned in the furnace can be estimated from

$$\text{volume of gas burned} = \left[\frac{(152 \times 10^6 \text{ Btu/year})}{0.94}\right]\left[\frac{1}{1,000 \text{ Btu/ft}^3}\right]$$

$$= 161,700 \text{ ft}^3/\text{year}$$

Think about the amount of gas that we need to burn to keep one home warm during the winter months in Minnesota. Also, think about how much extra energy it took to extract the gas and transport it to the home!

---

**EXAMPLE 5.10**

**Replacing windows**

Let us consider an older home with six 3 × 4 ft windows with U-values of $1.4\frac{\text{Btu}}{\text{h}\cdot\text{ft}^2\cdot\text{°F}}$. What if we were to replace these windows with more energy-efficient, newer models with $U = 0.2\frac{\text{Btu}}{\text{h}\cdot\text{ft}^2\cdot\text{°F}}$? How much energy would we save? The home is located in Massachusetts with annual degree days of 5,634 and a design temperature difference of 62 degrees Fahrenheit (68°F indoor and 6°F outdoor).

Solve this problem using Equations (5.8), (5.10), and (5.12). For the six old windows,

$$q = UA\,\Delta T = (6)\,(1.4\frac{\text{Btu}}{\text{h}\cdot\text{ft}^2\cdot\text{°F}})(12 \text{ ft}^2)(68 - 6)\text{°F} = 6,250\frac{\text{Btu}}{\text{h}}$$

For the new windows,

$$q = UA\,\Delta T = (6)\,(0.2\frac{\text{Btu}}{\text{h}\cdot\text{ft}^2\cdot\text{°F}})(12 \text{ ft}^2)(68 - 6)\text{°F} = 893\frac{\text{Btu}}{\text{h}}$$

The reduction in heat transfer that translates to energy savings is

$$q = 6,250\frac{\text{Btu}}{\text{h}} - 893\frac{\text{Btu}}{\text{h}} = 5,357\frac{\text{Btu}}{\text{h}}$$

$$Q_{\text{DD}} = \frac{5,357\left(\frac{\text{Btu}}{\text{h}}\right) \times 24 \text{ hrs}}{(68 - 6)(\text{°F})} = \frac{5,357\left(\frac{\text{Btu}}{\text{h}}\right) \times 24 \text{ hrs}}{62 \,(\text{°F})} = 2,074 \text{ Btu/DD}$$

$$Q_{\text{yearly}} = 2,074\left(\frac{\text{Btu}}{\text{DD}}\right)(5,634 \text{ DD}) = 11.7 \times 10^6 \text{ Btu/year}$$

Assuming the fuel oil used in Massachusetts has a heating value of 139,000 Btu per gallon and the furnace has an efficiency of 90%, the amount of fuel oil saved can be estimated from

$$\text{volume of fuel oil burned} = \left[\frac{(11.7 \times 10^6 \text{ Btu/year})}{0.90}\right]\left[\frac{1}{139{,}000 \text{ Btu/gallon}}\right]$$

$$= 94 \text{ gallons/year}$$

Therefore, by replacing the six windows, we can save 94 gallons of fuel oil per year! Think about it! Then, depending on the cost of both the fuel oil and the replacement windows, we can determine how long it would take for the dollar savings to take effect.

## Before You Go On

Answer the following questions to test your understanding of the preceding section:

1. What do we mean by daylighting?
2. What are the typical U-values for windows?
3. What does the air leakage (AL) rating of a window represent?
4. What is a degree day?

*Vocabulary—State the meaning of the following terms:*

Visible transmittance

SHGC

Condensation resistance

Degree days

## SUMMARY

### LO¹ Temperature Difference and Heat Transfer

Thermal energy transfer occurs whenever there exists a temperature difference within an object or a temperature difference between two things, such as the air inside a room and the air outside the building. This form of energy is called *heat transfer*. Additionally, heat always flows from a high-temperature region to a low-temperature region. There are three different mechanisms by which energy is transferred from a high-temperature region to a low-temperature region. These mechanisms are referred to as the *modes* of heat transfer. Moreover, there are three units that are commonly used to quantify thermal energy: (1) the British thermal unit (Btu), (2) the calorie, and (3) the joule (J).

## LO² Modes of Heat Transfer

The three modes of heat transfer are conduction, convection, and radiation. *Conduction* refers to the mode of heat transfer that occurs when a temperature difference exists *in* a medium. The energy is transported within the medium from the region with more energetic molecules (high-temperature region) to the region with less energetic molecules (low-temperature region). The rate of heat transfer by conduction is given by *Fourier's law* according to

$$q = kA\left(\frac{T_1 - T_2}{L}\right)$$

In this equation, $q$ = heat transfer rate in watts (W) or British thermal unit per hour (Btu/h), $k$ = thermal conductivity $\left(\dfrac{W}{m \cdot °C} \text{ or } \dfrac{Btu}{h \cdot ft \cdot °F}\right)$, $A$ = cross-sectional area normal to heat flow (m² or ft²), and $T_1 - T_2$ = temperature difference (°C or °F) across the material of $L$ thickness (m or ft).

The R-value of a material provides a measure of resistance to heat flow: The higher the value, the more resistance the material offers to heat flow. The thermal resistance for a piece of material is defined as $R_{thermal} = \dfrac{L}{kA}$. $R_{thermal}$ has the units of °C/W or °F·h/Btu. When this equation is expressed per unit area of the material, it is referred to as the R-value or the R-factor $R = \dfrac{L}{k}$, where $R$ has the units of $\left(\dfrac{m^2 \cdot °C}{W} \text{ or } \dfrac{ft^2 \cdot °F}{\frac{Btu}{h}}\right)$. Note that neither $R_{thermal}$ nor $R$ is dimensionless, and sometimes the manufacturers of insulating materials express the R-value of their products per unit thickness.

### Convection Heat Transfer

*Convection* heat transfer occurs when a fluid (a gas or a liquid) in motion comes into contact with a solid surface whose temperature differs from the moving fluid. There are two broad areas of convection heat transfer: *forced convection* and *free (natural) convection*. Forced convection refers to situations where the flow of fluid is forced by a fan or a pump. Free convection, on the other hand, refers to situations where the flow of fluid occurs naturally due to density variation in the fluid. For both the forced and the free convection situations, the overall heat transfer rate between the fluid and the surface is governed by Newton's law of cooling, which is given by $q = hA(T_s - T_f)$ where $h$ is the heat transfer coefficient in W/m²·°C or Btu/h·ft²·°F, $A$ is the area of the exposed surface in m² or ft², $T_s$ is the surface temperature in degrees (°C or °F), and $T_f$ represents the temperature of the fluid in degrees (°C or °F). The value of the heat transfer coefficient for a particular situation is determined from experiments. It is also common to define a resistance term for the convection process, similar to the R-value in conduction. The thermal convection resistance is defined as $R_{thermal} = \dfrac{1}{hA}$. This equation is commonly expressed per unit area of solid surface exposure and is called *film resistance* or the *film coefficient* $R = \dfrac{1}{h}$, where $R$ has units of $\dfrac{m^2 \cdot °C}{W}$ or $\dfrac{ft^2 \cdot °F}{\frac{Btu}{h}}$.

### Radiation Heat Transfer

All matter emits thermal *radiation*. The higher the temperature of the surface of the object, the more thermal energy is emitted by the object, and unlike the conduction and convection modes, heat transfer by radiation can occur in a vacuum. The amount of radiant energy emitted by a surface is given by $q = \epsilon\sigma A T_s^4$; $q$ represents the rate of thermal energy per unit time emitted by the surface in watts; $\epsilon$ is the emissivity of the surface where $0 < \epsilon < 1$; $\sigma$ is the Stefan-Boltzmann constant where $\sigma = 5.67 \times 10^{-8}$ W/m²·K⁴; $A$ represents the area of the surface in m², and $T_s$ is the surface temperature of the object expressed in Kelvin.

## LO³ Daylighting

Daylighting refers to the use of windows and skylights to bring natural light into a building. The National Fenestration Rating Council (NFRC) defines the ratings for window and skylight energy performance.

Factors that are considered when selecting windows and skylights include the U-factor, solar heat gain coefficient (SHGC), air leakage (AL), sunlight transmittance and visible transmittance (VT).

### LO⁴ Degree Days and Energy Estimation

A degree day (DD) is the difference between 65°F (typically) and the average temperature of the outside air during a 24-hour period. In practice, historical degree-day values (based on the average of data over many years) are used to estimate monthly and annual energy consumptions to heat buildings from the following relationships:

$$Q_{DD} = \frac{\text{heat loss}\left(\frac{\text{Btu}}{\text{h}}\right) \times 24 \text{ h}}{\text{design temperture difference (°F)}}$$

$$Q_{monthly} = (Q_{DD})(\text{monthly degree days})$$

$$Q_{yearly} = (Q_{DD})(\text{yearly degree days})$$

## KEY TERMS

Condensation Resistance 148
Conduction Heat Transfer 133
Convection Heat Transfer 140
Daylighting 146
Degree Days 149
Emissivity 144

Film Resistance 141
Heat Transfer 132
Heat Transfer Coefficient 141
Overall Heat Transfer
Coefficient 143
Radiation Heat Transfer 144

Solar Heat Gain Coefficient
(SHGC) 147
Thermal Conductivity 134
Thermal Resistance 136
U-factor 144
Visible Transmittance (VT) 147

## *Apply What You Have Learned*

You are to investigate the size of the furnace in your own, your parents' or a friend's home and the degree-day values for your city. Estimate the monthly and annual costs to heat the home. State all your assumptions. Suggest ways to reduce the heat loss for the home (e.g., by replacing older windows with ENERGY STAR® windows with lower U-values or by adding more insulation to the walls or the roof) and estimate the associated savings.

Elena Elisseeva / Shutterstock.com

# PROBLEMS

*Problems that promote life-long learning are denoted by* 🔑

**5.1** Calculate the R-value for the following materials: (a) 4-in.-thick brick and (b) 12-in.-thick concrete slab.

**5.2** Calculate the R value (film resistance) for a situation when the wind is blowing over a wall, with $h = 4.0$ Btu/h·ft²·°F.

**5.3** Calculate the overall U-value for a 12-in.-thick concrete wall with the wind blowing over its outside surface ($h = 5.88$ Btu/h·ft²·°F), and still air over its inside surface ($h = 1.47$ Btu/h·ft²·°F).

**5.4** If a window has a total R-value of 1.7, what is its U-value?

**5.5** Calculate the heat transfer rate from a 1,000 ft², 6-in.-thick concrete wall with inside and outside surface temperatures of 70°F and 40°F, respectively.

**5.6** Calculate the heat transfer rate from a 1,000 ft², 9-in.-thick concrete wall with inside and outside surface temperatures of 70°F and 40°F, respectively.

**5.7** Calculate the heat transfer rate from a 2,000-ft² ceiling of a house with a total R-value of 41.0 h·ft²·°F/Btu. Assume an inside room temperature of 68°F and an attic air temperature of 10°F.

**5.8** Calculate the heat loss through the walls of a building with a net surface area of 1,400 ft² and a total R-value of 25 h·ft²·°F/Btu. Assume an inside room temperature of 68°F and an outside air temperature of 5°F.

**5.9** Calculate the heat transfer rate through a 24-ft² door with $U = 0.73 \dfrac{\text{Btu}}{\text{h} \cdot \text{ft}^2 \cdot °\text{F}}$. The indoor and outdoor temperatures are 68°F and 10°F, respectively.

**5.10** Calculate the heat transfer rate through a 12-ft² window with $U = 0.8 \dfrac{\text{Btu}}{\text{h} \cdot \text{ft}^2 \cdot °\text{F}}$. The indoor and outdoor temperatures are 68°F and 10°F, respectively.

**5.11** Calculate the heat transfer rate through a 12-ft² window with $U = 0.2 \dfrac{\text{Btu}}{\text{h} \cdot \text{ft}^2 \cdot °\text{F}}$. The indoor and outdoor temperatures are 68°F and 10°F, respectively. Compare the results of this exercise with the results of Problem 5.10. How much energy will be saved if three windows with $U = 0.8 \dfrac{\text{Btu}}{\text{h} \cdot \text{ft}^2 \cdot °\text{F}}$ are replaced with three windows with $U = 0.2 \dfrac{\text{Btu}}{\text{h} \cdot \text{ft}^2 \cdot °\text{F}}$ for a home in a location with annual degree-days of 6,200?

**5.12** A typical exterior masonry wall of a house consists of the items in the table shown in Figure P5.12. Assume an inside room temperature of 68°F, an outside air temperature of 10°F, and an exposed area of 150 ft². Calculate the heat loss through the wall.

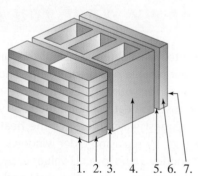

1. 2. 3.    4.    5. 6. 7.

| Items | Resistance (h · ft² · °F/Btu) |
|---|---|
| 1. Outside film resistance (winter, 15 mph wind) | 0.17 |
| 2. Face brick (4 in.) | 0.44 |
| 3. Cement mortar (1/2 in.) | 0.1 |
| 4. Cinder block (8 in.) | 1.72 |
| 5. Air space (3/4 in.) | 1.28 |
| 6. Gypsum wallboard (1/2 in.) | 0.45 |
| 7. Inside film resistance (winter) | 0.68 |

**FIGURE P5.12**

**5.13** In order to increase the thermal resistance of a typical exterior frame wall, it is customary today to use 2 × 6 studs instead of 2 × 4 studs to allow for the placement of more insulation within the wall cavity. A typical exterior (2 × 6) frame wall of a house consists of the materials shown in Figure P5.13. Assume an inside room temperature of 68°F, an outside air temperature of 20°F, and an exposed area of 150 ft². Determine the heat loss through this wall.

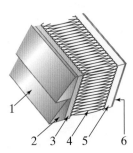

**Items** | **Resistance (h · ft² · °F/Btu)**
---|---
1. Outside film resistance (winter, 15 mph wind) | 0.17
2. Siding, wood | 0.81
3. Sheathing (1/2 in. regular) | 1.32
4. Insulation | 19.0
5. Gypsum wallboard (1/2 in.) | 0.45
6. Inside film resistance (winter) | 0.68

**FIGURE P5.13**

**5.14** A typical ceiling of a house consists of items shown in Figure P5.14. Assume an inside room temperature of 70°F, an attic air temperature of 15°F, and an exposed area of 1,000 ft². Calculate the heat loss through the ceiling.

**5.15** What would be the reduction in heat loss through the ceiling of a house if the insulation R-value of 19 is increased to 40 by adding more insulation? The ceiling area is 2,000 ft².

**5.16** Look up the low and high daily temperature values for a location and a month assigned to you by your instructor. Calculate the degree days for the given month.

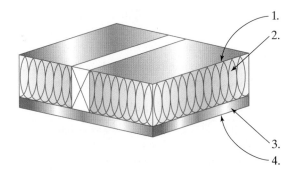

**Items** | **Resistance (h · ft² · °F/Btu)**
---|---
1. Inside attic film resistance | 0.68
2. Insulation | 19.0
3. Gypsum wallboard (1/2 in.) | 0.45
4. Inside film resistance (winter) | 0.68

**FIGURE P5.14**

**5.17** Calculate the heat loss from a double-pane glass window consisting of two pieces of glass, each having a thickness of 10 mm with a thermal conductivity of $k = 1.3 \dfrac{W}{m \cdot K}$. The two glass panes are separated by an air gap of 7 mm. Assume the thermal conductivity of air to be $k = 0.022 \dfrac{W}{m \cdot K}$. Also, express the total R- and U-values.

**5.18** A building is located in Baltimore, Maryland where the annual heating degree days are 4,654. The building has a heating load (heat loss) of 30,000 Btu/h, and a design temperature difference of 52°F (68°F indoor and 16°F outdoor). Estimate the building's annual energy consumption. If the building is heated with a furnace with an efficiency of 92%, how much gas is burned to keep the home at 68°F? State your assumptions.

**5.19** Nine old, 12-ft² windows with a U-value of $U = 1.2 \dfrac{Btu}{h \cdot ft^2 \cdot °F}$ were replaced with new windows having $U = 0.3 \dfrac{Btu}{h \cdot ft^2 \cdot °F}$. Calculate the energy savings on a day during a five-hour period when $T_{in} = 68°F$ and $T_{outside} = 10°F$.

**5.20**  For Problem 5.18, calculate the savings in cubic feet of natural gas. Assume the furnace has an efficiency of 98%.

**5.21**  Calculate the annual degree days for Boston, Massachusetts using the following given monthly values.

| Month | Degree-Days |
|---|---|
| January | 1,088 |
| February | 972 |
| March | 846 |
| April | 513 |
| May | 208 |
| June | 36 |
| July | 0 |
| August | 9 |
| September | 60 |
| October | 316 |
| November | 603 |
| December | 983 |

**5.22**  Visit a hardware store or go online to look up the U-values for some common windows and doors. Create a table that shows their U-values and cost. Based on the data collected, comment on how you would decide to change some old windows and doors. State all your assumptions.

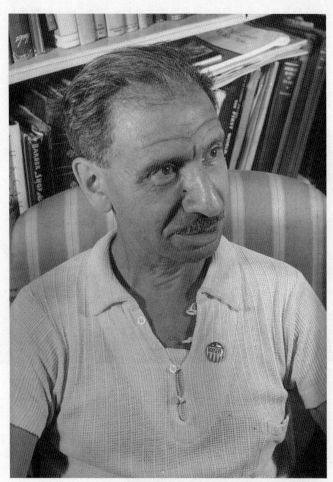

Photo by Gabriel Benzur/The LIFE Picture Collection/Getty Images

*"I find that a great part of information I have was acquired by looking up something and finding something else on the way."* —Franklin P. Adams (1881–1960)

CHAPTER

# 6

# Energy Consumption Rates and Non-Renewable Energy Sources

## LEARNING OBJECTIVES

**LO¹** World Energy Consumption Rates: describe how much energy is consumed in the world

**LO²** United States Energy Consumption Rates: describe how much energy is consumed in the United States in buildings, transportation, and industry

**LO³** Fossil Fuels: know about the production and consumption of gasoline, diesel, fuel oil, natural gas, propane, and coal

**LO⁴** Nuclear Energy: describe nuclear fuel and nuclear energy

# Discussion Starter

According to a report recently released by the U.S. Energy Information Administration (EIA): "... the world energy consumption is expected to increase from 524 quadrillion British thermal units (Btu) in 2010 to 820 quadrillion Btu in 2040. The increase in world energy use is largely in the developing world, where growth is driven by strong, long-term economic growth. Half of the total world increase in energy consumption is attributed to China and India. Fossil fuels are expected to continue supplying much of the energy used worldwide. Although petroleum and other liquids remain the largest source of energy, the liquid fuels share of world marketed energy consumption falls from 34 percent in 2010 to 28 percent in 2040. Renewable energy and nuclear power are the world's fastest-growing energy sources, each increasing by 2.5 percent per year; however, fossil fuels continue to supply almost 80 percent of world energy use through 2040."

**China and India account for half of the world increase in energy consumption through 2040**

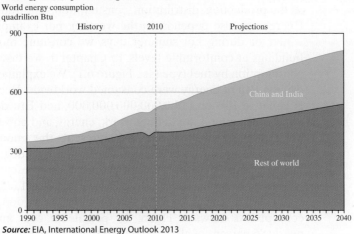

*Source:* EIA, International Energy Outlook 2013

**Renewable energy and nuclear power are the fastest growing sources of energy consumption**

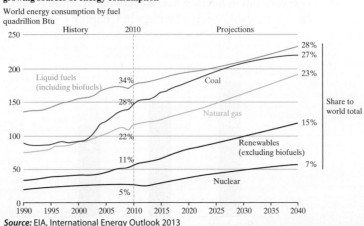

*Source:* EIA, International Energy Outlook 2013

 **To the Students:** Do you expect an increase in your own energy consumption during the next 20 years? Take a few minutes to think about it. Provide assumptions when explaining your answers.

# LO¹  6.1  World Energy Consumption Rates

We have been emphasizing that there are certain concepts that all well-educated citizens should know. Earlier, we discussed the importance of energy in our everyday lives and said that *without energy, we cannot do anything*! We need energy to keep our homes comfortable, to make goods, and to provide services that allow us to enjoy a high standard of living. We use energy in our homes for heating and cooling our living space, providing hot water and lighting systems, operating appliances, and powering electronic devices. We also use energy in our vehicles for private and business travel. In addition to our personal energy requirements, we need energy for industry to make and transport all kinds of products and food, to make building materials, to erect buildings, and to construct and maintain our infrastructure (roads, bridges, railroad systems, airports, etc.).

Personal energy consumption depends on an individual standards of living, and industrial energy consumption depends on economic activities such as the production, distribution, consumption, and trade of goods and services. Energy use also depends on the weather. For example, in an unusually cold winter or during hot summer days, we consume more energy to keep our buildings at comfortable levels. In Chapter 1, we discussed the world energy consumption by fuel type (see Figure 6.1). We explained that in 2011, 519 quadrillion Btu of energy were consumed worldwide. Recall that one quadrillion is equal to $10^{15}$ or 1,000,000,000,000,000, and Btu denotes British Thermal units. In Chapter 3, we defined work, energy, and power and stated their units. For example, recall that one Btu represents the amount of energy needed to raise the temperature of one pound of water by 1 degree Fahrenheit (°F), and kilowatt-hour (kWh) is a unit of energy that represents the amount of energy consumed during 1 hour (h) by a device that uses 1,000 watts (W) or one kilowatt (kW) of power. By now, you should have a good understanding of these concepts. As shown in Figure 6.1, petroleum, coal, and natural gas made up nearly 86 percent of all the fuel used to generate energy in 2011. Because fossil fuels comprise the majority of energy sources worldwide, we discuss them in great detail in Section 6.3. Moreover, we discuss nuclear energy in Section 6.4 and renewable energy sources in Chapter 7.

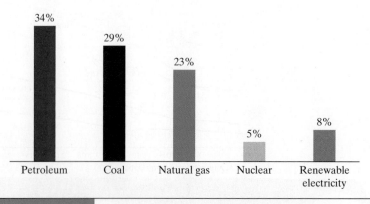

| | | |
|---|---|---|
| FIGURE 6.1 | World energy consumption by fuel type, 2011 (519 quadrillion Btu) | |

*Source:* Data from U.S. EIA

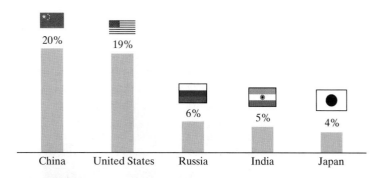

**FIGURE 6.2**

World energy consumption by top five countries, 2011 (519 quadrillion Btu)
*Source:* Data from U.S. EIA

In 2011, the top five countries with the largest energy consumption were China, the United States, Russia, India, and Japan (see Figure 6.2). The per capita consumption for these countries is shown in Figure 6.3. Note that the United States has the largest per capita energy consumption in the world with the value of 313 million Btu.

Also, recall from our discussion in Chapter 1 that in order to keep track of how energy is consumed in society, the U.S. Energy Information Administration (EIA) classifies the energy consumption rates by major sectors of the economy. These sectors are organized into *industrial, transportation, residential*, and *commercial*.

The *industrial sector* accounts for the share of total energy needed for all the facilities and equipment for construction, mining, agriculture, and manufacturing. The *transportation sector* includes energy use by all types of vehicles (motorcycles, cars, trucks, buses, trains, subways, aircraft, boats, barges, ships, etc.) used to transport people and goods. The *residential sector* accounts for energy use in homes and apartments, and the *commercial sector* represents energy consumption in malls, stores, offices, educational campuses (e.g., high schools, colleges), hospitals, and hotels. The share of the total world energy consumption

> The United States Energy Information Administration (EIA) classifies the energy consumption rates by major sectors of the economy: industrial, transportation, residential, and commercial.

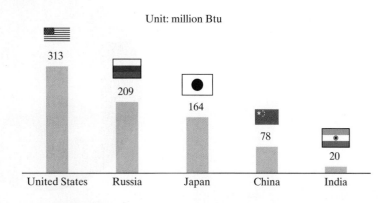

**FIGURE 6.3**

Per capita energy consumption of selected countries, 2011
*Source:* Data from U.S. EIA

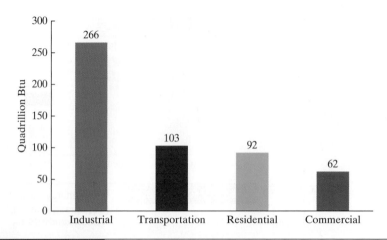

FIGURE 6.4 World energy consumption by sector
*Source:* Data from U.S. EIA

by various sectors is shown in Figure 6.4. The percentage of the total by each sector is also shown in Figure 6.5. From examining these figures, it becomes clear that half of the energy consumed worldwide is for industrial activities such as construction, mining, agriculture, and manufacturing. The transportation sector accounted for 20 percent of the total world energy, followed by residential with 18 percent and commercial with 12 percent.

In order to better understand how much energy we have been consuming in recent decades, see Figure 6.6 for the trend in the world annual energy consumption from 1980 to 2012. During this period, our annual energy consumption rate increased from 283 quadrillion Btu in 1980 to 524 quadrillion Btu in 2012. Furthermore, as mentioned in this chapter's *Discussion Starter*, this value is projected to increase to 820 quadrillion Btu by 2040.

Now let us identify the countries with the most energy consumption. The trend in energy consumption by the top five countries in the world is shown in Figure 6.7. In recent years, manufacturers of goods have moved to China

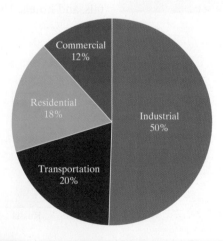

FIGURE 6.5 The percentage of world energy consumption by sector
*Source:* Data from U.S. EIA

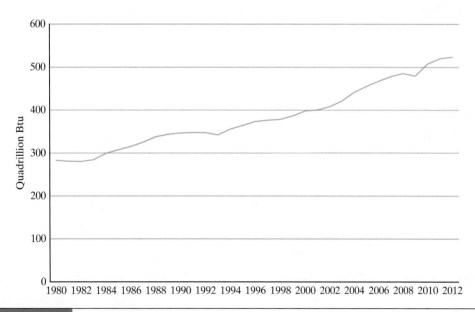

**FIGURE 6.6**   World annual total primary energy consumption
*Source:* Data from U.S. EIA

to take advantage of low labor costs. This activity is reflected in the steep rise in energy consumption in China during the past 20 years shown in Figure 6.7. Note that in 2012, China's energy consumption surpassed that of the United States.

Finally, the estimated world's non-renewable energy reserve by country is shown in Table 6.1; the United States has the largest coal reserve, while Saudi Arabia and Russia have the largest crude oil and gas reserves in the world.

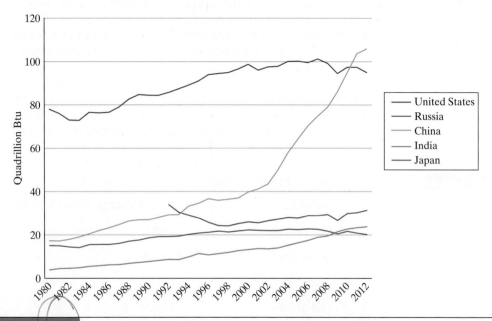

**FIGURE 6.7**   World annual total primary energy consumption—top five countries
*Source:* Data from U.S. EIA

| TABLE 6.1 | Estimated World Non-Renewable Energy Reserves by Country |

| Coal | | Crude Oil | | Gas | |
|---|---|---|---|---|---|
| United States | 27.5% | Saudi Arabia | 17.7% | Russia | 25.2% |
| Russia | 18.3% | Venezuela | 14.4% | Iran | 15.7% |
| China | 13.3% | Canada | 11.9% | Qatar | 13.4% |
| Other Non-OECD* Europe and Eurasia | 10.7% | Iran | 9.3% | Saudi Arabia | 4.1% |
| Australia and New Zealand | 8.9% | Iraq | 7.8% | United States | 4.1% |
| India | 7.0% | Kuwait | 6.9% | Turkmenistan | 4.0% |
| OECD Europe | 6.5% | United Arab Emirates | 6.7% | United Arab Emirates | 3.4% |
| Africa | 3.7% | Russia | 4.1% | Nigeria | 2.8% |
| Other Central and South America | 0.9% | Libya | 3.2% | Venezuela | 2.7% |
| Rest of World | 3.2% | Rest of World | 18.2% | Rest of World | 24.7% |
| Total | 100% | Total | 100% | Total | 100% |

*OECD: Organization for Economic Co-operation and Development.
*Source:* U.S. Energy Information Administration, *International Energy Outlook*, September 2011, Tables 5, 7, and 10.

## Heating Values of Fossil Fuels (Revisited)

By now, you should have noticed that the energy consumption rates are reported using the British thermal unit (Btu). The reason for using the Btu is that most of the energy consumed is generated by burning fossil fuels. In Chapter 3, we explained that we use fuels such as gasoline, diesel, fuel oil, and natural gas to generate energy for consumption. We also explained that the heating value of a fuel quantifies the amount of energy that is released when a unit mass (e.g., pound or kilogram) or a unit volume (cubic foot or cubic meter) of the fuel is burned. The average energy contents of common fossil fuels are shown in Table 6.2. Finally, recall that when you burn a fuel to generate energy, not all of its energy content is converted to useful energy; some of it is lost due to the inefficiency of the system generating and transmitting the energy.

| TABLE 6.2 | Average Energy Content of Common Fossil Fuels |

| Fuel | Quantity | Average Energy Content in Btu |
|---|---|---|
| Coal | One pound | 10,000 |
| Diesel | One gallon | 139,000 |
| Fuel oil (home heating oil) | One gallon | 139,000 |
| Gasoline | One gallon | 124,000 |
| Natural gas | One cubic foot | 1,000 |

**EXAMPLE 6.1**

As shown in Figure 6.1, coal provided 29 percent of the world's energy consumption of 519 quadrillion Btu in 2011. Assuming coal has an average energy content of 10,000 Btu per pound, how much coal was consumed?

amount of coal consumed in pounds

$$= (0.29)(519 \times 10^{15} \, \text{Btu})\left(\frac{1 \text{ pound}}{10,000 \, \text{Btu}}\right)$$

$$= 1.5051 \times 10^{13} \text{ pounds}$$

amount of coal consumed in tons

$$= (0.29)(519 \times 10^{15} \, \text{Btu})\left(\frac{1 \text{ pound}}{10,000 \, \text{Btu}}\right)\left(\frac{1 \text{ ton}}{2,000 \text{ pound}}\right)$$

$$= 7,525,500,000 \text{ tons} \approx 7.5 \text{ billion tons}$$

Take a moment and think about this value! How much of this coal was due to your energy consumption?

ArturNyk / Shutterstock.com

**EXAMPLE 6.2**

As shown in Figure 6.1, natural gas provided 23 percent of the world's energy consumption of 519 quadrillion Btu in 2011. Assuming that natural gas has an average energy content of 1,000 Btu per cubic foot, how much natural gas was consumed in 2011?

amount of natural gas consumed in ft$^3$

$$= (0.23)(519 \times 10^{15} \, \text{Btu})\left(\frac{1 \text{ ft}^3}{1,000 \, \text{Btu}}\right)$$

$$= 1.1937 \times 10^{14} \text{ ft}^3 \approx 119 \text{ trillion cubic feet}$$

Let's visualize how much natural gas this represents. An average bedroom in the United States has a volume of 1,800 ft$^3$(15 ft × 15 ft × 8 ft). It would take approximately 66 billion bedrooms to hold this much natural gas!

muratart / Shutterstock.com

Next, we turn our attention to some of our energy-consuming daily activities. The following examples show how many Btu it takes to enjoy our current life style. As you study these problems, reflect on your daily energy-consuming activities and habits, and think about ways to reduce your energy consumption rates.

**EXAMPLE 6.3**

nikkytok / Shutterstock.com

As shown in Figure 6.3, the United States per capita energy consumption was 313 million Btu in 2011. In Chapter 3, Example 3.6, we showed that you need to expend 8,340 Btu to heat up 20 gallons of water from room temperature at 70°F to 120°F to produce enough hot water to take a nice long shower. Let's project this daily value to an annual figure.

total annual energy needed to take a shower

$$= \left( 8,340 \frac{\text{Btu}}{\cancel{\text{day}}} \right) \left( \frac{365 \; \cancel{\text{days}}}{1 \; \text{year}} \right)$$

$$= 3,044,100 \; \frac{\text{Btu}}{\text{year}} \approx 3 \; \text{million} \; \frac{\text{Btu}}{\text{year}}$$

Therefore, each year you could spend nearly 3 million Btu to heat up water to take your daily shower.

---

**EXAMPLE 6.4**

Maria Dryfhout / Shutterstock.com

In Chapter 5, Example 5.9, we estimated the annual heating energy consumption of 152 million Btu for a house located in Minneapolis, Minnesota. Recall that we assumed a heating load (heat loss) of 62,000 Btu/h, indoor and outdoor design temperatures of 68°F and −14°F, and annual heating degree days of 8,382. Let's now assume that two people live in the house, and determine what percentage of the 2011 United States per capita energy consumption of 313 million Btu the annual heating load represents.

percentage of the total per capita consumption

$$= \left( \frac{\dfrac{152 \; \cancel{\text{million Btu}} \; \text{for heating the house}}{2 \; \cancel{\text{persons}}}}{313 \; \dfrac{\cancel{\text{million Btu}}}{\cancel{\text{person}}}} \right) (100\%)$$

$$= 24.3\%$$

---

**EXAMPLE 6.5**

According to the U.S. Federal Highway Administration's recent data, on average we drive nearly 13,500 miles per year. Let us now calculate the amount of energy consumed by driving our vehicles, assuming an average fuel economy rating of 25 miles per gallon (mpg), and compare it to the 2011 United States per capita energy consumption of 313 million Btu. Note that gasoline has an average energy content of 124,000 Btu per gallon, as shown in Table 6.2.

amount of energy consumed

$$= \left(13{,}500\ \frac{\text{miles}}{\text{year}}\right)\left(\frac{1\ \text{gallon}}{25\ \text{miles}}\right)\left(\frac{124{,}000\ \text{Btu}}{1\ \text{gallon}}\right)$$

$$= 66{,}960{,}000\ \frac{\text{Btu}}{\text{year}}$$

$$\approx 67\ \frac{\text{million Btu}}{\text{year}}$$

Mino Surkala / Shutterstock.com

percentage of the total per capita consumption

$$= \left(\frac{\dfrac{67\ \text{million Btu for driving a car}}{\text{person}}}{\dfrac{313\ \text{million Btu}}{\text{person}}}\right)(100\%) = 21.4\%$$

Note that, if two persons were to carpool, the answer changes to 10.7 percent!

---

**EXAMPLE 6.6**

In Chapter 4, we discussed electricity and examined how much electricity we consume due to our daily activities. You may recall that a United States household could consume as much as 10,000 kWh in a year. Let us now convert the kilowatt-hour unit to Btu, noting that 1 kWh = 3,412 Btu.

annual electricity consumption

$$= \left(10{,}000\ \frac{\text{kWh}}{\text{year}}\right)\left(\frac{3{,}412\ \text{Btu}}{1\ \text{kWh}}\right) = 34{,}120{,}000\ \frac{\text{Btu}}{\text{year}}$$

$$\approx 34\ \frac{\text{million Btu}}{\text{year}}$$

As you think about the result, recall that a typical power plant has an efficiency of 36 percent, and an additional 6 percent of the energy produced at the plant is lost in the power transmission lines. Therefore, the power plant must burn enough fuel to produce 113 million Btu in order for you to have access to 34 (113 × 0.3 = 34) million Btu or 10,000 kWh of electricity!

Africa Studio / Shutterstock.com

## Before You Go On

Answer the following questions to test your understanding of the preceding section:

1.  Which country in the world consumes the largest amount of energy?

2.  Name the three countries with the highest per capita energy consumption.

3.  What are the major sectors of the economy as defined by the EIA?

4.  Which sector of the world economy consumes the largest amount of energy?

*Vocabulary—State the meaning of the following terms:*

Quadrillion

Btu

EIA

kWh

## LO² 6.2 United States Energy Consumption Rates

The U.S. Energy Information Administration provides reliable statistics and detailed analyses of energy consumption for different sectors of the economy. In this section, we use the United States data as a means to convey important information; however, realize that the rising energy consumption is a global concern that requires a broader understanding and a collective global effort to address. In 2014, 98.3 quadrillion British thermal units were consumed by major sectors of the United States economy, including transportation, industrial, residential and commercial, along with electric power generation (see Figure 6.8). Non-renewable energy sources, such as petroleum, natural gas, and coal, made up 81 percent, while renewable and nuclear sources provided only 10 and 8 percent of the total energy consumed, respectively.

The United States primary energy consumption by source and sector is shown in Figure 6.9. When studying Figure 6.9, note that there are two numbers on each arrow. The numbers at the beginning of arrows, originating from an energy source, add up to 100. Also, numbers at the end of arrows leading to a sector add up to 100. Moreover, the numbers originating from an energy source represent the percentage of the source delivered to a sector, while the numbers shown close to a sector signify the percentage of the energy source used in that sector. For example, in 2014, 71 percent of all the petroleum used in the United States was consumed in the transportation sector, and 92 percent of all energy sources consumed in the transportation sector came from petroleum. The remaining 8 percent came from natural gas (3 percent) and renewable energy (5 percent).

In order to have a better understanding of how we have been consuming energy over the years, trends in the energy consumption

The United States has the largest per capita energy consumption in the world.

by different sectors during the 1949 to 2014 period are shown in Figure 6.10. As you can see from examining Figure 6.10, the energy consumed by all sectors is generally on the rise. Also, note brief declines in energy consumption in the industry and transportation sectors during certain periods. These declines were due to the economic crises of those periods.

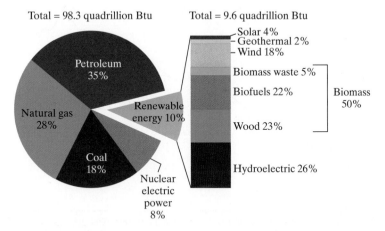

Note: Sum of components may not equal 100% as a result of independent rounding.

| **FIGURE 6.8** | The United States energy consumption by energy source, 2014 |
|---|---|
| | *Source:* U.S. Energy Information Administration, *Monthly Energy Review*, Table 1.3 and 10.1 (March 2015), preliminary data. |

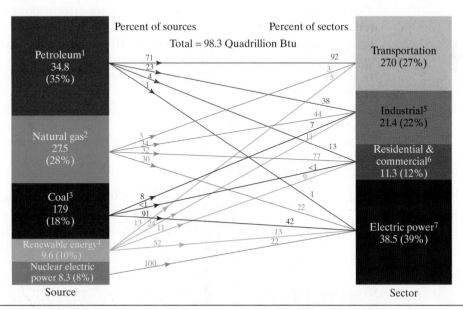

[1] Does not include biofuels that have been blended with petroleum—biofuels are included in "Renewable Energy."
[2] Excludes supplemental gaseous fuels.
[3] Includes less than –0.1 quadrillion Btu of coal coke net imports.
[4] Conventional hydroelectric power, geothermal, solar/photovoltaic, wind and biomass.
[5] Includes industrial combined-heat-and-power (CHP) and industrial electricity-only plants.
[6] Includes commercial combined-heat-and-power (CHP) and commercial electricity-only plants.

[7] Electricity-only and combined-heat-and-power (CHP) plants whose primary business is to sell electricity, or electricity and heat, to the public, includes 0.2 quadrillion Btu of electricity net imports not shown under "Source."
  Notes: Primary energy in the form that it is first accounted for in a statistical energy balance, before any transformation to secondary or tertiary forms of energy (for example, coal is used to generate electricity). • Sum of components may not equal total due to independent rounding.

| **FIGURE 6.9** | The United States energy consumption by source and sector. |
|---|---|
| | *Source:* U.S. Energy Information Administration, *Monthly Energy Review*, (March 2015). Tables 1.3, 2.1–2.6. |

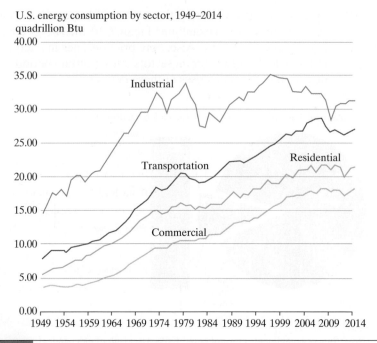

U.S. energy consumption by sector, 1949–2014
quadrillion Btu

| FIGURE 6.10 | The United States energy consumption trends by sector |

*Source:* U.S. Energy Information Administration, *Monthly Energy Review*, Table 2.1 (March 2015), preliminary data for 2014

**EXAMPLE 6.7**

In 2014, which sector of the United States economy used the largest amount of coal? What percentage of the energy sources for electric power generation came from coal?

Use Figure 6.9 to answer these questions. It should be clear from examining Figure 6.9 that 91 percent of coal was consumed in producing electricity, and 42 percent of all the energy sources consumed in generating electric power came from coal.

Next, we look at the energy consumption by each sector in more detail.

## Transportation Sector

As shown in Figure 6.9, 27 percent of United States energy consumption was due to transporting people and goods in 2014. Think about all of the cars, buses, trains, planes and subway systems that are used to transport people. Also, consider all of the trucks, trains, barges, and planes that are used to carry goods every day. Most of the energy consumption in the transportation sector is by automobiles and light trucks. Gasoline and diesel fuel account for nearly 85 percent of energy consumed by these vehicles. According to the EIA, there are over 250 million vehicles on the roads in the United States, and vehicles travel more than 3 trillion (3,000,000,000,000) miles each year, with automobiles accounting for nearly 56 percent of the miles travelled. Currently, a small percentage of vehicles use electricity, natural gas, or ethanol. The energy use by each type of vehicle is shown in Figure 6.11. Moreover, **gasoline** is the main transportation fuel, as shown in Figure 6.12. The fuel consumption of motor vehicles during the 1966 through 2013 period is shown in Figure 6.13, which gives you some idea of the trends in motor-vehicle fuel consumption.

Transportation energy use by type

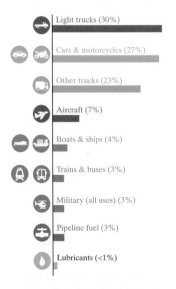

## FIGURE 6.11

The percentage of energy consumption by different modes of transportation
*Source:* U.S. Energy Information Administration, *Annual Energy Outlook 2015*, Reference case, Table 36, estimates for 2014

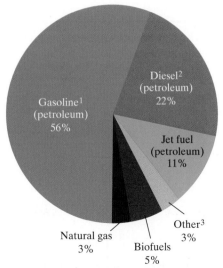

[1] Motor gasoline and aviation gas; excludes ethanol
[2] Excludes biodiesel
[3] Electricity, liquid petroleum gas, lubricants, residual fuel oil and other fuels

Note: Due to rounding, data may not sum to exactly 100%.

## FIGURE 6.12

Fuel used for United States transportation
*Source:* U.S. Energy Information Administration, *Monthly Energy Review*, (March 2015), Table 2.5 and 3.8c, preliminary data

**Motor vehicle fuel consumption, 1966-2013**

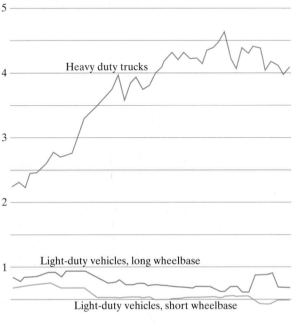

## FIGURE 6.13

The United States petroleum consumption trends by transportation mode
*Source:* U.S. Energy Information Administration (EIA), *Monthly Energy Review*, (March 2015), Table 1.8

**EXAMPLE 6.8**

The U.S. Federal Highway Administration's recent data regarding the average annual miles driven per driver by age group is shown in the accompanying table.

| Age | Male | Female | Average |
|---|---|---|---|
| 16–19 | 8,206 | 6,873 | 7,624 |
| 20–34 | 17,976 | 12,004 | 15,098 |
| 35–54 | 18,858 | 11,464 | 15,291 |
| 55–64 | 15,859 | 7,780 | 11,972 |
| 65+ | 10,304 | 4,785 | 7,646 |
| Average | 16,550 | 10,142 | 13,476 |

Let's calculate how many gallons of gasoline a woman who is between the age of 20 and 34 would consume on average if she were to drive a car with an average fuel economy rating of 25 miles per gallon (mpg) for the next ten years.

amount of gasoline to be consumed in the next 10 years

$$= \left(12,004 \frac{\text{miles}}{\text{year}}\right)\left(\frac{1 \text{ gallon}}{25 \text{ miles}}\right)(10 \text{ years}) = 4,802 \text{ gallons}$$

**EXAMPLE 6.9**

In Example 6.8, how many gallons of gasoline would be saved if she were to drive a more efficient car with an average fuel economy of 30 miles per gallon (mpg) for the next ten years?

amount of gasoline to be consumed in the next 10 years

$$= \left(12,004 \frac{\text{miles}}{\text{year}}\right)\left(\frac{1 \text{ gallon}}{30 \text{ miles}}\right)(10 \text{ years}) = 4,001 \text{ gallons}$$

Now compare the two cars by calculating the amount of gasoline to be saved over a 10-year period:

comparing a car with 25 mpg to

one with 30 mpg = 4,802 gallons − 4,001 gallons = 801 gallons

Imagine the fuel savings if the fuel economy of one million or more cars was increased from 25 to 30 mpg!

| EXAMPLE 6.10 |
|---|

From Example 6.9, how much money would the driver save if the gasoline prices were to fluctuate between $2.00 and $4.00 during the next 10 years?

In the previous example, we showed that 801 gallons of gasoline could be saved through increased fuel-consumption efficiency. Let's now investigate the savings in increments of 25-cent increases in the gasoline price. This type of examination is called a *what-if-scenario* or *sensitivity* (to price changes) *analysis*.

| Gasoline Cost (in dollars) | Savings (in dollars) |
|---|---|
| 2.00 | 1,602.00 |
| 2.25 | 1,802.25 |
| 2.50 | 2,002.50 |
| 2.75 | 2,202.75 |
| 3.00 | 2,403.00 |
| 3.25 | 2,603.25 |
| 3.50 | 2,803.50 |
| 3.75 | 3,003.75 |
| 4.00 | 3,204.00 |

As you can see, based on the price of gas, the average savings for the 10-year interval could range from $1,602.00 (or $160.20 per year) to $3,204.00 (or $320.40 per year). In addition to monetary gains, a car with a superior gas mileage produces less carbon dioxide—a gas that is contributing to global warming. In Chapter 8, we show that a gallon of gasoline can produce 20 pounds of carbon dioxide.

## Residential and Commercial Sector

The residential sector accounts for the energy consumption in homes and apartments, and the commercial sector represents the energy use in schools, lodging such as hotels, retail buildings such as malls, and health care buildings such as hospitals and clinics. As shown in Figure 6.9, the residential and commercial sectors together accounted for nearly 12 percent of the total primary energy consumption in the economy in 2014.

It is important to note here that electricity consumption is not included in the 12 percent figure, as electricity is considered a secondary source of energy; electricity is produced using primary energy sources such as fossil fuels, nuclear materials, or renewable sources. According to the EIA, electricity's share of household consumption has grown significantly during the past decades. Think about the space heating and cooling equipment, lighting systems, electronic devices, and appliances such as refrigerators, freezers, ovens, washers, and dryers that are used in our homes every day.

The results of a recent survey of energy use in homes are shown in Figure 6.14, with space heating accounting for 42 percent, lighting and other appliances 30 percent, water heating 18 percent, air conditioning 6 percent, and refrigeration 5 percent. The survey results also show that the energy-efficiency gains in recent years are offset by the use of more electronic devices and appliances (see Figures 6.15 and 6.16). For example, today over 20 percent of homes in the United States have a second refrigerator.

Since space heating constitutes the largest portion of energy use in homes, an additional survey was conducted to determine the type of primary energy

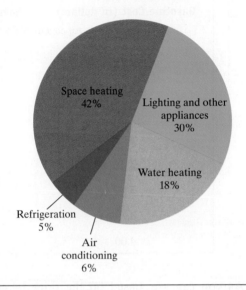

**FIGURE 6.14**    How we use energy in our homes
*Source:* U.S. Energy Information Administration, *Residential Energy Consumption Survey (RECS) 2009.*

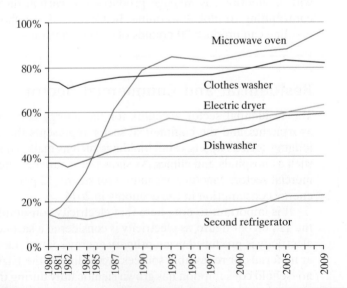

**FIGURE 6.15**    Share of homes with appliances
*Source:* U.S. Energy Information Administration, *Residential Energy Consumption Survey 1980, 1981, 1982, 1984, 1987, 1990, 1993, 1997, 2001, 2005 and 2009.*

source used for space heating. The result of this survey, shown in Figure 6.17, indicates that natural gas and electricity account for 82 percent of the energy sources in American homes. Natural gas and propane are used primarily for space heating and cooking purposes. In parts of the country with milder winters, electricity is also used for heating.

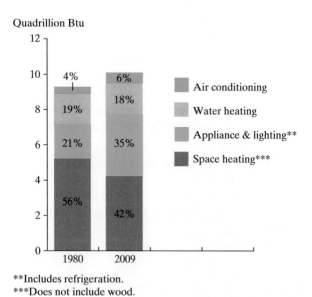

**Includes refrigeration.
***Does not include wood.

**FIGURE 6.16**   A comparison of home energy end-use between 1980 and 2009
*Source:* U.S. Energy Information Administration, *Residential Energy Consumption Survey 1980 and 2009.*

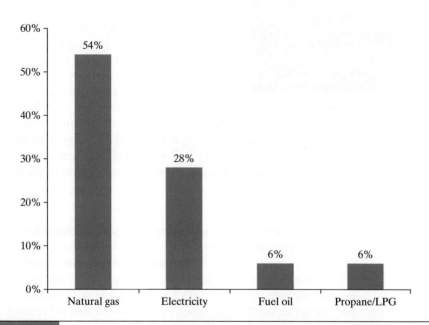

**FIGURE 6.17**   The percent of primary source of energy used in space heating in single-family homes
*Source of data:* U.S. Energy Information Administration

## Commercial Sector

According to the U.S. Energy Information Administration, the top five energy-consuming commercial building categories are:

1. Retail and Service (20 percent of total energy consumption)
   - Malls and stores
   - Car dealerships
   - Dry cleaners
   - Gas stations
2. Office (17 percent of energy consumption)
   - Professional and government offices
   - Banks
3. Education (13 percent of energy consumption)
   - Elementary, middle, and high schools
   - Colleges
4. Health Care (9 percent of energy consumption)
   - Hospitals
   - Medical offices
5. Lodging (8 percent of energy consumption)
   - Hotels
   - Dormitories
   - Nursing homes

hacohob / Shutterstock.com

As you can see from this list, the EIA classifies schools and colleges as commercial buildings and groups them with the commercial sector of the economy. Even though each commercial building category—because of its activities—may have different energy needs, in general, commercial buildings use more than half of their energy for space heating and lighting.

In the past, lighting systems accounted for a major portion of electricity consumption in commercial buildings. However, as more lighting systems switch to LED lights, the share of electricity consumption for lighting systems will decline. Furthermore, as you would expect, computers, copiers, and other office equipment also contribute to the consumption of electricity. According to the EIA, electricity and natural gas are the most common energy sources used in commercial buildings. Now that you have some understanding of how energy is consumed in the transportation, residential, and commercial sectors, let us look at energy consumption in the industrial sector.

## Industrial Sector

As shown in Figure 6.9, the *industrial sector* accounted for 22 percent of the total energy use in the United States in 2014. This value represents the share of total energy consumed by all facilities, activities, and equipment for construction, mining, agriculture, and manufacturing. The sources of energy used in industry include natural gas, electricity, propane, coal, fuel oil, and other sources such as agricultural waste, wood residues from mill processing, and wood and paper-related refuse (see Figure 6.18). The petroleum refining, chemical, paper, and metal industries are among the largest consumers of energy in this sector.

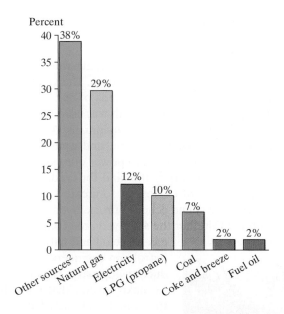

¹Includes all use of energy and fuels; excludes shipments of energy sources produced onsite.

²Other sources include steam, pulping liquor from paper making, agricultural waste, tree wood, wood residues from mill processing, and wood-related and paper-related refuse.

| FIGURE 6.18 | Sources of energy used for industry and manufacturing, 2010[1] |
|---|---|

*Source:* U.S. Energy Information Administration, *Manufacturing Energy Consumption Survey 2010.* Table 1.2 (March 2013)

## Before You Go On

Answer the following questions to test your understanding of the preceding section:

1.   What percentage of the United States total energy consumption is due to the residential sector?

2.   What are the major commercial building categories?

3.   The transportation sector accounts approximately for what percentage of the United States total energy use?

4.   What does the industrial sector represent?

*Vocabulary—State the meaning of the following terms:*

Commercial sector

Industrial sector

## LO³  6.3  Fossil Fuels

Oil, natural gas, and coal are commonly referred to as fossil fuels because they were formed millions of years ago. They make up over 80 percent of the fuel used to address the energy needs of our modern society. Historically, oil found near the surface of the Earth was used for thousands of years by different civilizations, including the Babylonians, Assyrians, Egyptians, and Native North Americans, as fuel, medicine, and waterproofing agents. Today we use fossil fuels to heat our homes, to power our cars, and to generate electricity.

Today, geologists survey the land for crude oil using vibrating devices or with small amounts of explosives to create sonic waves to listen for echoes. From studying the sound echoes of rock layers below the Earth's surface, they then determine whether the surveyed area contain deposits of oil and natural

Yarygin / Shutterstock.com

gas. Once oil and natural gas deposits are found, the petroleum engineers decide on the drilling and production approach. They first drill an exploratory well to make certain that enough oil exists in the area that can be extracted economically. If the results are favorable, they proceed to drill development wells. The oil may flow to the surface naturally or be forced out of the ground using different technologies. The newest technology, which is called hydraulic fracturing, refers to situations wherein water, chemicals, sands, and other agents are forced underground to break up geological formations such as shale, sandstone, and rock that contain oil and natural gas.

**World Crude Oil Production** According to the EIA, nearly one hundred countries produced crude oil in 2013; Russia (13 percent), Saudi Arabia (13 percent), United States (10 percent), China (5 percent), and Canada (4 percent) were among the largest producers (see Figure 6.19). The values shown in parentheses

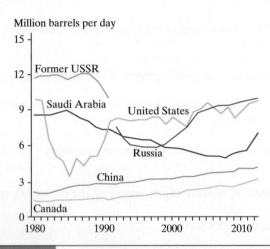

| FIGURE 6.19 | The top five oil producing countries from 1980 to 2013 |

*Source:* U.S. Energy Information Administration, International Energy Statistics (as of February 2015)

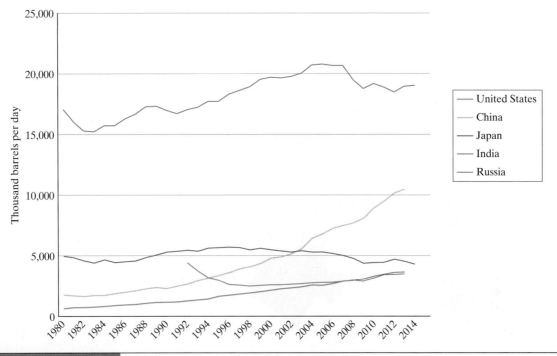

The top five oil consuming countries
*Source:* U.S. Energy Information Administration

represent the share of total world production. During this period, the world consumed about 90.4 million barrels of oil per day. In 2013, the United States alone consumed nearly 19 million barrels per day. The trend in crude oil consumption in the world during the 1980 to 2014 period is shown in Figure 6.20; China's consumption is on the rise due to its increased manufacturing activities and to more people owning automobiles in China.

**United States Crude Oil Production** In the United States, crude oil is recovered in 31 States and from offshore wells in the Gulf of Mexico. In 2014, more than two-thirds of the United States crude oil production came from Texas (37 percent), North Dakota (13 percent), California (6 percent), Alaska (6 percent), Oklahoma (4 percent), and the Gulf of Mexico (16 percent) (see Figure 6.21). Automobiles are the most common mode of transportation in the United States, and as we discussed earlier, most of these vehicles are fueled by either gasoline or diesel. The liquid fuel consumption rates are projected to increase to 16.1 million barrels per day by 2035.

The process that takes the ***crude oil*** from a source and produces ***gasoline*** and ***diesel*** fuel is shown in Figure 6.22. The imported and domestic crude oil is first sent to refineries via ships and pipelines. After the crude oil is refined into gasoline, diesel, or fuel oil, it is sent to refinery storage tanks and from there to bulk storage terminals near consuming areas. The oil is then distributed to gas stations via tanker trucks. There are nearly 162,000 gas stations in the United States, of which approximately 50,000 are stations with no affiliation with any oil companies. As a result, these gas stations may be selling any brand of gasoline. Moreover, because gasoline is sent through pipelines that are shared, some mixing of products may occur; therefore, at a given gas station, it would be difficult to track down the origin of a product from a specific refinery. Additionally, because of cost, refineries use a mixture of domestic and foreign crude oil to make petroleum products. This fact would make it even more difficult to pinpoint the origin of a refined product.

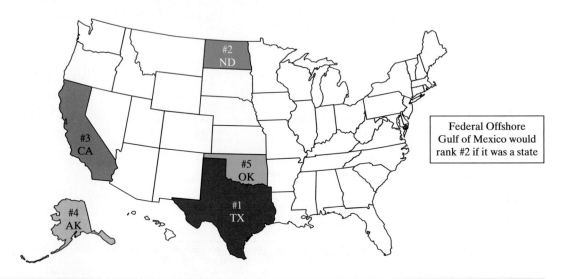

Federal Offshore Gulf of Mexico would rank #2 if it was a state

**FIGURE 6.21**  The top oil producing states in the United States
*Source:* U.S. Energy Information Administration, *Petroleum Supply Monthly*, Table 26 (February 2015), preliminary data for 2014

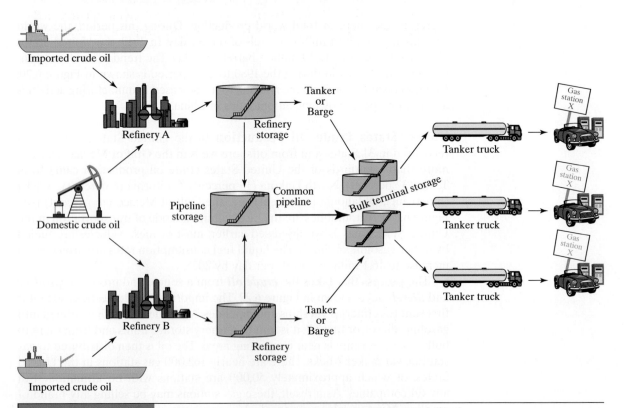

**FIGURE 6.22**  The process of converting crude oil to gasoline and diesel
*Source:* U.S. Energy Information Administration

In a refinery, from each barrel (42 gallons) of oil, 19 gallons of gasoline are made. The remaining 23 gallons are turned into diesel, heating oil, jet fuel, and other petroleum-based products. One barrel of oil is equal to 42 gallons or 159 liters. To provide a means of understanding how much gasoline we use each day in the United States, the top five gasoline-consuming states are shown in Table 6.3. As you might expect, California leads this category.

| TABLE 6.3 | Top Five Gasoline Consuming States, 2013 | |
|---|---|---|
| **State** | **Million Gallons/Day (approximate values)** | **Share of Total U.S. Consumption** |
| California | 39 | 11% |
| Texas | 36 | 10% |
| Florida | 20 | 6% |
| New York | 14 | 4% |
| Ohio | 13 | 4% |

*Source:* U.S. Energy Information Administration

**EXAMPLE 6.11**

In 2014, 375 million gallons of gasoline were consumed each day in the United States. What is the per capita gasoline consumption per day and per year, given that the population of the United States in that year was 319 million?

per capita consumption per day

$$= \left( \frac{\frac{375 \text{ million gallons}}{\text{day}}}{319 \text{ million persons}} \right)$$

$$= 1.175 \text{ gallons per person per day}$$

$$\approx 1.2 \text{ gallon per person per day}$$

per capita consumption per year

$$= \left( \frac{\frac{375 \text{ million gallons}}{\text{day}}}{319 \text{ million persons}} \right) \left( \frac{365 \text{ days}}{1 \text{ year}} \right)$$

$$= 429 \text{ gallons per person per year}$$

**EXAMPLE 6.12**

From Example 6.11, assuming that on average most cars have a fuel economy of 25 miles per gallon, what is the distance travelled per capita per year?

distance travelled per capita per year

$$= \left( \frac{429 \; \cancel{\text{gallons}}}{\dfrac{\text{person}}{\text{year}}} \right) \left( \frac{25 \text{ miles}}{1 \; \cancel{\text{gallon}}} \right)$$

$$= 10,725 \text{ miles per person per year} \approx 11,000 \text{ miles per person per year}$$

How far do you drive each year?

## Diesel Fuel

We use diesel fuel to power automobiles, public and private busses, trucks, farm equipment and tractors, construction machinery, and boats. Many of these vehicles play important roles in our daily lives in building infrastructure, moving or lifting things, farming, fishing, and transporting people, goods, and food. Diesel fuel is also used in military vehicles and tanks. *Diesel* fuel accounts for nearly one-fifth of the total transportation fuel consumption in the United States. It is also used in remote and emergency electricity generators. The energy content of diesel is greater than many other fuels. For example, on average, the energy content of diesel fuel per gallon is approximately 12 percent more than gasoline (see Table 6.1).

The past and projected future trends of liquid fuel consumption in the transportation sector are shown in Figure 6.23.

Dmitry Kalinovsky / Shutterstock.com

EPG_EuroPhotoGraphics / Shutterstock.com

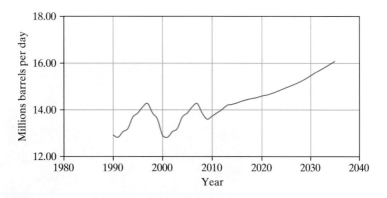

**FIGURE  6.23**  The liquid fuels consumption trends in transportation
*Source:* U.S. Energy Information Administration

## Fuel Oil (Heating Oil)

***Fuel oil*** is a petroleum product used to heat homes in America—particularly in the northeast—and to produce electricity. ***Heating oil*** and diesel fuel are similar in composition; the main difference between the two fuels is their sulfur content. Heating oil has more sulfur than diesel fuel does. From each barrel of crude oil (42 gallons), approximately 10 gallons of diesel fuel and 2 gallons of fuel oil are produced. In addition, because heating oil is tax-exempt and cannot be used legally to fuel cars and trucks on highways, the U.S. Internal Revenue Service requires heating oil to be dyed red. The red color makes it clear that the product is tax-exempt and cannot be used legally as highway diesel.

In the northeastern section of the United States, nearly 6.4 million homes rely on fuel oil to heat their homes during winter months. According to the Energy Information Administration, in 2009, these homes purchased nearly 3.7 billion gallons of heating oil, which accounted for 84 percent of total residential fuel oil consumption in the United States that year.

**Top five heating oil consuming states in 2009**

*Source:* U.S. Energy Information Administration, *Fuel and Kerosene Sales 2009* (February 2011).

**Sales of residential heating oil by region, 2009**

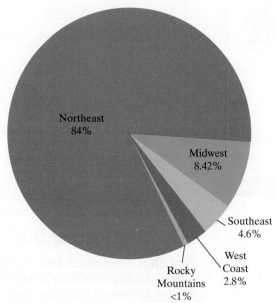

Northeast
84%

Midwest
8.42%

Southeast
4.6%

West
Coast
2.8%

Rocky
Mountains
<1%

*Source:* U.S. Energy Information Administration, *Fuel and Kerosene Sales 2009* (February 2011).

As you know, when petroleum products are burned, they produce pollutants such as carbon dioxide ($CO_2$), carbon monoxide (CO), sulfur dioxide ($SO_2$), and nitrogen oxides. These pollutants cause acid rains, global warming, and hazy conditions in cities. In order to reduce the emission of sulfur dioxide from buses and trucks, the U.S. Environmental Protection Agency (EPA) began reinforcing new emission standards in 2006 that require an 87 percent reduction in the sulfur content of diesel fuel. This type of fuel is commonly referred to as ultra-low sulfur diesel (ULSD). Starting in December of 2010, all diesel fuels used for trucks and buses were ULSD, and by 2014, all diesel fuel produced met the ULSD standards.

## Before You Go On

Answer the following questions to test your understanding of the preceding section:

1. Which countries are among the top five producers of crude oil?

2. What are the top five gasoline-consuming states in the United States?

3. Which fuel has more energy content per gallon: gasoline or diesel?

4. What is the difference between diesel fuel and heating oil?

*Vocabulary—State the meaning of the following term:*

ULSD

isak55 / Shutterstock.com

## Natural Gas

In 2013, the world consumed nearly 121 trillion cubic feet of *natural gas*. In 2014, according to the U.S. Energy Administration, 26,819 billion cubic feet of natural gas were consumed to generate electricity (30 percent) for industrial (29 percent), residential (19 percent), commercial (13 percent), and other activities (9 percent). The United States natural gas transportation network consists of about 1.5 million miles of mainline and secondary pipelines, which connect the production areas to the consumers as shown in Figure 6.24. Salt caverns, depleted oil reservoirs, or aquifer reservoirs serve as underground storage facilities to store natural gas as a seasonal backup supply. Aboveground storage facilities for liquefied natural gas are also used. There are approximately 400 active storage fields. The major gas transportation pipelines in the United States are shown in Figure 6.24, and the percentage of natural gas transmission pipeline mileage in each state is shown in Figure 6.25.

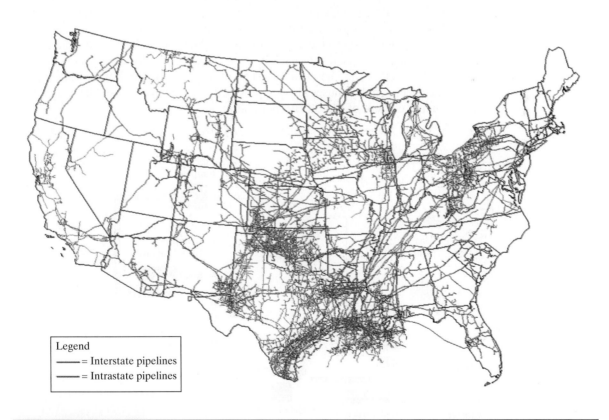

Legend
——— = Interstate pipelines
——— = Intrastate pipelines

**FIGURE 6.24**   United States natural gas transportation network: 1.5 million miles of mainline and other pipelines that link production areas and markets
*Source:* U.S. Energy Information Administration

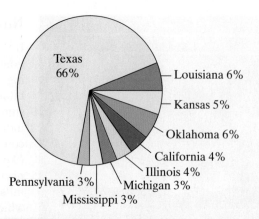

**FIGURE 6.25**   Percent of United States natural gas transmission pipeline mileage in each state (2008)
*Source:* Energy Information Administration, Natural Gas Transportation Information System, Natural Gas Pipeline Maps Database (December 2008).

## Propane

Refineries and natural-gas processing plants also make millions of barrels per day of *liquefied petroleum gases (LPGs)*, such as propane. A gas such as *propane* is referred to as a liquid petroleum gas because it is stored in a tank under relatively high pressures, which keeps it in a liquid state. The LPGs become gas once released from the pressurized tank. The process for making LPGs is depicted in Figure 6.26. In the northeast, liquid petroleum gases, such as propane, are used for space heating, cooking, and heating water. Propane is also used in the chemical industry to make plastics and other materials. In many cities, public buses also are powered by propane. Propane has an average energy content of 91,600 Btu per gallon.

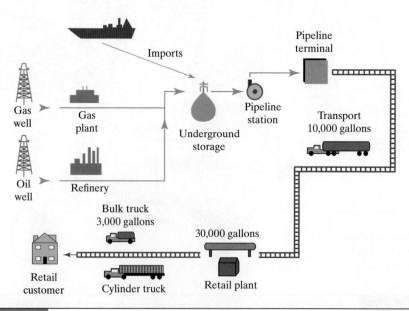

**FIGURE 6.26**   Propane production and distribution system: the process for making propane
*Source:* U.S. Energy Information Administration

## Coal

michael sheehan / Shutterstock.com

Coal is relatively inexpensive to mine and use as a fuel source to generate electricity. In 2013, the world consumed about 8.5 billion tons of coal. During the same period, we consumed about 925 million tons of coal in the United States, and 93 percent of this amount was burned in power plants to generate electricity. The rest of the coal was used in other industries, including steel, cement, and paper, to process materials. Figure 6.27 shows the major regions where coal is mined in the United States. According to the EIA, five countries have 73 percent of the world's coal reserves: the United States (26 percent), Russia (18 percent), China (13 percent), Australia (9 percent), and India (7 percent). In the United States, as shown in Figure 6.28, the top five coal-producing mines were located in Wyoming (39 percent), West Virginia (12 percent), Kentucky (8 percent), Illinois (5 percent), and Pennsylvania (5 percent) in 2013.

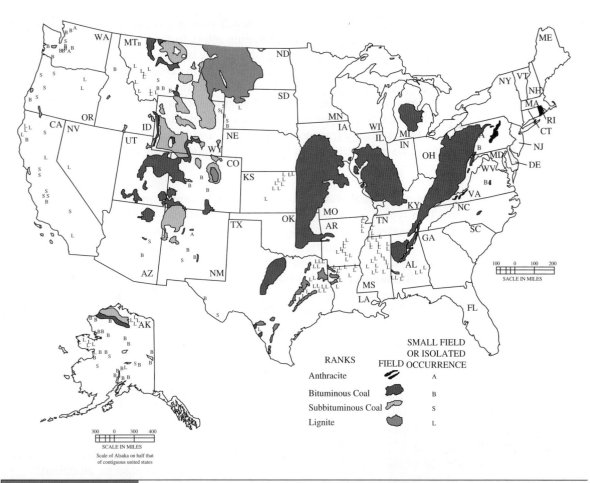

**FIGURE 6.27**    Regions where coal is mined in the United States
*Source:* U.S. Energy Information Administration

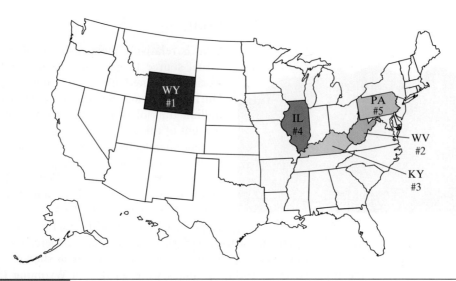

**FIGURE 6.28**

Top coal-producing states, 2013
*Source:* U.S. Energy Information Administration, *Annual Coal Report* (January 2015)

There are two methods by which coal is extracted: surface mining and underground mining. Surface mining refers to the process where coal resides less than 200 feet below ground and can be extracted by first removing the top soil and rock layers to gain access to the coal beneath. When the coal is located several hundred feet below the Earth's surface, underground mining techniques are used. After coal is mined, to increase its heating value, it is processed to remove dirt and other unwanted materials. The processed coal is then typically transported via trains and barges to power plants. Coal also may be transported (pumped) through pipelines by first crushing it and then mixing it with water.

**Types of Coal** Based on its carbon and energy content, coal is grouped into four types: *anthracite, bituminous, subbituminous*, and *lignite*.

***Anthracite*** has the highest heating and carbon content values (86 to 97 percent). Found mostly in Pennsylvania, this coal makes up less than 2 percent of the total reserve in the United States.

***Bituminous coal*** is ranked second in terms of carbon content and heating value. It contains 45 to 86 percent carbon and is produced from mines in West Virginia, Kentucky, and Pennsylvania. Bituminous coal makes up nearly 50 percent of the total reserve. It is mostly used in power plants to generate electricity and in production facilities to make steel.

***Subbituminous coal*** is ranked third in terms of carbon content and heating value. It contains 25 to 35 percent carbon and is mined in Wyoming and Montana. It makes up nearly 37 percent of the total reserve. Nearly half of the coal mined in the United States is subbituminous.

***Lignite***, which has the lowest carbon content and heating value of the coals, is mined in North Dakota, West Virginia, Kentucky, and Pennsylvania. Lignite is also used to produce electricity.

We discuss the environmental impact of fossil fuels in Chapters 8 and 9.

> Based on its carbon and energy contents, coal is grouped into four types: anthracite, bituminous, subbituminous, and lignite.

## *Before You Go On*

Answer the following questions to test your understanding of the preceding section:

1.   What is a liquid petroleum gas?

2.   How much of the United States electricity is generated by coal?

3.   Which states are among the largest producers of coal?

4.   What are different types of coal?

***Vocabulary—State the meaning of the following terms:***

Propane

LPGs

Anthracite

Bituminous coal

Subbituminous coal

Lignite

## LO⁴   6.4  **Nuclear Energy**

Nuclear energy represents 5 percent of the energy sources in the world and 8 percent in the United States. As we explained in Chapter 4, there are two processes by which nuclear energy is harnessed: nuclear fission and nuclear fusion.

Kletr / Shutterstock.com

Nuclear power plants use nuclear fission to heat water to create steam to turn the turbines that run the generators that produce electricity. During the nuclear fission process, to release energy, atoms of uranium are bombarded by small particles called neutrons. This process splits the atoms of uranium and releases more neutrons and energy in the form of heat and radiation. The additional neutrons go on to bombard other uranium atoms, and the process keeps repeating itself, leading to a chain reaction. The fuel most widely used by nuclear power plants is Uranium 235 or simply *U-235*. U-235 is relatively rare and must be processed from the uranium that is mined. After it is processed, the uranium fuel is made into ceramic pellets that are stacked end-to-end to form *fuel rods*. The fuel

FooTToo / Shutterstock.com

rods are then bundled together to create *fuel assemblies*, which are then used in the reactor core of a nuclear power plant. Depending on a reactor design, a reactor could hold as many as 193 fuel assemblies, and each fuel assembly could hold up to 264 fuel rods. In the United States there are currently 61 nuclear power plants with 99 reactors that produce about 800 billion kWh of electricity annually. The amount of electricity generated by nuclear fuel from 1980 through 2014 is shown in Figure 6.29. Presently, there are 30 countries in the world that have nuclear power plants. The top 10 countries that generated nearly 1,977 billion kWh of electricity in 2012 are shown in Figure 6.30.

According to the EIA, the owners and operators of United States civilian nuclear power reactors purchased the equivalent of 53.3 million pounds of uranium during 2014. Only 6 percent of this uranium came from the United States, and the remaining 94 percent was of international origin (38 percent from Australia and Canada, 39 percent originated in Kazakhstan, Russia, and Uzbekistan, 16 percent came from Namibia, Niger, and South Africa, and 2 percent from various other countries).

Finally, because spent fuel assemblies are highly radioactive, they must be stored in underwater pools for several years and then moved to dry cask concrete or steel storage containers that are cooled by air. Eventually, the spent fuel assemblies are moved from interim storage sites to permanent underground storage facilities.

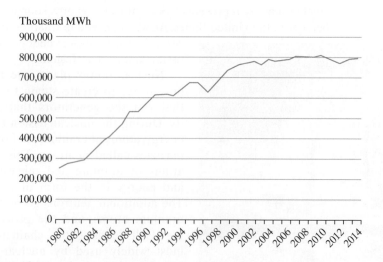

Thousand MWh

**FIGURE 6.29** Electricity generated by nuclear fuel from 1980 through 2014.
*Source:* U.S. Energy Information Administration, *Electric Power Monthly*, Tables 1.1 (February 2015), preliminary 2014 data

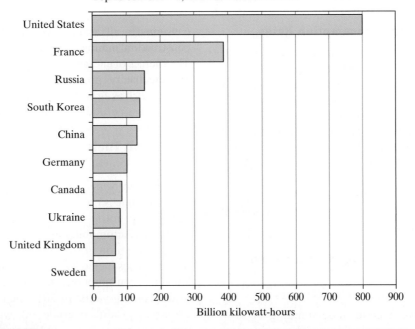

**FIGURE 6.30**   The top ten countries with nuclear energy generation
***Source:*** U.S. Energy Information Administration

# Before You Go On

Answer the following questions to test your understanding of the preceding section:

1.  In your own words, explain how electricity is generated in a nuclear power plant.

2.  Currently, how much of the world's electricity is generated using nuclear energy?

3.  What is a fuel rod?

4.  What is a fuel assembly?

*Vocabulary—State the meaning of the following terms:*

Fission process

Fuel rod

Fuel assembly

Spent fuel

# SUMMARY

## LO¹ World Energy Consumption Rates

Your personal energy consumption depends on your standard of living, while industrial energy consumption depends on economic activities such as production, distribution, consumption, and trade of goods and services. Energy use also depends on the weather. In 2011, 519 quadrillion British thermal units (Btu) of energy were consumed worldwide. In 2011, petroleum, coal, and natural gas made up nearly 86 percent of all the fuel used in generating energy. The top five countries with the largest energy consumptions were China, the United States, Russia, India, and Japan.

## LO² United States Energy Consumption Rates

### Residential and Commercial Sectors

You should have a good understanding of how we consume energy in buildings. The residential sector, which accounts for energy use in homes and apartments, represents the third largest portion of energy consumption in our society. The residential energy is spent for space heating, lighting, appliances, water heating, air conditioning, and refrigeration. Natural gas and electricity provide nearly 86 percent of the energy used in American homes. Commercial buildings include retail and service, malls and stores, car dealerships, dry cleaners, gas stations, professional and government offices, banks, schools and colleges, hospitals, and hotels. Electricity and natural gas are the most common energy sources used in commercial buildings as well.

### Transportation Sector

Nearly twenty-seven percent of the United States energy consumption is spent transporting people and products. Most of the transportation energy is consumed by automobiles and light trucks; gasoline and diesel fuel account for nearly 85 percent of energy consumed by vehicles.

### Industrial Sector

The industrial sector accounts for about twenty-two percent of the total energy use in the United States. This value represents the share of total energy consumed by all facilities, activities, and equipment for construction, mining, agriculture, and manufacturing. The sources of energy used in industry include natural gas, electricity, propane, coal, fuel oil, and other sources such as agricultural waste, wood residues from mill processing, and wood and paper-related refuse.

## LO³ Fossil Fuels

### Gasoline, Diesel, and Fuel Oil

As a well-informed citizen, you should know how much gasoline is processed from a barrel of crude oil, and which countries are among the world's top five oil producers. Automobiles are the most common mode of transportation in the United States, with most of these vehicles fueled by either gasoline or diesel. The liquid fuel consumption rates are expected to increase in the coming years. In 2013, nearly one hundred countries produced crude oil, with the top five producers being: Russia (13 percent), Saudi Arabia (13 percent), United States (10 percent), China (5 percent), and Canada (4 percent). In a refinery, each barrel of oil—equal to 42 gallons—makes 19 gallons of gasoline. The remaining 23 gallons of oil are turned into diesel, heating oil, jet fuel, and other petroleum-based products.

Diesel fuel accounts for nearly one-fifth of the total transportation fuel consumption in the United States. Fuel oil is a petroleum product used to heat homes in America—especially in the northeast. Heating oil and diesel fuel are similar in composition; the main difference between the two fuels is their sulfur content. Heating oil has more sulfur than diesel fuel does. In 2006, in order to reduce emissions of sulfur dioxide from buses and trucks, the EPA began reinforcing new emission standards that required an 87 percent reduction in the sulfur content of diesel fuel. This type of fuel is commonly referred to as ultra-low sulfur diesel (ULSD). Starting in December 2010, all diesel fuels used for trucks and buses were ULSD, and by 2014, all diesel fuel was ULSD.

### Natural Gas and Propane

The natural gas transportation network in the United States is made up of nearly 1.5 million miles of mainline and secondary pipelines. You should be able to explain the process for making and distributing both natural gas and liquefied petroleum gases (LPGs). Refineries and natural gas processing plants also make millions of barrels per day of LPGs, such as propane. Propane is referred to as a liquid petroleum gas because it is stored in a tank under relatively high pressures, keeping it in a liquid state. The LPGs change to gas once released from the pressurized tank. In the northeast, a liquid petroleum gas, such as propane, is used for cooking and to heat water and homes. Propane is also used in the chemical industry to make plastics and other materials.

## Coal

You should know that coal, based on its carbon and energy content, is classified into anthracite, bituminous, subbituminous, and lignite. Anthracite has the highest heating and carbon content values (86 to 97 percent). Bituminous coal is ranked second in terms of carbon content and heating value. Subbituminous coal is ranked third in terms of carbon content and heating value. Lignite has the lowest carbon content and heating value. According to the U.S. Department of Energy, most of the coal mined in the United States is used for generating electricity. The mines in Wyoming, West Virginia, Kentucky, Illinois, and Pennsylvania are among the largest producers of coal.

## LO⁴  Nuclear Energy

Currently, nuclear energy represents 5 and 8 percent of the world and U.S. energy sources, respectively.

There are two processes by which nuclear energy is harnessed: nuclear fission and nuclear fusion. Nuclear power plants use nuclear fission to heat water to create steam to turn the turbines that in turn run the generators that produce electricity. In nuclear fission, to release energy, atoms of uranium are bombarded by neutrons. This process splits the atoms of uranium and releases more neutrons and energy in the form of heat and radiation. The additional neutrons go on to bombard other uranium atoms, and the process keeps repeating itself, leading to a chain reaction. The fuel most widely used by nuclear power plants is Uranium 235 or simply U-235. U-235 is relatively rare and must be processed from the uranium that is mined. After it is processed, the uranium fuel is made into ceramic pellets that are stacked end-to-end to form fuel rods. The fuel rods are then bundled together to create fuel assemblies, which are then used in the reactor core of a nuclear power plant. In the United States there are currently 61 nuclear power plants with 99 reactors that produce about 800 billion kWh of electricity annually. Today, there are 30 countries in the world that have nuclear power plants. The top 10 countries generate nearly 1,977 billion kWh of electricity.

The spent fuel assemblies are highly radioactive and must be stored in pools underwater for several years, then moved to dry cask concrete or steel storage containers that are cooled by air. Eventually, the spent fuel assemblies are moved from interim storage sites to permanent underground storage facilities.

## KEY TERMS

Anthracite  190
Bituminous Coal  190
Commercial Sector  163
Crude Oil  181
Diesel  181
Fuel Assembly  192
Fuel Oil  185

Fuel Rod  191
Gasoline  181
Industrial Sector  163
Heating Oil  185
Lignite  190
Liquid Petroleum Gases (LPG) 188

Natural Gas  187
Propane  188
Residential Sector  163
Subbituminous Coal  190
Transportation Sector  163
U-235  191

# Apply What You Have Learned

For this project, you are to work as a group to determine your primary annual energy resource footprint. Each group (the size to be determined by your instructor) is to estimate how much fuel they consumed last year. You need to consider fuels such as gasoline, diesel, fuel oil, natural gas, propane, and coal. Don't forget about your electricity consumption, and remember that even though electricity is considered a secondary energy source, it is generated in power plants mostly by burning fossil fuels. State all your assumptions, compile your findings into a single

Thanapun / Shutterstock.com

report, and present it to the class. How many Btu of energy were consumed by the group? What is your per capita energy consumption value, and how does it compare to the values in Figure 6.3?

# PROBLEMS

*Problems that promote life-long learning are denoted by* 🔑

**6.1** Estimate how much energy you consume 🔑 by driving your vehicle. State all your assumptions and express your answer using Btu. Suggest ways to reduce your consumption by 10%.

**6.2** Estimate how much electricity you consume 🔑 in a year. Express your answer in kWh and Btu. State all your assumptions, and suggest ways to reduce your consumption by 10%.

**6.3** Estimate how much energy you consume for 🔑 showering and bathing activities in a year. State your assumptions and express your answer in Btu.

**6.4** How many gallons of gasoline would be saved during the next 10 years if a driver were to change her existing car with a 25-mpg fuel efficiency to a more efficient car with an average fuel economy of 40 mpg? Assume she drives her car about 12,000 miles per year.

**6.5** Perform a sensitivity cost saving analysis for Problem 6.4. See Example 6.10.

**6.6** How much energy in Btu would be saved during the next five years if a household were to reduce its annual electricity consumption rates from 10,000 kWh to 7,000 kWh? Perform a sensitivity cost saving analysis, assuming the electric utility company could charge between 10 to 20 cents per kWh. Use increments of 2-cent change for your analysis.

**6.7** As shown in Figure 6.1, in 2011, petroleum provided 34% of the world's energy consumption of 519 quadrillion Btu. Assuming petroleum has an average energy content of 130,000 Btu/gallon, how many barrels of petroleum were consumed? Also express your answer in gallons.

**6.8** How many gallons of gasoline are consumed annually on average by a 25-year-old male driver (see Example 6.8) if he drives a car with a fuel economy of 20 mpg?

**6.9** Assume the annual heating energy consumption of a house is 122 million Btu.

How many cubic feet of natural gas does it take to keep the house warm during the cold months?

**6.10** The annual heating energy consumption of a house located in the northeastern part of the United States is 140 million Btu. How many gallons of fuel oil does it take to keep the house warm during the cold months?

**6.11** The annual electricity consumption of a household is 9,800 kWh. How many pounds of coal must be burned in a power plant to address this need? Assume a combined overall efficiency of 30% for the power plant and the loss in transmission lines.

**6.12** In a recent year, California consumed 41 million gallons/day of gasoline. How many Btu of energy were consumed each day? What is California's gasoline consumption per capita? Assume an approximate population of 37 million.

**6.13** In a recent year, 23 trillion ft$^3$ of natural gas was delivered to 70 million customers in the United States. How many rooms with dimensions of 15 ft $\times$ 15 ft $\times$ 10 ft could be filled with 23 trillion ft$^3$? This problem is intended to give you a visual image of how much natural gas we consume.

**6.14** In 2010, United States coal mines produced 1,805.3 million tons of coal. What was the coal consumption per capita for the United States? Assume an approximate population of 308 million for that year.

**6.15** In the United States, wood and wood waste could account for 2% of energy use. How many Btus of energy are generated from wood and wood waste? State your assumptions.

**6.16** How many gallons per day would be saved if we increased the fuel efficiency of gas-consuming vehicles by 10% in California? State all your assumptions.

**6.17** How many pounds of coal would be saved in the United States if the efficiency of power plants that use coal were to increase by 5%?

**6.18** As Americans, we enjoy barbequing, especially when the weather is nice. You have seen propane tanks that are used with barbeque grills. Investigate how much propane is consumed annually for outdoor barbequing. State your assumptions and present your findings in a brief report.

**6.19** Investigate the rise in coal consumption in power plants if one hundred million automobiles were to become electric. State your assumptions and present your findings in a brief report.

**6.20** In a recent year, the United States consumed 29,000,000,000,000 kWh of energy. Express this value in MWh, GWh, TWh, and PWh. See Table 2.2 for the list of decimal multiples and prefixes used with SI base units.

Phillip Harrington / Alamy Stock Photo

*"Our culture runs on coffee and gasoline, the first often tasting like the second."*

—EDWARD ABBEY (1927–1989)

CHAPTER

7

# Renewable Energy

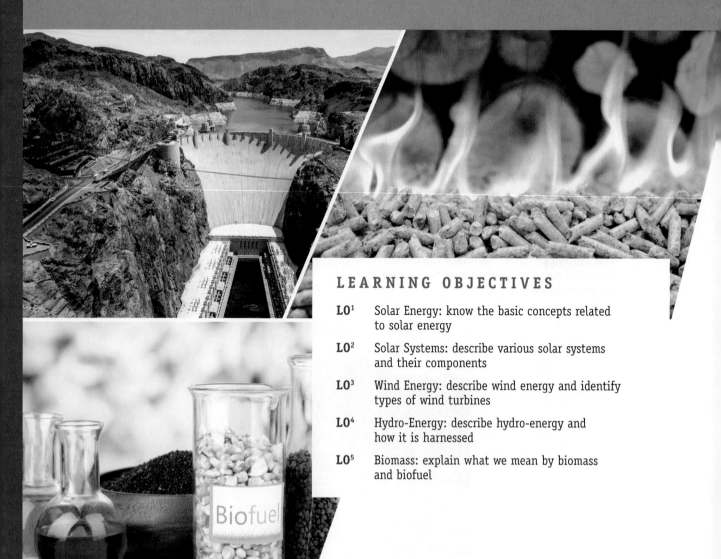

## LEARNING OBJECTIVES

**LO¹** Solar Energy: know the basic concepts related to solar energy

**LO²** Solar Systems: describe various solar systems and their components

**LO³** Wind Energy: describe wind energy and identify types of wind turbines

**LO⁴** Hydro-Energy: describe hydro-energy and how it is harnessed

**LO⁵** Biomass: explain what we mean by biomass and biofuel

# *Discussion Starter*

**U.S. energy consumption by energy source, 2014**

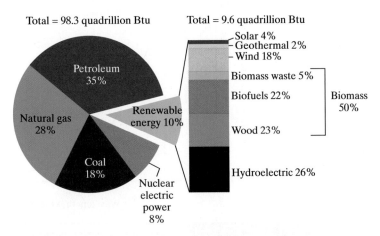

Note: Sum of components may not equal 100% as a result of independent rounding.

Source: U.S. Energy Information Administration, *Monthly Energy Review,* Table 1.3 and 10.1 (March 2015), preliminary data

E nergy obtained from the Sun's rays wind moving water the Earth's interior heat and wood, grain, and plant by-products are commonly referred to as renewable energy. According to the U.S. Energy Information Administration, renewable energy accounted for about 10 percent of total energy consumption in the United States (approximately 9.6 quadrillion Btu) in 2014. Moreover, the world's renewable energy sources including solar, wind, hydropower, and biomass are anticipated to account for only 15 percent of total world energy consumption by year 2040.

**To the Students:** What do you think are the reasons for renewable energy being such a relatively small percentage of energy consumption? Take a few moments and think about it. To get you started, think about factors such as cost, efficiency, the location of the renewable source sites relative to the whereabouts of the energy demand (e.g., where windy sites are and where high electricity demand regions are), availability, and reliability (e.g., cloudy days or days without wind or lack of flowing water due to droughts). How might these and other concerns be addressed in the future?

## LO¹ 7.1 Solar Energy

Solar energy starts with the Sun at an average distance of 93 million miles (~150 million kilometers) from the Earth. The Sun is a nuclear fusion reactor, with its surface temperature at approximately 10,000 degrees Fahrenheit (°F) or 5,500 degrees Celsius (°C). *Solar energy* that reaches the Earth is in the form of electromagnetic radiation, consisting of a wide spectrum of wavelengths

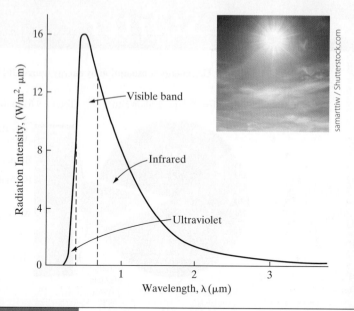

**FIGURE 7.1**    The solar radiation bands.

and energy intensities. Almost half of the solar energy received on the Earth is in the band of visible light. Solar radiation can be divided into three bands: *ultraviolet*, *visible*, and *infrared*, as shown in Figure 7.1. Many of you have had firsthand experience with the ultraviolet band that causes sunburn. The visible band comprises about 48 percent of useful radiation for heating, and the infrared makes up the rest.

As you may know, the Earth's orbit around the Sun is elliptical (see Figure 7.2). When the Sun is closer to the Earth, the Earth's surface receives a little more solar energy. The Earth is closer to the Sun when it is winter in the northern hemisphere. However, because the Earth is tilted away from the Sun, the winter months are colder, and the 23.5 degree tilt in the Earth's axis of rotation is the factor that dictates the amount of solar radiation striking the Earth at a particular location (see Figure 7.3). Moreover, because of the tilt of the Earth, the days are longer in the northern hemisphere from the spring (vernal equinox) to the fall (autumnal equinox). The opposite is true in the southern hemisphere; that is, the longer days occur during the other six months.

The distance from the Earth to the Sun changes during the year so that the energy reaching the outer atmosphere of the Earth varies from 410 to 440 Btu/ft² · h. *At the average Earth to Sun distance, the intensity of solar energy is 428 Btu/ft² · h or 1,350 W/m² out in space at the edge of the Earth's atmosphere.* The amount of radiation available at a place depends on many factors, including geographical location, season, local landscape, weather, and time of day.

As solar energy passes through the Earth's atmosphere, some of the energy is absorbed, some of it is scattered, and some of it is reflected by clouds, dust, pollutants, forest fires, volcanoes, and water vapor. The solar radiation that reaches the Earth's surface

> As solar energy passes through the Earth's atmosphere, some of it is absorbed, some of it is scattered, and some of it is reflected by clouds, dust, etc.

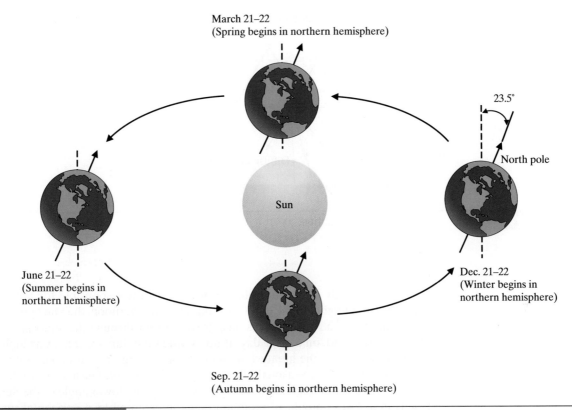

March 21–22
(Spring begins in northern hemisphere)

23.5°

North pole

Sun

June 21–22
(Summer begins in
northern hemisphere)

Dec. 21–22
(Winter begins in
northern hemisphere)

Sep. 21–22
(Autumn begins in northern hemisphere)

FIGURE 7.2    The orbit of the Earth around the Sun in respect to the seasons.

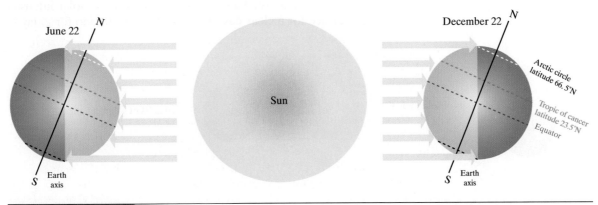

June 22

N

December 22

N

Arctic circle
latitude 66.5°N

Tropic of cancer
latitude 23.5°N

Equator

Sun

S   Earth
axis

S   Earth
axis

FIGURE 7.3    The Northern Hemisphere is tilted away from the Sun in winter and tilted toward the
Sun in summer.

without being diffused is called *direct beam* solar radiation. Atmospheric
conditions can reduce direct beam radiation by 10 percent on clear, dry
days and by 100 percent during thick, cloudy days. This process is shown in
Figure 7.4.

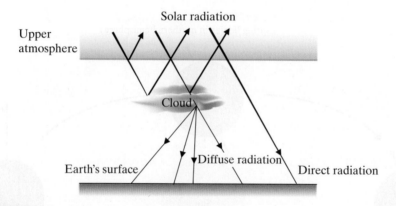

**FIGURE 7.4**    Direct and diffuse radiation.

As you already know, the rotation of the Earth is responsible for hourly variations in sunlight. In the early morning and late afternoon, the Sun appears low in the sky; as a result, its rays must travel further through the atmosphere. On the other hand, on a clear day, at noon, when the Sun appears at its highest point in the sky, the greatest amount of solar energy reaches a horizontal surface on the Earth. Seasonal effects are also important. During the winter, the Sun is at a lower angle than it is in the summer. The lower angle of the Sun results in a lower amount of radiation being intercepted by a horizontal surface. As shown in Figure 7.5, the amount of energy intercepted by a one-foot or one-meter width of a horizontal surface when the Sun is at a low angle in the winter is smaller than when the Sun is at a high angle during summer months. As depicted, more radiation is intercepted by a horizontal surface during June and July than in December and January.

To provide an additional visual aid, the hourly variation in solar intensity on a horizontal surface for a clear day at a location in Colorado for a day in

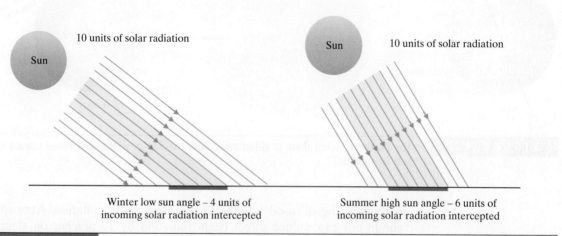

**FIGURE 7.5**    The amount of energy intercepted by a one-foot or one-meter width of a horizontal surface in winter versus summer.

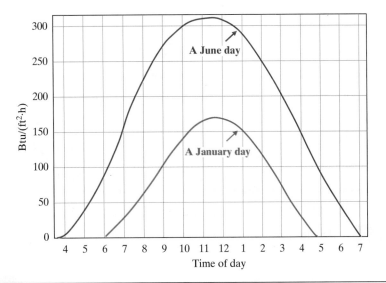

FIGURE 7.6 The hourly variations in solar intensity on a horizontal surface for a clear day at a location in Colorado for a day in January and in June.

January and one in June is shown in Figure 7.6. As shown, the hourly peak values occur at solar noon, when the Sun is at the highest angle and its rays are passing through the minimum thickness of the atmosphere. Also note from examining Figure 7.6 that the higher intensity of solar radiation occurs during a day in June when compared to a day in January (the red curve has higher values than the blue curve). Moreover, since winter days are shorter than summer days, the period during which solar energy can be collected is less in the winter. In other words, the amount of solar energy that can be collected varies with the season. To further demonstrate this point, the monthly variation in solar intensity on a horizontal surface for a clear day at a location in Colorado is shown in Figure 7.7.

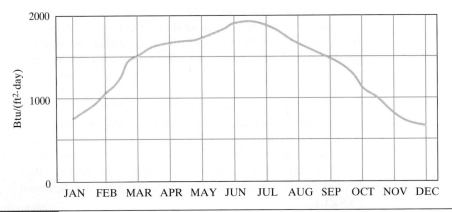

FIGURE 7.7 The monthly variations in solar intensity on a horizontal surface for a clear day at a location in Colorado.

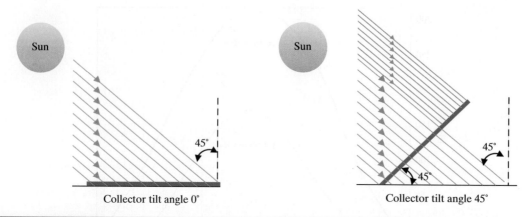

Collector tilt angle 0°    Collector tilt angle 45°

FIGURE 7.8    The effect of a solar collector's tilt angle.

You also may have noticed that in the northern hemisphere, most solar collectors face directly south and are tilted. When installing a solar collector to receive energy from the Sun, it is important to tilt the collector so that its receiving surface is nearly perpendicular to the Sun's rays. As shown in Figure 7.8, when the tilt angle of a solar collector is changed from zero (a horizontal position) to an angle equal to the incoming rays, more energy is intercepted by the collector. Then, in order to maximize solar energy collection throughout the day and throughout the year, the collector's surface must track the Sun across the sky so that the rays always remain perpendicular to its receiving surface. This requirement involves continuous movement of the collector from east to west as well as continuous change in the angle of tilt. Even though the tracking is technologically possible, it is too costly to be economically feasible for home-scale solar systems. Considering the mechanical and economical constraints for hot water systems where both winter and summer solar collection is desired, a good compromise is to tilt the collector to an angle equal to the latitude of the location.

However, if the solar system is used predominately for space-heating purposes, maximum collection is typically required during the coldest months from approximately October until March. During this period, the angle of the Sun's rays varies from about 5 degrees in October to about 23 degrees in December below that of the angle on September 21, as shown in Figure 7.9. So to maximize solar energy collection during the space-heating period, a good compromise is to tilt the collector at an angle equal to about latitude plus 15 degrees, as shown. Finally, as mentioned previously, in the northern hemisphere the collector should face south; however, note that in the southern hemisphere, the solar collector must face north.

In many countries, solar radiation data are represented as kilowatt-hours per square meter ($kWh/m^2$). The average daily solar radiation for the months of January and July in the United States are shown in Figures 7.10 and 7.11, respectively. The average solar data for the other months in the United States are available through the National Renewable Energy Lab (NREL).

Example 7.1 shows how you can use the information given in Figures 7.10 and 7.11.

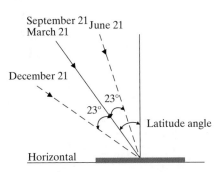

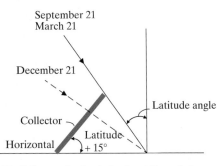

(A) December 21, sun 23° below latitude angle from perpendicular.
June 21, sun 23° above latitude angle from perpendicular.
September 21 and March 21, sun at latitude angle from perpendicular.

(B) Collector tilted at latitude + 15° maximizes winter collection.

**FIGURE 7.9**     The effect of latitude on solar energy interception.

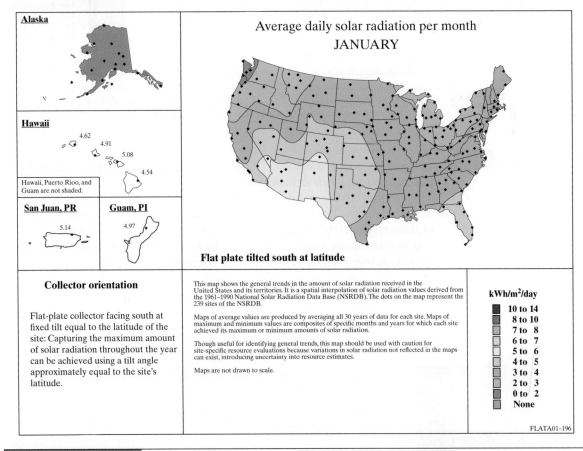

**FIGURE 7.10**     The United States average daily solar radiation for the month of January.
*Source:* National Renewable Energy Laboratory Resource Assessment Program.

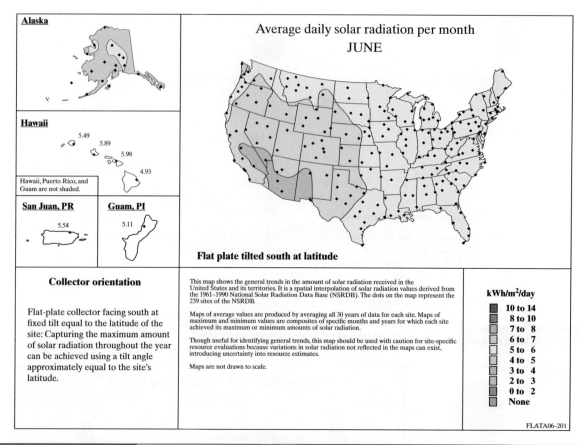

---

**FIGURE 7.11**    The United States average daily solar radiation for the month of June.
*Source:* National Renewable Energy Laboratory Resource Assessment Program.

---

**EXAMPLE 7.1**

On average, how much solar radiation is intercepted in Alaska by four flat plate collectors (with dimensions 1 m × 1.5 m) that are tilted at an angle equal to the latitude in the month of January as compared to that in June?

Using Figure 7.10, we note that in Alaska, based on a particular location in January, from 0 to 2 kWh/m²/day of solar radiation could strike a surface that is tilted at an angle of the given latitude, while from Figure 7.11, the value in June could vary from 4 to 6 kWh/m²/day.

total area of the collectors = (4)(1.0 m)(1.5 m) = 6 m²

Then for the month of January,

total solar energy intercepted by the four collectors

$$= (6 \text{ m}^2) \left( \frac{2 \text{ kWh}}{\text{m}^2 \cdot \text{day}} \right) (31 \text{ days in January}) = 372 \text{ kWh or}$$

$$= (372 \text{ kWh}) \left( \frac{3,412 \text{ Btu}}{1 \text{ kWh}} \right) = 1,269,264 \text{ Btu} \approx 1.27 \text{ million Btu}$$

Note we used the conversion factor 1 kWh = 3,412 Btu to convert the result into Btu units as well. For the month of June, assuming 4 kWh/m²/day, we have

total solar energy intercepted by the four collectors

$$= (6 \text{ m}^2) \left( \frac{4 \text{ kWh}}{\text{m}^2 \cdot \text{day}} \right) (30 \text{ days in June}) = 720 \text{ kWh or}$$

$$= (720 \text{ kWh}) \left( \frac{3,412 \text{ Btu}}{1 \text{ kWh}} \right) = 2,456,640 \text{ Btu} \approx 2.45 \text{ million Btu}$$

And assuming 6 kWh/m²/day, we have

total solar energy intercepted by the four collectors

$$= (6 \text{ m}^2) \left( \frac{6 \text{ kWh}}{\text{m}^2 \cdot \text{day}} \right) (30 \text{ days in June})$$

$$= 1,080 \text{ kWh} = (1,080 \text{ kWh}) \left( \frac{3,412 \text{ Btu}}{1 \text{ kWh}} \right) = 3,684,960 \text{ Btu}$$

$$\approx 3.68 \text{ million Btu}$$

Therefore, the total amount of solar energy intercepted by the collectors for the month of January could vary from 0 to 372 kWh (1.27 million Btu), whereas during the month of June, between 720 kWh (2.45 million Btu) and 1,080 kWh (3.68 million Btu) could be intercepted.

In the next section, we discuss different types of solar systems.

## Before You Go On

Answer the following questions to test your understanding of the preceding section:

1. What is solar energy?
2. Into how many bands can solar radiation be divided, and what are they?
3. What are the factors that define how much solar energy is available at a given location?
4. What is the difference between direct and diffuse radiation?
5. What is the effect of a lower angle of the Sun during winter months?
6. Why are solar collectors tilted?

*Vocabulary—State the meaning of the following terms:*

Solar energy

Sun angle

Direct solar radiation

Diffuse solar radiation

# LO² 7.2 Solar Systems

The economic feasibility of solar systems depends on the amount of solar radiation available at a geographic location. Moreover, using various technologies, solar radiation can be converted into useful applications such as heating water or space (air) or generating electricity. In general, solar systems can be categorized into *active*, *passive*, and *photovoltaic* systems.

## Active Solar Systems

There are two basic types of **active solar systems** used for heating: liquid and air. The *liquid flat-panel systems* make use of water or water–antifreeze mixture (in cold climates) to collect solar energy. In such systems, the liquid is heated in a solar collector (Figure 7.12) and then transferred to a storage system. In contrast, *air systems* heat the air in "air collectors" and transport it to a storage or a designated space using blowers. Most active solar systems cannot provide adequate heating for both space and hot water needs. Consequently, an auxiliary or back-up heating system is needed.

An active solar system makes use of mechanical components such as a collector, pumps, and a storage tank to collect and store solar energy.

The main components of an active liquid hot-water solar system are shown in Figure 7.13. The solar energy is captured in the flat panel solar collector and then pumped through a heat exchanger where the energy collected is transferred to the domestic hot water (DHW) pre-heater tank and eventually to the auxiliary hot water tank as needed. A home in Golden, Colorado, using a flat-plate solar system is shown in Figure 7.14.

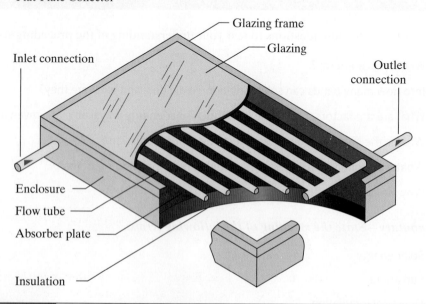

Flat-Plate Collector

Glazing frame

Glazing

Inlet connection

Outlet connection

Enclosure

Flow tube

Absorber plate

Insulation

**FIGURE 7.12**    A schematic of a solar collector.
*Source:* U.S. Department of Energy

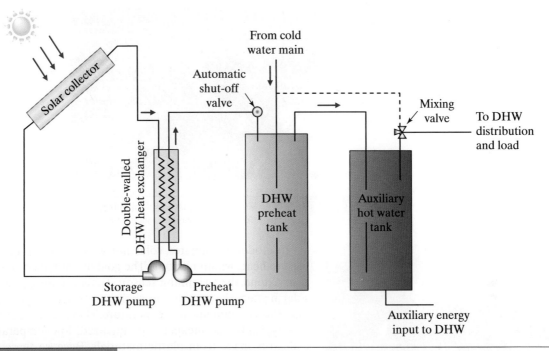

FIGURE 7.13     A schematic of a solar hot water system.

FIGURE 7.14     This home in Golden, Colorado, uses a liquid-based solar system for space and water heating.
*Source:* Courtesy of DOE/NREL

Evacuated tubes (see Figure 7.15) are another type of hot-water solar collector that are more expensive and operate at higher temperatures than flat-plate collectors. For these types of collectors, the vacuum inside the tubes minimizes the heat loss from the collector to the surrounding air.

In moderate climates, solar hot-water systems are also used to heat swimming pools. The goal of this type of system is to extend the swimming season. The swimming pool has solar heaters that operate at slightly warmer temperatures than the surrounding air temperature. These types of collectors typically use inexpensive, unglazed, low-temperature collectors made from plastic materials. Because these systems are not insulated, they require large collector areas, approximately 50 to 100 percent of the pool area.

When you travel to the southwestern part of the United States or abroad, you may see parabolic-shaped (U-shaped) collectors with mirror-like surfaces (see Figure 7.16). These collectors are used to generate electric power in power plants. The parabolic concentrating systems make use of tracking devices to follow the Sun during the day. In these systems, the rays of the Sun are reflected against solar collectors, which are basically U-shaped mirrors that are connected together inline, to concentrate all the reflected energy onto a receiving pipe that is filled with fluids that have a high heat capacity, such as oil or molten salt. The concentrated solar energy heats up the fluid, and the energy collected by the fluid in the pipe is then transferred to water to create steam in a heat exchanger. Similar to the conventional steam power plant that we discussed in Chapter 4, the steam then runs through a turbine that turns a generator to create electricity.

**EXAMPLE 7.2**

Assume that the solar collector system of Example 7.1 has an average efficiency of 60 percent during the month of June and is located at a location in Alaska, where 4 kWh/m²/day of solar energy is intercepted by the system. On average, how many gallons of water at 60°F could be heated to 110°F by the system each day during the month of June?

$$\text{total area of the collectors} = (4)(1.0 \text{ m})(1.5 \text{ m}) = 6 \text{ m}^2$$

total solar energy intercepted by the four collectors each day

$$= (6 \text{ m}^2)\left(\frac{4 \text{ kWh}}{\text{m}^2 \cdot \text{day}}\right)$$

$$= 24\ \frac{\text{kWh}}{\text{day}} \text{ or } \left(24\ \frac{\text{kWh}}{\text{day}}\right)\left(\frac{3{,}412 \text{ Btu}}{1 \text{ kWh}}\right) = 81{,}888\ \frac{\text{Btu}}{\text{day}}$$

And considering the efficiency of the system, the total available energy becomes

$$\text{total available energy} = (0.6)\left(81{,}888\ \frac{\text{Btu}}{\text{day}}\right) = 49{,}133\ \frac{\text{Btu}}{\text{day}}$$

Recall from Chapter 3, Example 3.6, that each gallon of water has a mass of 8.34 pounds, and that one Btu represents the amount of thermal energy needed to raise the temperature of one pound mass (lbm) of water by one degree Fahrenheit (°F). Realizing these facts, we can now solve for the unknown gallons ($x$) of water in the following manner:

$$\text{total available energy} = 49{,}133\ \frac{\text{Btu}}{\text{day}}$$

$$= \left(x\ \frac{\text{gallons of water}}{\text{day}}\right)\left(\frac{8.34 \text{ lbm}}{1 \text{ gallon of water}}\right)\left(\frac{1 \text{ Btu}}{(1 \text{ lbm})(1°\text{F})}\right)\left(\overset{50}{\overline{(110-60)}}°\text{F}\right)$$

Solving for $x$, we get

$$x \approx 118\ \frac{\text{gallons of water}}{\text{day}}$$

Therefore, each day the given system can provide about 118 gallons of hot water for activities such as bathing or showering.

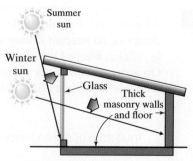

Direct gain through south-facing wall

| FIGURE 7.17 | A schematic of a direct passive solar system. |

## Passive Solar Systems

***Passive solar systems*** do not make use of any mechanical components such as collectors, pumps, blowers, or fans to collect, transport, or distribute solar heat to various parts of a building. Instead, a *direct passive solar* system uses large glass areas on the south wall of a building and a thermal mass to collect the solar energy. The solar energy is stored in the interior thick masonry walls and floors during the day and is released at night. In cold climates, passive systems also use insulated curtains at night to cover the glass areas to reduce the heat loss. Another feature of a passive solar system is an overhang to shade the windows during summer, as shown in Figure 7.17.

*Indirect-gain passive* designs use a storage mass placed between the glass wall and the heated space. As the air between the glass and masonry wall is heated, it rises and enters the room through a vent at the top of the wall and is replaced by the cooler room air that enters the lower vent. Not all of the solar heat is transferred to the air; some is stored in the masonry wall or floor (see Figure 7.18).

Another common type of passive solar system is a sunspace. The space may be used as a greenhouse, atrium, sun porch, or sun room. Masonry or concrete floors and walls, water containers, or covered pools of water may serve as thermal storage. A photograph of an interior section of a house with a sunspace is shown in Figure 7.19.

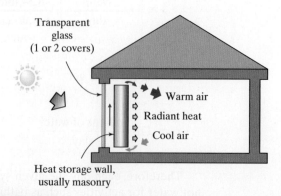

| FIGURE 7.18 | A schematic of an indirect passive solar system. |

**FIGURE 7.19**

The interior (sunspace) of a passive solar house.
*Source:* Tom Grundy / Shutterstock.com

## Photovoltaic Systems

A *photovoltaic system* converts light energy directly into electricity. These systems come in all sizes and shapes, as shown in Figure 7.20. You have also seen small photovoltaic cells that provide power for a calculator. A larger system that produces power for a home, however, often consists of a photovoltaic array, batteries, a charge controller, and an inverter.

A *photovoltaic (PV) cell* is the backbone of any photovoltaic system. Photovoltaic cell materials include crystalline, polycrystalline, and amorphous silicon. The crystalline and polycrystalline silicon PV cells have high efficiencies; however, they are expensive to produce. On the other hand, the amorphous silicon cells have lower efficiencies, are less expensive to produce, and are easier to work with. Thin-film amorphous silicon solar cells can be affixed directly to a metal roof of a building. The efficiencies of various solar cells and the improvements that have been made since 1976 are shown in Figure 7.21.

The manufacturers of photovoltaic systems combine cells to form a module, and then the modules are combined to form what is known as a *photovoltaic array* (see Figure 7.22). Photovoltaic systems are classified as *stand-alone*, *hybrid*, or *grid-tied*. The systems that are not connected to a utility grid are called **stand-alone** and require batteries to store the electrical energy to be used at nights and during cloudy days. *Hybrid systems* are those which use a combination of photovoltaic arrays and some other form of energy, such as diesel generation. As the name implies, the *grid-tied systems* are connected to a utility grid. A grid-tied system does not need a battery bank to store energy. A schematic diagram for a grid-tied system is shown in Figure 7.23.

A typical photovoltaic system consists of a photovoltaic array, batteries, a charge controller, and an inverter.

Courtesy of DOE/NREL

Courtesy of DOE/NREL

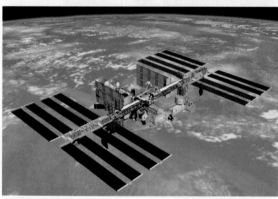
Courtesy NASA, ID: jsc2006e43519

Courtesy of DOE/NREL

Courtesy of DOE/NREL

Courtesy of DOE/NREL

**FIGURE 7.20** Examples of photovoltaic systems, top row: parking rooftop, solar bike; middle row: space station, a building rooftop; and bottom row: photovoltaic roof shingles, a remote communication facility.
*Source:* Courtesy of DOE/NREL

The electricity that is generated by photovoltaic panels is *direct current* (DC). Every photovoltaic system has an inverter. An *inverter* is a device that converts direct current into alternating current that is used in homes. As shown in Figure 7.23, the line coming out of the inverter goes to the main utility breaker panel in the house, and from there it goes to the utility meter and the electricity grid. Systems that use batteries to store electricity for cloudy days

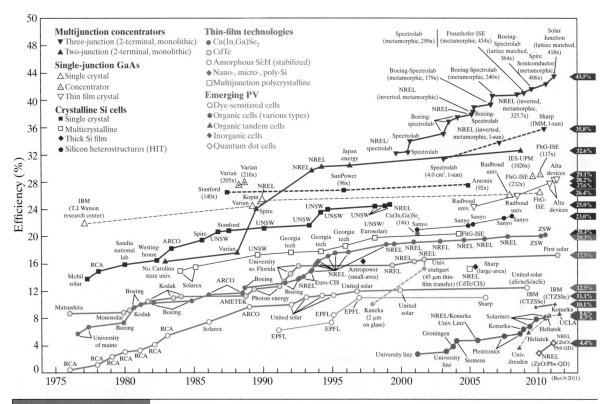

The efficiency of solar cells and how improvements have been made over time.
*Source:* NREL.

also make use of a charge controller. A ***charge controller*** protects the batteries from overcharging. When the batteries are fully charged, the charge controller disconnects them from the PV array.

Photovoltaic systems are also used in photovoltaic power plants, which represent large-scale commercial systems that produce electricity. One of the largest grid-tied photovoltaic power plants in the United States is the Alamosa photovoltaic plant, which is located in an area of 82 acres in south central Colorado. It went online in 2007 and generates about 8.2 megawatt (MW) of power.

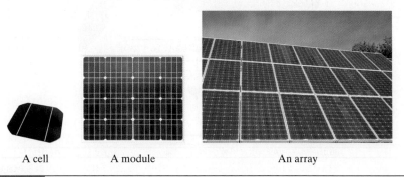

A cell                A module                         An array

A photovoltaic cell, module, and array.
*Source:* Winbjörk / Shutterstock.com, Mrs_ya / Shutterstock.com, CJimenez / Shutterstock.com.

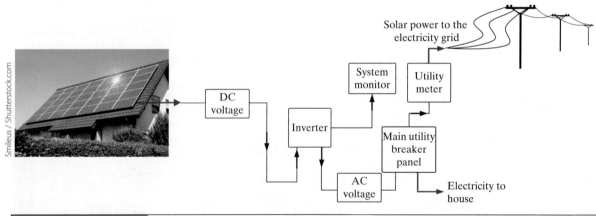

**FIGURE 7.23**    An example and a schematic drawing of a grid-tied PV system.

The U.S. solar data for sizing photovoltaic systems are shown in Figure 7.24. Examples 7.4 and 7.5 show you how to use this information.

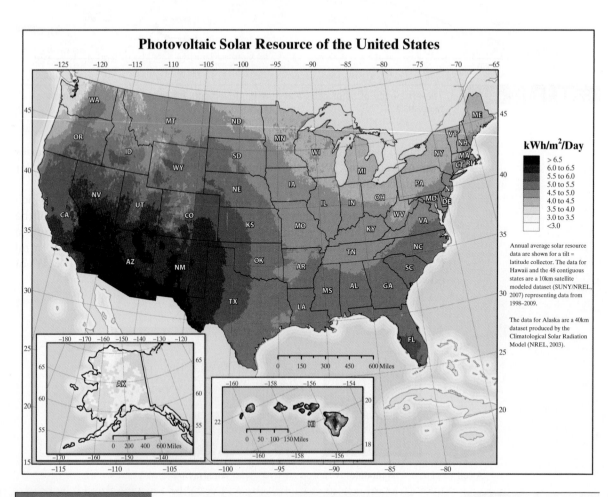

**FIGURE 7.24**    Photovoltaic resources of the United States.
*Source:* National Renewable Energy Laboratory.

**EXAMPLE 7.3**

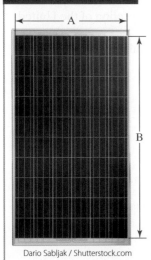

Dario Sabljak / Shutterstock.com

A manufacturer of photovoltaic systems provides the following specifications for one of its modules:

maximum power = 250 W (at illumination of 1 kW/m²)

A = 960 mm

B = 1,600 mm

What is the efficiency of this module?

We discussed efficiency in Chapter 3. Recall that efficiency = $\dfrac{\text{output}}{\text{input}}$, and note that for this example, the input is 1 kWh/m² or 1,000 W/m² and the output is 250 W for the entire module. Also note that

area of the module = (0.9 m)(1.6 m) = 1.44 m²

Then,

$$\text{efficiency} = \frac{\text{output}}{\text{input}} = \frac{250\ \text{W}}{\left(\dfrac{1{,}000\ \text{W}}{\text{m}^2}\right)(1.44\ \text{m}^2)} = 0.17 \text{ or } 17\%$$

This result represents the maximum possible efficiency of the module under ideal laboratory test conditions. It is important to note that, under real outdoor conditions, the efficiency of the module is less—closer to 13 or 14 percent.

**EXAMPLE 7.4**

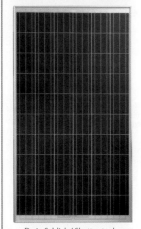

Dario Sabljak / Shutterstock.com

As shown in Figure 7.24, the average solar radiation available for photovoltaic systems for the southern part of Arizona is about 6.5 kWh/m²/day. If a photovoltaic array consists of ten modules (the 250-watt modules from Example 7.3), each module has the dimensions of 0.9 m × 1.6 m, and assuming an efficiency of 14 percent, how much electricity is generated by this system each year?

total area of the photovoltaic system = (10)(0.9 m)(1.6 m)

= 14.4 m²

amount of electricity generated

$$= (0.14)(14.4\ \text{m}^2)\left(\frac{6.5\ \text{kWh}}{\text{m}^2 \cdot \text{day}}\right)\left(\frac{365\ \text{days}}{\text{year}}\right)$$

$$= 4{,}783\ \frac{\text{kWh}}{\text{year}} \approx 4{,}800\ \frac{\text{kWh}}{\text{year}}$$

Therefore, this system will generate about 4,800 kWh/year. Note that there are additional losses in the system, depending on the wiring and the inverter, so the amount of electricity that could be consumed would be less than 4,800 kWh/year.

**EXAMPLE 7.5**

How much electricity would be generated if the photovoltaic system of Example 7.4 is located in Vermont in the northeastern section of the United States?

From Figure 7.24, the average solar radiation available for Vermont is about 4 kWh/m$^2$/day. Then

amount of electricity generated

$$= (0.14)(14.4 \text{ m}^2)\left(\frac{4 \text{ kWh}}{\text{m}^2 \cdot \text{day}}\right)\left(\frac{365 \text{ days}}{\text{year}}\right)$$

$$= 2{,}943 \ \frac{\text{kWh}}{\text{year}}$$

$$\approx 2{,}900 \ \frac{\text{kWh}}{\text{year}}$$

Note that the same system in Vermont produces approximately 39 percent less electricity than it would in Arizona.

## Before You Go On

Answer the following questions to test your understanding of the preceding section:

1. Describe ways by which we convert solar energy into useful forms of energy.

2. Describe how an active solar system works.

3. What are the main components of an active liquid solar system?

4. Describe how a passive solar system works.

5. How does a photovoltaic system work, and what are its main components?

*Vocabulary—State the meaning of the following terms:*

Passive system

Solar collector

Inverter

Photovoltaic array

Grid-tied system

## LO³  7.3  **Wind Energy**

Wind energy is a form of solar energy. As you may know, because of the Earth's tilt and orbit, the Sun heats the Earth and its atmosphere at different rates. You also know that hot air rises and cold air sinks to replace it. As the air moves, it has kinetic energy (we explained kinetic energy in Chapter 3). Part of this kinetic energy can be converted into mechanical energy and electricity. Historically, the Persians, Chinese, and Egyptians were among the first civilizations to harness wind energy to grind grains, pump water, and sail boats. As was the case with new technologies of that era, the methods for harnessing wind energy found its way to Europe, and the Europeans brought this technology to the Americas when they settled in the New World.

> Two types of wind turbines are used to extract energy from the wind: vertical axis turbines and horizontal axis turbines.

As you know, you can expect to experience windy days with different wind intensities during certain times of the year. Therefore, it should be self evident that the amount of wind energy that can be harnessed depends on how fast and how often the wind blows in an area. Thus the potential for harnessing wind energy and generating electricity varies by geogrpahical location. The United States wind resource map in Figure 7.25 shows annual average wind speeds at a

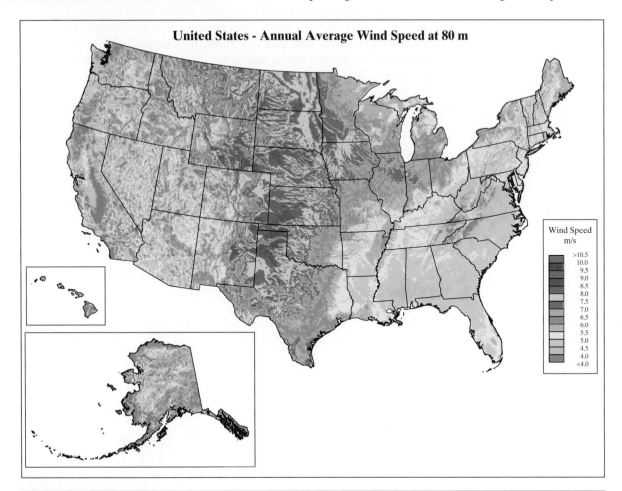

**United States - Annual Average Wind Speed at 80 m**

Wind Speed
m/s

>10.5
10.0
9.5
9.0
8.5
8.0
7.5
7.0
6.5
6.0
5.5
5.0
4.5
4.0
<4.0

**FIGURE 7.25**   United States wind resource map.
***Source:*** National Renewable Energy Laboratory. Wind resource estimates developed by AWS Truepower, LLC for wind Navigator@, Web: http://www.windnavigator.com|http: //www.awstruepower.com. Spatial resolution of wind resource data: 2.5km. Projection: Albers Equal Area WGS84.

height of 80 m above ground.  The rotors of wind turbines are usually mounted on tall towers. This is done because the wind speed increases with the vertical distance from the ground.  On a windy day, air at a higher elevation moves faster than the air near ground, as shown in Figure 7.26.

Two types of wind turbines are used to extract energy from the wind: *vertical axis* and *hoizontal axis*. Schematic diagrams of vertical axis and horizontal axis turbines are shown in Figure 7.27. The **vertical axis turbine** can accept wind from any angle, requires light-weight towers, and is easy to service. The main disadvantage of the vertical axis turbine is that, because the rotors are near the ground where the wind speeds are relatively low, it has poor performance. Most wind turbines in use in  the United States are the horizontal axis type. As the name implies, the rotor blades of a **horizontal axis turbine** rotate about an axis that is horizontal (see Figure 7.27).Wind turbines are typically classified as small (< 100 kW), intermediate (< 250 kW), and large (250 kW to 8 MW).

Here are some wind turbine terms that you will find useful.

- The *blades* and *hub* are called **rotors**. Most horizontal axis turbines have either two or three blades. The blades are typically made from wood, steel, aluminum, or fiberglass. Wooden blades are strong, light-weight, inexpensive, and flexible, whereas blades made from steel are strong, but they are also heavy and expensive. The newer turbines use fiberglass blades because they are strong, lightweight, and inexpensive. Aluminum blades are strong and lighter than steel, but are also expensive. Depending on the size of the system, the blades could be as long as 100 feet or more.
- The *gear box* connects the low-speed shaft attached to the rotor to the high-speed shaft that is attached to the generator to increase the rotational speed.
- The *yaw motor* runs the yaw drive to keep the blades facing into the wind as the wind direction changes.
- A *controller* starts the wind turbine at speeds of about 8 to 16 mile per hour (mph) and stops the turbine at relatively high speeds to prevent damage to the blades and components. An *anemometer* measures the wind speed and transmits the data to the controller.
- A *brake* stops the rotor in emergencies or high wind speeds. The brake is applied mechanically, electrically, or hydraulically.

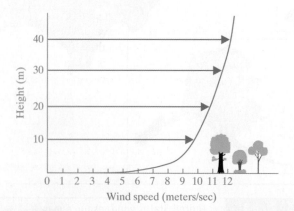

**FIGURE 7.26**    An example of wind speeds near the ground.

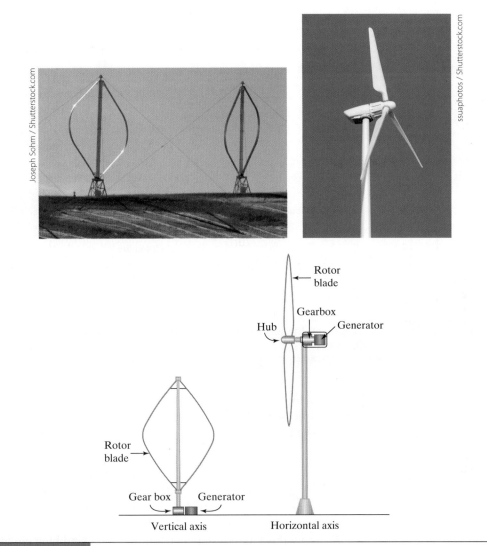

**FIGURE 7.27**     Vertical axis and horizontal axis wind turbines.

- The *sweep area* of the blades is shown in Figure 7.28. The sweep area is equal to the area of the circle through which wind moves; sweep area $= \pi(\text{blade length})^2$.
- ***Rotor solidity*** is the ratio of the total rotor platform area to the total sweep area. Low solidity results in high speed and low torque, whereas high solidity (values greater than 0.8) results in low speed and high torque (see Figure 7.28).

Another important principle that you should know is the ***Betz limit***, which states that not all wind power can be captured. Think about it; if all the wind energy was captured, the air behind the rotor will have a zero speed, which would mean that no air is flowing over the blades. The theoretical limit for rotor efficiency is 59 percent, with most current wind turbines having an efficiency in the range of 25 to 45 percent.

The Betz limit states that not all wind power can be captured.

High solidity

---

**FIGURE 7.28**  The turbine on the left has a lower solidity (3a/A) (low) than the turbine on the right.
*Source:* Deyan Georgiev / Shutterstock.com.

Let us now look at how we might estimate the amount of energy that could be extracted from wind. Recall from our discussion in Chapter 3 that an object having a known mass $m$ and moving with a speed $V$ has a kinetic energy that is equal to

$$\text{kinetic energy} = \left(\frac{1}{2}\right)(\text{mass})(\text{speed})^2 = \left(\frac{1}{2}\right)mV^2$$

We can apply this kinetic energy equation to the wind blowing over a turbine by noting that, in this case, $m$ represents the mass of the moving air and $V$ is the wind speed. Next, recall the definition of power as

$$\text{power} = \frac{\text{energy}}{\text{time}}$$

Then the amount of power that could be extracted from the moving air (wind) is given by

$$\text{power} = \frac{\text{energy}}{\text{time}} = \frac{\left(\frac{1}{2}\right)(\text{mass})(\text{speed})^2}{\text{time}}$$

$$= \left(\frac{1}{2}\right)(\text{mass flow rate})(\text{speed})^2$$

7.1

In Equation (7.1), the quantity $\dfrac{\text{mass}}{\text{time}}$ is called the mass flow rate; how much air per unit of time (for example, per second) is moving through the wind turbine's sweep area. The mass flow rate is related to the density of the air and the volume of the moving air as shown in the following equation. Moreover, the volume of air moving through the wind turbine is related to the area (i.e., sweep area) and the distance travelled by air. Recognition of these facts then results in

$$\text{mass flow rate} = \frac{\text{mass}}{\text{time}} = \frac{\overbrace{(\text{density})(\text{volume})}^{\text{mass}}}{\text{time}}$$

$$= \frac{(\text{density})\overbrace{(\text{area})(\text{distance travelled})}^{\text{volume}}}{\text{time}}$$

$$= \frac{(\text{density})(\text{area})(\text{distance travelled})}{\text{time}}$$

$$= \frac{(\text{density})(\text{area})(\text{speed})(\overbrace{\cancel{\text{time}}}^{\text{distance travelled}})}{\cancel{\text{time}}}$$

$$\text{mass flow rate} = (\text{density})(\text{area})(\text{speed}) \qquad \textbf{7.2}$$

Substituting Equation (7.2) into Equation (7.1), we get

$$\text{power} = \left(\frac{1}{2}\right)(\text{mass flow rate})(\text{speed})^2$$

$$= \left(\frac{1}{2}\right)\overbrace{(\text{density})(\text{area})(\text{speed})}^{\text{mass flow rate}}(\text{speed})^2$$

or

$$\text{power} = \left(\frac{1}{2}\right)(\text{density})(\text{area})(\text{speed})^3 \qquad \textbf{7.3}$$

Next, we must account for the Betz limit and the efficiency of the wind turbine. This last step yields a relationship for ***wind power*** in terms of turbine efficiency, air density, sweep area, and wind speed according to

$$\text{wind power} = (\text{efficiency})\left(\frac{1}{2}\right)(\text{density})(\text{sweep area})(\text{speed})^3 \qquad \textbf{7.4}$$

Let us now look at some examples where we apply Equation (7.4).

**EXAMPLE 7.6**

A wind turbine manufacturer states that one of its largest systems with a blade length of 35.25 meters (m) can generate 1.5 megawatts (MW) of electricity when the wind speed is 12 meters per second (m/s) or ~27 miles per hour (mph). The manufacturer does not mention anything about the efficiency of its system, so let us calculate the efficiency of this system. *Note*: The density of air is 1.2 kg/m³.

$$\text{wind power} = (\text{efficiency})\left(\frac{1}{2}\right)(\text{density})(\text{sweep area})(\text{speed})^3$$

$$\text{sweep area} = \pi(\text{blade length})^2 = \pi(35.25 \text{ m})^2 = 3{,}904 \text{ m}^2$$

$$1.5 \times 10^6 \text{ W} = (\text{efficiency})\left(\frac{1}{2}\right)\left(1.2\,\frac{\text{kg}}{\text{m}^3}\right)(3{,}904 \text{ m}^2)\left(12\,\frac{\text{m}}{\text{s}}\right)^3$$

$$\text{efficiency} = \frac{1.5 \times 10^6 \text{ watts}}{\left(\frac{1}{2}\right)\left(1.2\,\frac{\text{kg}}{\text{m}^3}\right)(3{,}904 \text{ m}^2)\left(12\,\frac{\text{m}}{\text{s}}\right)^3}$$

$$= 0.37 \text{ or } 37\%$$

Tyler Olson / Shutterstock.com

**EXAMPLE 7.7**

Estimate the power generated by the wind turbine of Example 7.6 for wind speeds of 6 m/s (13.4 mph), 8 m/s (17.9 mph), and 10 m/s (22.4 mph), assuming the same efficiency of 37 percent at all of the given wind speeds. Again, the density of air is 1.2 kg/m³.

$$\text{wind power} = (\text{efficiency})\left(\frac{1}{2}\right)(\text{density})(\text{sweep area})(\text{speed})^3$$

For wind speed = 6 m/s,

$$\text{power} = (0.37)\left(\frac{1}{2}\right)\left(1.2\,\frac{\text{kg}}{\text{m}^3}\right)(3{,}904 \text{ m}^2)\left(6\,\frac{\text{m}}{\text{s}}\right)^3$$

$$= 187{,}205 \text{ watts} \approx 187 \text{ kW}$$

For wind speed = 8 m/s,

$$\text{power} = (0.37)\left(\frac{1}{2}\right)\left(1.2\,\frac{\text{kg}}{\text{m}^3}\right)(3{,}904 \text{ m}^2)\left(8\,\frac{\text{m}}{\text{s}}\right)^3$$

$$= 443{,}744 \text{ watts} \approx 444 \text{ kW}$$

For wind speed = 10 m/s,

$$\text{power} = (0.37)\left(\frac{1}{2}\right)\left(1.2\,\frac{\text{kg}}{\text{m}^3}\right)(3{,}904 \text{ m}^2)\left(10\,\frac{\text{m}}{\text{s}}\right)^3$$

$$= 866{,}688 \text{ watts} \approx 867 \text{ kW}$$

Note the power generated is proportional to the wind speed cubed. Therefore, a relatively small increase in wind speed could result in a large increase in power generation. The results are summarized in the following table.

| Wind Speed (m/s) | Power Generated (kW) |
|:---:|:---:|
| 6 | 187 |
| 8 | 444 |
| 10 | 867 |
| 12 | 1,500 |

Today, the United States, China, Germany, Spain, and India are among the countries with the largest amount of electricity generated from wind. In the United States, the top five states with the largest electricity generation from wind were Texas, Iowa, California, Oklahoma, and Kansas in 2014 (refer to Figure 7.29). In fact, one of the largest wind farms in the United States is located in Texas, with 430 turbines that together produce 735 MW of electricity.

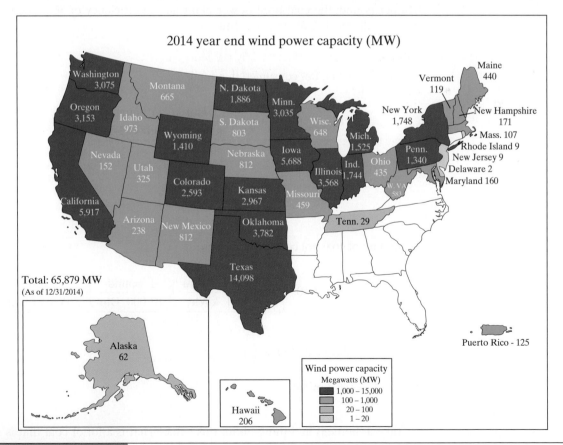

**FIGURE 7.29**   The top wind energy-producing states.

*Source:* National Renewable Energy Laboratory. Data is from the American Wind Energy Association Fourth Quarter 2014 Market Report: http://www.awea.org

**EXAMPLE 7.8**

In 2014, 182 billion kilowatt-hours (kWh) of electricity was generated from wind energy in the United States. Assuming an annual household electricity consumption of 10,000 kWh, how many households could be supplied with electricity from wind energy?

$$\text{number of households} = (182 \times 10^9 \ \cancel{\text{kWh}})\left(\frac{1 \text{ household}}{10,000 \ \cancel{\text{kWh}}}\right)$$

$$= 182 \times 10^5$$

$$\approx 18 \text{ million}$$

Note that for this result we did not account for any losses in transmission lines.

**EXAMPLE 7.9**

For Example 7.8, how much coal was saved (not consumed) because of wind energy generation? Assume coal has an average energy content of 10,000 Btu per pound; the coal-fired power plant has an efficiency of 36 percent. Also, recall that 1 kWh = 3,412 Btu.

amount of coal not consumed in pounds

$$= \frac{(182 \times 10^9 \ \cancel{\text{kWh}})\left(\dfrac{3,412 \ \cancel{\text{Btu}}}{1 \ \cancel{\text{kWh}}}\right)\left(\dfrac{1 \text{ pound}}{10,000 \ \cancel{\text{Btu}}}\right)}{0.36}$$

$$= 172.495 \times 10^9 \text{ pounds}$$

amount of coal not consumed in tons

$$= \frac{(182 \times 10^9 \ \cancel{\text{kWh}})\left(\dfrac{3,412 \ \cancel{\text{Btu}}}{1 \ \cancel{\text{kWh}}}\right)\left(\dfrac{1 \ \cancel{\text{pound}}}{10,000 \ \cancel{\text{Btu}}}\right)\left(\dfrac{1 \text{ ton}}{2,000 \ \cancel{\text{pound}}}\right)}{0.36}$$

$$= 86,247,777 \text{ tons}$$

$$\approx 86.3 \text{ million tons}$$

Take a moment and think about this value! That is a lot of coal that is not being consumed and does not add to pollution and $CO_2$ emissions.

## Before You Go On

Answer the following questions to test your understanding of the preceding section:

1. Which regions of the United States have the least potential for producing wind energy?

2. What are the two types of wind turbines?

3. Describe the main components of a wind turbine.

4. What are wind turbine blades typically made from?

5. Explain the Betz limit.

***Vocabulary—State the meaning of the following terms:***

Horizontal axis turbine

Vertical axis turbine

Rotor

Rotor solidity

## LO⁴  7.4  Hydro-Energy

Electricity is also generated by flowing water stored behind dams. Naturally, water flows from higher elevation to lower elevation. Throughout history, we have captured the power of moving water in rivers to perform various tasks such as cutting lumber or grinding flour. Today, the power of moving water is converted into electricity using water turbines that are connected to generators. In 2014, hydropower accounted for 7 percent of the total United States electricity generation or 26 percent of renewable energy generated. The power of moving water is converted into electricity using a number of techniques including *impoundment, diversion*, and *pumped storage hydropower*.

### Impoundment

The *impoundment approach* makes use of dams to store water. As shown in Figure 7.30, as water is released through the dam, it is guided into water turbines located in hydroelectric power plants housed within the dam to generate electricity. The potential energy due to the height of the water stored behind the dam is converted to kinetic energy (moving energy), and as the water flows through the turbine it spins the turbine. The turbine connected to a generator then turns the generator.

Today in the United States there are nearly 80,000 dams, of which only 3 percent produce electricity. Therefore, there still exists untapped potential for electricity generation using hydropower. Many of the dams were built to control flooding and to irrigate crops. The U.S. Department of Energy is currently working

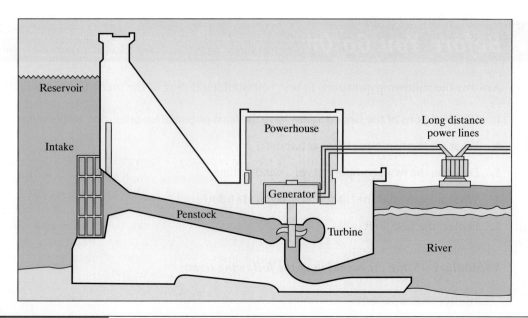

**FIGURE 7.30**    A schematic of a hydropower plant.
*Source:* Based on Tennessee Valley Authority

on modifying many of these dams to produce electricity as well. Research is also being conducted to design more fish friendly turbines and ladders to accommodate the movement of fish. Today, most of the hydroelectric facilities in the United States are located in western states such as Washington, Oregon, and California with the Grand Coulee Dam being the largest. The hydroelectric capacity of conventional facilities in the United States is shown in Figure 7.31.

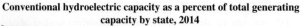

**Conventional hydroelectric capacity as a percent of total generating capacity by state, 2014**

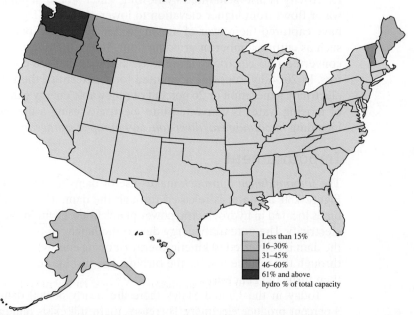

Less than 15%
16–30%
31–45%
46–60%
61% and above
hydro % of total capacity

**FIGURE 7.31**    The United States conventional hydroelectric capacity.
*Source:* U.S. Energy Information Administration, Electric Power Monthly, Tables 6.2 and 6.2B (February 2015)

## Diversion

Let us next consider the *diversion technology* used in harnessing hydropower. As the name implies, with this approach part of the water in a river is diverted to run through turbines. This technology does not require a large dam and makes use of the natural flow of water.

## Pumped Storage Hydropower

The *pumped storage technique* pumps the water from a lower elevation and stores it in a higher elevation at night when the energy demand is low. During the daytime when the energy use is high, the water is released from the higher elevation storage to the lower elevation to produce electricity.

Next, we estimate the amount of energy that could be extracted from the water running from a dam through the penstock (and eventually through the turbine) to the river downstream of the dam. Recall from our discussion in Chapter 3 the potential energy of an object is equal to

$$\text{potential energy} = \frac{(\text{mass})(\text{acceleration due to gravity}) \times}{(\text{change in its elevation})}$$

$$= mgh$$

We can then apply the potential energy equation to the water flowing through the dam through the turbine. Note that, in this case, $m$ represents the mass of the moving water and $h$ is the change in the elevation of water between the water level stored behind the dam and the discharge side on the downstream river. We again use the definition of power $\left(\text{power} = \frac{\text{energy}}{\text{time}}\right)$ to determine how much power could be extracted from the flowing water in the following manner:

$$\text{power} = \frac{\text{energy}}{\text{time}}$$

$$= \frac{(\text{mass})(\text{acceleration due to gravity}) \times (\text{change in elevation})}{\text{time}}$$

$$= (\text{mass flow rate})(\text{acceleration due to gravity}) \times (\text{change in elevation})$$

**7.5**

Again in Equation (7.5), the quantity $\frac{\text{mass}}{\text{time}}$ is called mass flow rate; in this case, how much water per unit time is moving through the turbine.

We also need to account for losses due to friction in the penstock and piping system, as well as the efficiency of the turbine. This last step yields a relationship for water power generation in terms of an overall efficiency of the system, mass flow rate, acceleration due to gravity, and change in elevation of water according to

$$\text{power} = \frac{(\text{overall efficiency})(\text{mass flow rate}) \times}{(\text{acceleration due to gravity})(\text{change in elevation})}$$

7.6

In Equation (7.6), note that power generated is directly proportional to the water mass flow rate and the change in elevation of the water. Next, we apply this relationship to an actual hydropower plant.

---

**EXAMPLE 7.10**

In a recent year, a hydropower station in Japan commissioned a new turbine and a generator with the following specifications:

$$\text{mass flow rate} = 207,000 \text{ kg/s}$$

$$\text{head (elevation change)} = 48 \text{ m}$$

$$\text{overall efficiency} = 89\%$$

What is the power output of this system?

$$\text{power} = \frac{(\text{overall efficiency})(\text{mass flow rate}) \times}{(\text{acceleration due to gravity})(\text{change in height})}$$

$$= (0.89)\left(207,000 \ \frac{\text{kg}}{\text{s}}\right)\left(9.81 \ \frac{\text{m}}{\text{s}^2}\right)(48 \text{ m}) = 86,750,222 \text{ watts}$$

$$\approx 87,000 \text{ kW}$$

Next, let us look at one of the largest hydropower plants in the U.S.

# *Hoover Dam*

The Hoover Dam is one of the Bureau of Reclamation's multipurpose projects on the Colorado River. These projects control floods; store water for irrigation, municipal, and industrial use; and provide hydroelectric power, recreation, and fish and wildlife habitat. The Hoover Dam is a concrete arch–gravity type of dam in which the water load is carried by both gravity and horizontal arch actions.

### *The Reservoir*

At an elevation of 1221.4 feet, Lake Mead is the largest man-made lake in the United States and contains 28,537,000 acre-feet (an acre-foot is the amount of water required to cover 1 acre to a depth of 1 foot). This reservoir stores the entire average flow of the river for two years. That much water would cover the entire state of Pennsylvania to a depth of 1 foot.

Lake Mead extends approximately 110 miles upstream toward the Grand Canyon and approximately 35 miles up the Virgin River. The width of Lake Mead varies from several hundred feet in the canyons to a maximum of 8 miles. The reservoir covers about 157,900 acres or 247 square miles.

Recreation, although a byproduct of this project, constitutes a major use of the lake and the controlled flows created by the Hoover and other dams on the lower Colorado River today. Lake Mead is one of America's most popular recreational areas, with a 12-month season that attracts more than 9 million visitors each year for swimming, boating, waterskiing, and fishing. The lake and its surrounding area are administered by the National Park Service as part of the Lake Mead National Recreation Area, which also includes Lake Mohave downstream from the Hoover Dam.

### *The Power Plant*

There are 17 main turbines in the Hoover power plant. The original turbines were all replaced through an upgrading program between 1986 and 1993. With a rated capacity of 2,991,000 horsepower (hp) and two

milosk50 / Shutterstock.com

station-service units rated at 3,500 horsepower each for a plant total of 2,998,000 horsepower, the plant has a nameplate capacity of 2,074,000 kilowatts (kW). This includes the two station-service units that are rated at 2,400 kilowatts each. The Hoover Dam generates low-cost hydroelectric power for use in Nevada, Arizona, and California. The Hoover Dam alone generates more than 4 billion kilowatt-hours (kWh) a year. From 1939 to 1949, the Hoover power plant was the world's largest hydroelectric installation; with an installed capacity of 2.08 million kilowatts, it is still one of this country's largest.

The $165 million-dollar cost of the Hoover Dam has been repaid, with interest, to the federal treasury through the sale of its power. The Hoover Dam energy is marketed by the Western Area Power Administration to 15 entities in Arizona, California, and Nevada under contracts that expire in 2017. More than half, 56 percent, goes to southern California users; Arizona contractors receive 19 percent; and Nevada users get 25 percent. The revenues from the sale of this power now pay for the dam's operation and maintenance. The power contractors also paid for the upgrading of the power plant's nameplate capacity from 1.3 million to over 2.0 million kilowatts.

*Source:* U.S. Bureau of Reclamation.

---

**EXAMPLE 7.11**

According to the U.S. Census Bureau, there are approximately 213,000 households in Las Vegas. Assuming an annual household electricity consumption of 10,000 kWh, can the Hoover Dam power plant generate enough electricity for all of the households in Las Vegas?

As we mentioned previously, the Hoover power plant generates 4 billion kWh a year. Then,

$$\text{number of households} = (4 \times 10^9 \text{ kWh}) \left( \frac{1 \text{ household}}{10,000 \text{ kWh}} \right) = 400,000$$

Therefore, the answer is yes.

---

## Before You Go On

Answer the following questions to test your understanding of the preceding section:

1.  In your own words, explain different approaches used to generate electricity in a hydro-energy plant.

2.  What are the main components of a hydro-energy plant?

## LO⁵  7.5  Biomass

*Biomass* refers to organic materials such as forest and wood trimmings, plants, fast growing grasses and trees, crops, or algae grown specifically to be converted to produce biofuels using different processes. In 2014, biomass accounted for approximately 52 percent of the renewable energy use in the United States, with about half from wood and wood waste and the other half from ethanol and municipal waste. We discuss municipal waste and energy recovery from waste in Chapter 11.

### Wood

Throughout history, wood, because of its abundance in many parts of the world, has been a material of choice for many applications. Wood is a renewable source, and because of its ease of workability and its strength, it has been used to make many products. Today, wood is used in a variety of products ranging from telephone poles to toothpicks. Common examples of wood products include hardwood flooring, roof trusses, furniture, frames, wall supports, doors, decorative items, window frames, trimming in luxury cars, tongue depressors, clothespins, baseball bats, bowling pins, fishing rods, and wine barrels. Wood is also the main ingredient used to make various paper products. When burned, wood is considered to be a biomass fuel. Throughout our history, wood has been used as fuel in stoves and fireplaces. In fact, it was the main

Wood, ethanol, and biodiesel are called biomass fuels; their energy is derived from plants, crops, and animal fats.

Victor Soares / Shutterstock.com                                    Jerax / Shutterstock.com

source of energy in the world until the mid-1800s. Table 7.1 shows the timeline of how wood has been used since 1860. Today, wood is still a major source of energy for people in many developing countries.

Wood and wood by-products and waste (e.g., sawdust and scraps) make up nearly 2 percent of the United States' energy source. For example, in recent years, sawdust has been compressed to form pellets—commonly known as

| TABLE 7.1 | Wood as Fuel—Timeline |
|---|---|
| 1860 | Wood was the primary fuel for heating and cooking in homes and businesses and was used for steam in industries, trains, and boats. |
| 1890 | Coal had replaced much of the wood used in steam generation. |
| 1900 | Ethanol was competing with gasoline to be the fuel for cars. |
| 1910 | Most rural homes were still heated with wood. In urban areas, coal was displacing wood in homes. |
| 1930 | Over half of all Americans lived in cities in buildings heated by coal. Rural Americans still heated and cooked with wood. Diesel and gasoline were firmly established as the fuel for trucks and automobiles. Street cars ran on electricity. Railroads and boats used coal and diesel fuel. |
| 1950 | Electricity and natural gas had replaced wood heat in most homes and commercial buildings. |
| 1974 | Some Americans used more wood for heating because of higher energy costs. Some industries switched from coal to waste wood. The paper and pulp industry also began to install wood and black liquor boilers for steam and power, displacing fuel oil and coal. |
| 1984 | Burlington Electric (Vermont) built a 50-MW, wood-fired plant with electricity production as the primary purpose. This plant was the first of several built since 1984. |
| 1989 | Pilot trials of direct wood-fired gas turbine plants were conducted for the first time in Canada and in the United States. |
| 1990 | The capacity to generate electricity from biomass (not including municipal solid waste) reached 6 GW. Of 190 biomass-fired, electricity-generating facilities, 184 were nonutility generators, mostly wood and paper. |
| 1994 | Successful operation of several biomass gasification tests identified hot gas clean-up as key to widespread adoption of the technology. |

*Source:* U.S. Energy Information Administration.

wood pellets—that are burned in heating stoves. Approximately 20 percent of the wood and wood waste is used for heating homes and cooking. Also, it is worth noting that wood- and paper-product plants use their own wood waste as fuel to generate electricity.

## Algae

In recent years, much attention has been focused on algae as a biofuel. *Algae* are small aquatic organisms that convert sunlight to energy. There are over 100,000 different types of algae. They can be grown in algae farms that are basically large man-made ponds. Some algae store energy in the form of oil that is extracted by breaking down the cell structure using solvents or sound waves. The extracted oil is then further refined to serve as a biofuel. Algae require a great deal of carbon dioxide to grow, which makes them even more attractive as a renewable fuel source because they can serve as a source that also removes carbon dioxide from the atmosphere. For example, algae farms can be placed near power plants or other sources that produce a lot of carbon dioxide.

## Ethanol

*Ethanol* refers to an alcohol-based fuel that is made from sugar found in crops such as corn and sugarcane. Most of the ethanol produced in the United States is made from corn and is mixed with gasoline. In fact, most of the gasoline sold in the United States today has approximately 10 percent ethanol by volume and is marked by E10. In the midwestern section of the United States, at many gas stations you may have also seen fuel that is marked by E85. This designation represents a fuel that has 85 percent ethanol mixed with gasoline. Only certain types of vehicles with flexible-fuel engines can run on fuel that has an ethanol content greater than 10 percent. Gasoline mixed with ethanol burns cleaner and produces less pollution. According to the EIA, approximately 13 billion gallons of ethanol was mixed with gasoline in 2014. Brazil is the second largest producer of ethanol, but unlike in the United States, sugarcane and sugar beets are used to make ethanol.

---

**EXAMPLE 7.12**

According to the U.S. Department of Agriculture Forest Product Laboratory test results, premium wood pellets have a net heating value of 13.6 million Btu per ton. How much wood pellets should be ordered for the house in Example 5.9 with an annual heating load of $152 \times 10^6$ Btu/year and a pellet-burning furnace efficiency of 0.85?

The amount of wood pellets that should be ordered can be estimated from:

amount of wood pellets

$$= \left[ \frac{(152 \times 10^6 \text{ Btu/year})}{0.85} \right] \left[ \frac{1 \text{ ton of wood pellets}}{13.6 \times 10^6 \text{ Btu}} \right]$$

$$= 13.15 \text{ tons of wood pellets/year}$$

This is a lot of wood pellets! Think about it.

## Biodiesel

*Biodiesel* refers to fuel made from vegetable oils, animal fats, or recycled restaurant grease. Most of the biodiesel fuel made in the United States comes from soybean oil and is commonly mixed with diesel fuel derived from petroleum. For example, B20 fuel denotes a mixture of 20 percent biodiesel with 80 percent petroleum diesel by volume. The top five countries in the world with the greatest biodiesel consumption are the United States, Germany, Brazil, France, and Spain.

## Before You Go On

Answer the following questions to test your understanding of the preceding section:

1. How much of the United States energy use comes from biomass?

2. How much of the United States energy use comes from wood and wood waste?

3. How are wood pellets made ?

4. What is algae?

5. What is ethanol?

*Vocabulary—State the meaning of the following terms:*

Biomass

Biodiesel

Algae

Biofuel

## SUMMARY

### LO¹ Solar Energy

You should know the basic concepts related to solar energy: radiation band, direct and diffuse radiation, and factors that determine how much solar radiation is available at a location. Solar energy starts with the Sun at an average distance of 93 million miles (~150 million kilometers) from the Earth. Solar radiation can be divided into three bands: ultraviolet, visible, and infrared. The visible band comprises about 48 percent of useful radiation for heating, and the

infrared makes up the rest. The amount of radiation available at a specific place depends on many factors, including geographical location, season, local landscape and weather, and time of day. As solar energy passes through the Earth's atmosphere, some of it is absorbed, some of it is scattered, and some of it is reflected by clouds, dust, pollutants, forest fires, or water vapor. The solar radiation that reaches the Earth's surface without being diffused is called direct beam solar radiation. On a clear day at noon when the

Sun appears at its highest point in the sky, the greatest amount of solar energy reaches a horizontal surface on the Earth. Seasonal effects are also important. During the winter, the angle of the Sun is lower than it is in the summer, which results in a lower amount of radiation being intercepted by a horizontal surface.

## LO² Solar Systems

Solar systems can be categorized into active, passive, and photovoltaic. There are two basic types of active solar heating systems: liquid and air. The liquid systems make use of water, a water-antifreeze mixture, or other liquids to collect solar energy. In such systems, the liquid is heated in a solar collector and then transferred to a storage system. In contrast, air systems heat the air in "air collectors" and transport it to storage or a space using blowers. In moderate climates, solar hot water systems are also used to heat swimming pools with the goal of extending the swimming season.

Passive solar systems do not make use of any mechanical components such as collectors, pumps, blowers, or fans to collect, transport, or distribute solar heat to various parts of a building. Instead, a direct passive solar system uses large glass areas on the south wall of a building and a thermal mass to collect the solar energy. The solar energy is stored in interior thick masonry walls and floors of the building during the day and is released at night.

A photovoltaic system converts light energy directly into electricity. A photovoltaic (PV) cell is the backbone of any photovoltaic system. Photovoltaic cell materials include crystalline, polycrystalline, and amorphous silicon. The manufacturers of photovoltaic systems combine cells to form a module, and then the modules are combined to form an array. A photovoltaic system often consists of batteries, a charge controller, and an inverter. A charge controller protects the batteries from overcharging. An inverter is a device that converts direct current into alternating current. Photovoltaic systems are classified into stand-alone, hybrid, or grid-tied systems.

## LO³ Wind Energy

Wind energy is a form of solar energy. The wind speed increases with the vertical distance from the ground, and the power generated by wind is directly proportional to the speed of the wind cubed. Not all wind power can be captured. If that were to happen, the air behind the rotor would have a zero speed, which would mean that no air is flowing over the blades. Two types of wind turbines are used to extract the energy from the wind: vertical axis and horizontal axis. Wind turbines are typically classified as small (<100 kW), intermediate (<250 kW), and large (250 kW to 8 MW).

## LO⁴ Hydro-Energy

The power of moving water is converted into electricity using a number of techniques including impoundment, diversion, and pumped storage hydropower. The impoundment approach makes use of dams to store water. The water is guided into water turbines located in hydroelectric power plants housed within the dam to generate electricity. The diversion technology diverts part of the water running through a river through turbines. This technology does not require a large dam and makes use of the natural flow of water. The pumped storage technique pumps the water from a lower elevation and stores it in a higher elevation at night when the energy demand is low; then during the daytime when the energy use is high, the water is released from the higher elevation storage to the lower elevation to produce electricity.

## LO⁵ Biomass

Biomass refers to organic materials such as forest and wood trimmings, plants, fast-growing grasses and trees, crops, or algae grown specifically to be converted to produce biofuels using different processes.

Wood is considered a biomass fuel. Throughout our history, wood has been used as fuel in stoves and fireplaces. Today, wood is still a major source of energy for people in many developing countries. In recent years, sawdust has been compressed to form pellets—commonly known as wood pellets—that are burned in heating stoves. Also, wood and paper product plants use their wood waste as fuel to generate electricity.

Lately, much attention has also been focused on algae as a biofuel. Algae are small aquatic organisms that convert sunlight to energy. They can be grown in algae farms that are basically large man-made ponds. Some algae store energy in the form of oil that is extracted by breaking down the cell structure using solvents or sound waves. The extracted oil is then further refined to serve as biofuel.

Ethanol refers to alcohol-based fuel that is made from sugar found in crops such as corn and sugarcane, and biodiesel refers to fuel made from vegetable oils, animal fats, or recycled restaurant grease. Most of the ethanol produced in the United States is made from corn and is mixed with gasoline; most of the biodiesel fuel comes from soybean oil and is mixed with diesel fuel derived from petroleum.

# KEY TERMS

Active Solar System  208
Algae  234
Betz Limit  221
Biodiesel  235
Charge Controller  215
Ethanol  234
Grid-Tied System  213

Horizontal Axis Turbine  220
Hybrid System  213
Hydropower  227
Inverter  214
Passive Solar System  212
Photovoltaic Array  213
Photovoltaic (PV) Cell  213

Photovoltaic System  213
Rotor  220
Rotor Solidity  221
Solar Energy  199
Stand-Alone System  213
Vertical Axis Turbine  220
Wind Power  223

# Apply What You Have Learned

Visit a website of a photovoltaic module manufacturer such as SHARP and look up the following information for a new product: the wattage, dimensions, and weight. How many modules would be needed to provide 4,000 kWh of electricity per year where you live? How much would it cost for the array? Also investigate the cost and types of batteries used in photovoltaic systems to store energy to be used during cloudy days. Present your findings in a brief report.

Smileus / Shutterstock.com

# PROBLEMS

*Problems that promote life-long learning are denoted by* 🔑

**7.1** In January and June, how much solar radiation (in kWh/m²/day) on average is intercepted by a surface (with an effective area of 2 m²) that is tilted at an angle equal to the latitude of the location for the following states: Georgia, Michigan, and New Mexico? State your assumptions.

**7.2** In the southern region of Arizona, how much solar radiation is intercepted on average by two flat plate collectors (with dimensions 1 m × 1.5 m) that are tilted at an angle equal to the latitude of its location in the month of January as compared to June?

**7.3** Assume that the solar collector system of Problem 7.2 has an average efficiency of 60% during the month of January and 68% during the month of June. On average, how many gallons of supply water at 70°F could be heated to 120°F by the system each day during each month?

**7.4** For a solar system located in Colorado, how many flat-panel solar collectors with dimensions of 1 m × 1.5 m and an efficiency of 58% would be required to heat up 80 gallons of water from 65°F to 120°F during the month of January? State your assumptions.

**7.5** In cold climates, solar water-heating systems, which use liquids, need protection from freezing when ambient temperatures approach 32°F (0°C). Basically, there are two options: (a) use an antifreeze solution or (b) drain the water in the collector(s) and the piping (collector loop) either manually or automatically. Investigate how these systems operate and give your findings in a brief report.

**7.6** The masonry wall shown in Figure P7.6 is used in passive solar design and is commonly referred to as a "Trombe" wall in honor of Felix Trombe, who developed this concept. Investigate the factors that must be considered when sizing (determining the thickness of) the "trombe" wall. Write a brief report discussing your findings.

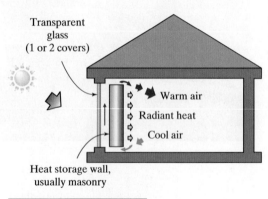

Transparent glass (1 or 2 covers)

Warm air

Radiant heat

Cool air

Heat storage wall, usually masonry

**FIGURE P7.6**

**7.7** What is the efficiency of a photovoltaic module (Figure P7.7) with the following specifications?

maximum power output
= 250 W (at illumination of 1 kW/m²)

A = 900 mm

B = 1,400 mm

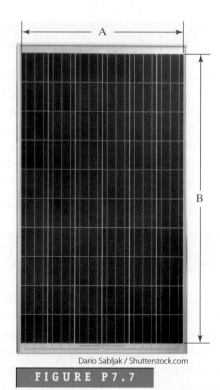

Dario Sabljak / Shutterstock.com

**FIGURE P7.7**

**7.8** How much electricity is generated by a photovoltaic system consisting of 14 modules? The system has an efficiency of 14%, and each module has an effective area of 1.4 m². The photovoltaic system is located in New York with an average solar radiation of 4.5 kWh/m²/day.

**7.9** Assume a photovoltaic system is located in Colorado with an average solar radiation of 6.0 kWh/m²/day. How much electricity is generated by a photovoltaic system consisting of 12 modules? The system has an efficiency of 13%, and each module has an effective area of 1.2 m².

**7.10** Investigate the current cost ($/kWh) of generating electricity using photovoltaic systems. Compare it to the cost of electricity provided by the electric company in your area.

**7.11** Solar shingles are placed on a roof (Figure P7.11) in the same way conventional roof shingles are. Investigate the pros and cons of solar shingles and their power outputs. Present your findings in a brief report.

© Bill Brooks / Alamy Stock Photo

**FIGURE P7.11**

**7.12** How much electricity is generated at wind speeds of 8 m/s, 10 m/s, 12 m/s, and 14 m/s by a wind turbine that has a blade length of 20 m? Assume an efficiency of 35% for the system and an air density of 1.2 kg/m³.

**7.13** A wind turbine manufacturer states that one of its systems with a blade length of 31 m can generate 1.3 MW of electricity when the wind speed is 14 m/s. What is the efficiency of this system? *Note*: The density of air is 1.2 kg/m³.

**7.14** How much electricity is generated at wind speeds of 8 m/s, 10 m/s, 12 m/s, and 14 m/s by a wind turbine that has a blade length of 50 m? Assume an efficiency of 37% for the system and an air density of 1.2 kg/m³.

**7.15** Investigate the current cost ($/kWh) of generating electricity using wind power and compare it to the cost of electricity provided by the electric company in your area.

**7.16** The Hoover Dam generates more than 4 billion kWh a year. How many 18.5 W LED light bulbs could be powered in a year by the Hoover Dam's power plant?

**7.17** How much coal must be burned in a steam power plant with a thermal efficiency of 34% to generate enough power to equal the 4 billion KWh a year generated by the Hoover Dam?

**7.18** As shown in the Discussion Starter of this chapter, 23% of the 9.6 quadrillion Btu of renewable energy in 2014 came from wood and wood by-products and waste (e.g., sawdust and scraps). Estimate the number of cords of wood that would have the equivalent energy content. State your assumptions.

**7.19** As shown in the Discussion Starter of this chapter, 4% of the 9.6 quadrillion Btu of renewable energy in 2014 came from solar systems. How much coal is saved (not consumed) because of the solar energy segment? Assume coal has an average energy content of 10,000 Btu per pound and a coal-fired power plant has an efficiency of 36% and a 6% loss in the transmission lines.

**7.20** As shown in the Discussion Starter of this chapter, 26% of the 9.6 quadrillion Btu of renewable energy in 2014 came from hydroelectric plants. How much coal is saved (not consumed) because of the hydroelectric plants? Assume coal has an average energy content of 10,000 Btu per pound and a coal-fired power plant has an efficiency of 36% and a 6% loss in the transmission lines.

Library of Congress Prints and Photographs Division Washington, D.C. 20540 USA

*"The future is not in the hands of fate but ours."* — JULES JUSSERAND (1855–1932)

# Environment

# 3

In Part Three of this book, we focus on the environment and introduce you to important concepts related to air, water, and natural resources; these include the rates of water and natural resources consumption and waste. We emphasize that our earth has finite resources. We provide general information about the atmosphere, weather, and climate, along with outdoor and indoor air quality standards. We also cover water resources, water quality standards, and water consumption rates in our homes, agriculture, and the industrial and manufacturing sectors of our society. We also provide a detailed understanding of common materials that are used to make products and structures in addition to discussing waste and recycling.

# Air and Air Quality Standards

## LEARNING OBJECTIVES

**LO¹** Atmosphere, Weather, and Climate: describe the Earth's atmosphere and its different layers; explain the difference between weather and climate

**LO²** Outdoor Air Quality Standards: understand the sources of outdoor air pollution and be familiar with the EPA's standards

**LO³** Indoor Air Quality Standards: understand the sources of indoor air pollution and be familiar with the EPA's standards

**LO⁴** Global Air Quality Issues: understand global air quality issues

# Discussion Starter

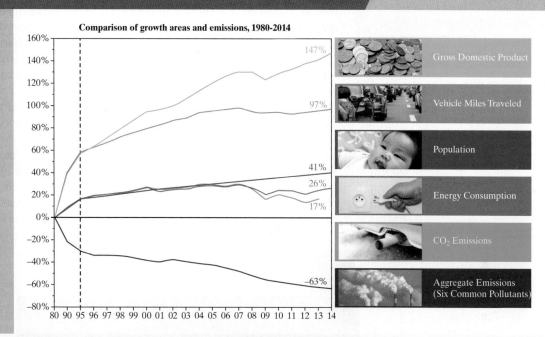

Comparison of growth areas and emissions, 1980-2014

Derek Hatfield / Shutterstock.com, disq / Shutterstock.com, thechatat / Shutterstock.com, mrstam / Shutterstock.com, Alexander Ishchenko / Shutterstock.com, Vadim Petrakov / Shutterstock.com

The U.S. Environmental Protection Agency (EPA) uses annual emission estimates to indicate the effectiveness of its programs. The EPA graph here shows the relationship between growth and pollution. Between 1980 and 2014, the United States gross domestic product increased 147 percent, vehicle miles traveled increased 97 percent, energy consumption increased 26 percent, and the population grew by 41 percent. During the same time period, total emissions of the six principal air pollutants (to be discussed in this chapter) dropped by 63 percent. The graph also shows that between 1980 and 2013, $CO_2$ emissions increased by 17 percent.

*Source:* EPA, Air Quality Trends, https://www.epa.gov/air-trends

**To the Students:** How do you see the relationship between population growth and air quality changing? Can we increase our population, standard of living, and energy consumption while reducing $CO_2$ emissions?

## LO[1] 8.1 Atmosphere, Weather, and Climate

We all need air and water to sustain life. The Earth's atmosphere, which is commonly referred to as air, is a mixture of approximately 78 percent nitrogen, 21 percent oxygen, and less than 1 percent argon. Small amounts of other gases are also present in the Earth's atmosphere, including carbon dioxide, sulfur dioxide, and nitrogen oxide. The atmosphere also contains water vapor. The concentration level of these gases depends on the altitude and geographical location. At higher altitudes (10 to 50 km), the Earth's atmosphere

Air is a mixture of mostly nitrogen, oxygen, and small amounts of other gases such as argon, carbon dioxide, sulfur dioxide, and nitrogen oxide.

also contains ozone. Even though these gases make up a small percentage of the Earth's atmosphere, they play a significant role in maintaining a thermally comfortable environment for us and other living species. For example, *ozone* absorbs most of the ultraviolet radiation arriving from the sun that could harm us. ***Carbon dioxide*** plays an important role in sustaining plant life; however, if the atmosphere contains too much carbon dioxide, it does not allow the Earth to cool down effectively by radiation. Water vapor in the atmosphere in the form of clouds allows for the transportation of water from the oceans to land in the form of rain and snow. We discuss the water cycle in greater detail in Chapter 9.

The air surrounding the Earth is divided into four distinct regions: *troposphere, stratosphere, mesosphere,* and *thermosphere* (see Figure 8.1). The layer of air closest to the Earth's surface is called the ***troposphere***, which plays an important role in shaping our weather. The radiation from the sun heats the Earth's surface which in turn heats the air near the surface. As the air heats up,

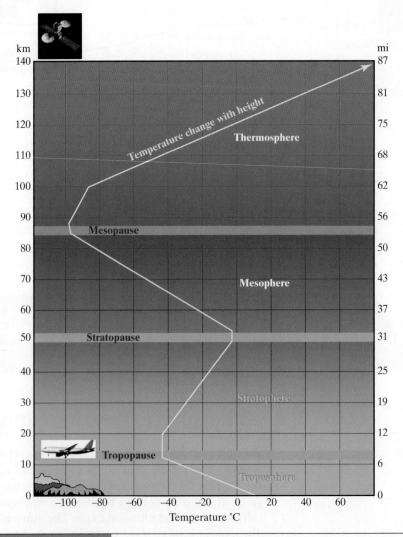

**FIGURE 8.1**    The different layers of the Earth's atmosphere

*Source:* Wire_man / Shutterstock.com, Yeko Photo Studio / Shutterstock.com

it moves away from the Earth's surface and cools down. As shown in Figure 8.1, the temperature of the troposphere decreases with altitude. The *stratosphere* starts at an altitude of about 20 kilometers (12 miles), and the air temperature in this region increases with altitude, as shown. The reason for the increase in temperature in the stratosphere is that the ozone in this layer absorbs ultraviolet (UV) radiation and as a result it warms up. The region above the stratosphere is called the *mesosphere*, which contains relatively small amounts of ozone; consequently, the air temperature decreases again, as shown in Figure 8.1. The last layer of air surrounding the Earth is called the *thermosphere*. The temperature in this layer increases again with altitude because of the absorption of solar radiation by oxygen molecules.

## The Difference between Weather and Climate

News reports often contain words such as *weather conditions* or *climate change*. What is the difference between weather and climate? Well, the word *weather* represents atmospheric conditions such as temperature, pressure, wind speed, and moisture level that could occur during a period of hours or days. For example, a weather report will list an approaching snowstorm or a thunderstorm with details about temperature, wind speed, and the amount of rain or snow based on the current conditions.

*Climate*, on the other hand, represents the average weather conditions over a long period of time. By "a long period of time," we mean many decades or centuries. For example, when we say "Chicago is cold and windy in winter," or "Houston is hot and humid in summer," then we are talking about the climate of these cities. Also, note that even though Chicago may experience a mild winter one year, we know from historical data that has been averaged over many years that the city is cold and windy in winter. Now, if Chicago were to experience many consecutive winters that were mild and calm, we could say that perhaps the climate of Chicago is changing. It should be clear by these examples that, when we talk about global climate change, we are talking about changes that are contrary to our expectations.

> Weather represents atmospheric conditions such as temperature that could occur during a period of hours or days. Climate, on the other hand, represents the average weather conditions over a long period of time (i.e., decades or centuries).

jessicakirsh/Shutterstock.com

Why is this distinction between weather and climate important to know? Well, when scientists warn us about global warming, they are talking about a warming trend—such as the average temperature of the Earth—that is on the rise. The trend is based on data collected and averaged over many decades. A warmer Earth temperature also indicates warmer oceans, which in turn means stronger storms and increased weather anomalies. Next, we discuss a possible cause of global warming—greenhouse gases—and how they affect the Earth's temperature.

## Greenhouse Gases

As mentioned previously, carbon dioxide plays an important role in sustaining plant life; however, if the atmosphere contains too much carbon dioxide, it will not allow the Earth to cool down effectively by radiation. When solar energy passes through the Earth's atmosphere, some of it is absorbed, some of it is scattered, and some of it is reflected by clouds, dust, pollutants, forest fires, volcanoes, and/or water vapor in the atmosphere. The solar energy that reaches the Earth's surface warms it, and eventually some of the absorbed energy re-radiates back toward outer space as the Earth's surface cools down in the evenings. Many gases present in the atmosphere trap some of this heat and as a result prevent the Earth's surface and its atmosphere from cooling (see Figure 8.2). The gradual warming of the Earth's atmosphere is commonly referred to as the *greenhouse effect*, and the gases that cause the warming are called **greenhouse gases**.

Len Green / Shutterstock.com

Some greenhouse gases occur naturally, while others are man-made due to our activities such as producing electricity, heating our homes, driving our cars, making goods, and so on. The major greenhouse gases that affect the warming of the atmosphere include:

- Carbon dioxide ($CO_2$)
- Methane ($CH_4$)
- Nitrous oxide ($N_2O$)

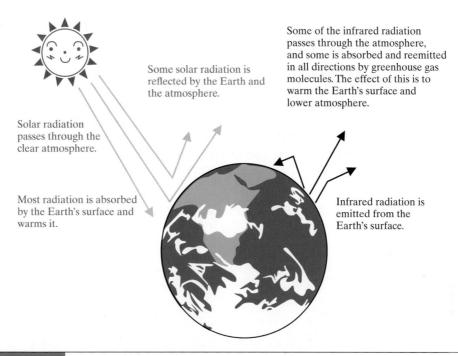

Solar radiation passes through the clear atmosphere.

Some solar radiation is reflected by the Earth and the atmosphere.

Some of the infrared radiation passes through the atmosphere, and some is absorbed and reemitted in all directions by greenhouse gas molecules. The effect of this is to warm the Earth's surface and lower atmosphere.

Most radiation is absorbed by the Earth's surface and warms it.

Infrared radiation is emitted from the Earth's surface.

**FIGURE 8.2**    The greenhouse gas effect
*Source:* U.S. Energy Information Administration

- Hydrofluorocarbons (HFCs)
- Perfluorocarbons (PFCs)
- Sulfur hexafluoride ($SF_6$)

Carbon dioxide accounts for nearly 80 percent of man-made greenhouse gas emissions. The emission of carbon dioxide results primarily from the combustion of gasoline, coal, and natural gas. These fuels are made up of hydrogen and carbon, and when burned, the carbon in them combines with oxygen in the air to create carbon dioxide.

Another important concept to understand is that when a fuel is burned, the amount of carbon dioxide created varies depending on the carbon content of the fuel. *Compared to gasoline and natural gas, coal creates the most carbon dioxide for each unit of energy it produces.* Burning fossil fuels also results in the emission of nitrous oxide, another greenhouse gas. The use of nitrogen fertilizers may also lead to the emission of nitrous oxide. In Chapter 6, we discussed energy consumption rates and the types of fuel used by each sector of our economy.  Here it is important to note that although the industrial sector accounts for a major share of the total energy consumed in the United States, the transportation sector (with nearly 27 percent of total energy consumption) emits more carbon dioxide because of its (almost) total reliance on gasoline. Think about this the next time you drive your car aimlessly!

Methane is another greenhouse gas that is generated by oil and natural gas operations. It may also be emitted from coal mines, landfills, and agricultural activities. Other greenhouse gases, such as hydrofluorocarbons (HFCs), perfluorocarbons (PFCs), and sulfur hexafluoride ($SF_6$), are released into the atmosphere due to various industrial activities. Some of these gases are also released into the atmosphere from slow leaks in refrigeration and air-conditioning units.

## A Comparison of the World and United States Carbon Dioxide Emissions

The annual world total carbon dioxide emissions from the consumption of energy are shown in Figure 8.3. The carbon dioxide emissions during the period shown increased from 18,200 million metric tons in 1980 to 32,310 million metric tons in 2012—a surge of 78 percent. Note that one metric ton is equal to 1,000 kilograms. Take a moment and think about the magnitude of these amounts. The five countries with the largest emissions are shown in Figure 8.4. In 2012, as expected, China led the world in $CO_2$ emissions due to

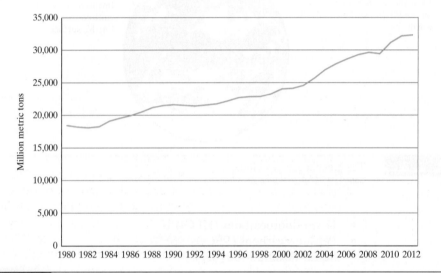

**FIGURE 8.3**    The annual world total carbon dioxide emissions from consumption of energy
*Source of data:* Data from U.S. Energy Information Administration

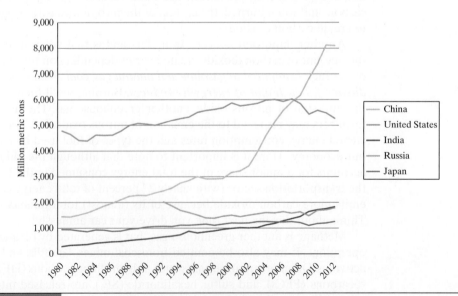

**FIGURE 8.4**    The annual total carbon dioxide emissions from consumption of energy—top 5 countries
*Source of data:* Data from U.S. Energy Information Administration

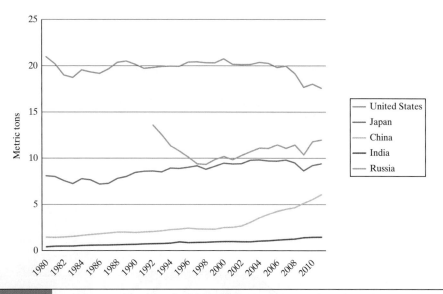

**FIGURE 8.5**   The annual carbon dioxide emissions per person for selected countries
*Source of data:* Data from U.S. Energy Information Administration

its recent rise in manufacturing activities. Also, note that most of the products manufactured in China are for consumption elsewhere! The annual carbon dioxide emissions per capita for selected countries are shown in Figure 8.5. Unsurprisingly, due to our life style, the United States leads in this category. Also, note that, due to improvements in the standard of living in China, their per capita carbon dioxide emissions are also on the rise. What do you think could be the future *global* consequences if China and India (with almost half of the world population) were to raise their standard of livings to the level of the United States?

Finally, the energy-consumption-related carbon dioxide emissions by sectors of the economy in the United States for 2013 (the latest data available) are shown in Figure 8.6. This bar chart shows the percent of contributions by residential and commercial, transportation, and industrial sectors to carbon dioxide emissions in the United States.

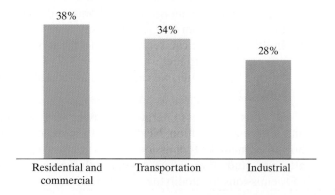

**FIGURE 8.6**   The energy-consumption-related carbon dioxide emissions in the United States for 2013
*Source of data:* Data from U.S. Energy Information Administration

Now let us look at the amount of carbon dioxide emitted due to one of our daily activities, driving. It is important to realize that each gallon of gasoline that you burn in your car creates nearly 20 pounds of $CO_2$ (see Example 8.1 for more detail). Moreover, assuming that you drive your car on average 12,000 miles annually and assuming a fuel economy rating of 25 miles per gallon for your car, then each year you are adding 9,600 pounds (4,354 kilograms) of $CO_2$ to the atmosphere.

Vehicles 51%

Appliances 26%

Heating and cooling 18%

U.S. Department of Energy, Reduce Climate Change, http://www.fueleconomy.gov/feg/climate.shtml

Fortunately, there are ways we can reduce our greenhouse-gas footprint.

- We can carpool or take public transportation when going to work or school.
- We should not leave our cars running idle for long periods of time, and we can remind others to consume less energy.
- We can conserve energy around home and school. For example, turn off the light in a room that is not in use.
- In the winter, we can set the thermostat at 65°F or slightly lower and wear a sweater to feel warm.
- During the summer, we can set the air-conditioning thermostat at 78°F or slightly higher.

By consuming less energy and driving less, we can help our environment and reduce air pollution. Next, we use Examples 8.1 through 8.4 to show you how to estimate $CO_2$ emissions due to your own personal activities. Hopefully, after you study these examples carefully, you will think about ways to reduce $CO_2$ emissions by modifying your behavior.

**EXAMPLE 8.1**

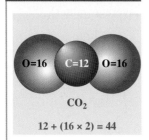

$CO_2$

$12 + (16 \times 2) = 44$

**A gallon of gasoline that weighs 6.3 pounds can produce 20 pounds of carbon dioxide ($CO_2$)**

The next time you get the urge to drive around with no purpose, think about the following. Every time you drive your car, 20 pounds of carbon dioxide ($CO_2$) are produced for every gallon of gasoline that you burn. A gallon of gasoline has a mass of approximately 6.3 pounds, and its chemical composition consists of 87 percent carbon and 13 percent hydrogen. Consequently, the carbon portion of a gallon of gasoline has a mass of 5.5 pounds (6.3 lbm × 0.87 = 5.5 lbm). When gasoline is burned, hydrogen and carbon separate and mix with oxygen and other constituents in the air; among other byproducts, they form water vapor and carbon dioxide. Carbon has an atomic weight of 12, while oxygen has an atomic weight of 16. When carbon mixes with oxygen to form $CO_2$, the carbon dioxide will have an atomic weight of 44 [12 (from carbon) + 2 × 16 (from oxygen) = 44].

Therefore, 27 percent (12/44 = 0.27) of the mass of $CO_2$ comes from carbon, while the other 73 percent (32/44 = 0.73) comes from oxygen. As a result, every gallon of gasoline that has 5.5 pounds of carbon produces 20 pounds of carbon dioxiode (5.5/0.27 = 20). Also remember that, for every gallon of gasoline you burn, you remove 14.5 pounds of oxygen from the air that you breathe. Of course, we rely on plants to absorb some of the carbon dioxide and replenish the lost oxygen!

In a similar fashion, we can show that when a gallon of diesel fuel is burned, approximately 22 pounds of carbon dioxide $CO_2$ are produced.

**EXAMPLE 8.2**

According to the U.S. Federal Highway Administration's recent data, the average annual miles driven by all age groups and genders is 13,476. Let us now calculate the amount (in pounds) of $CO_2$ released into the atmosphere by a car with an average fuel economy rating of 25 miles per gallon.

Maksim Toome / Shutterstock.com

amount of $CO_2$ released into atmosphere

$$= \left(13,476\,\frac{\text{miles}}{\text{year}}\right)\left(\frac{1\ \text{gallon}}{25\ \text{miles}}\right)\left(\frac{20\ \text{pounds of } CO_2}{1\ \text{gallon}}\right)$$

$$= 10,781 \text{ pounds of } CO_2 \text{ per year}$$

The $CO_2$ emissions produced by specific sources of energy are summarized in Table 8.1. You can use the data given in Table 8.1 to estimate your $CO_2$ footprint (see Examples 8.3 and 8.4).

| TABLE 8.1 | $CO_2$ Emissions by the Source of Energy |
|---|---|
| **Source of Energy** | **Pounds of $CO_2$ Emission (average values)** |
| 1 kWh of electricity | 1.7 |
| 1,000 cubic feet of natural gas | 120 |
| 1 gallon of gasoline | 20 |
| 1 gallon of diesel | 22 |
| 1 gallon of fuel oil | 22.5 |
| 1 gallon of propane | 12.4 |

**EXAMPLE 8.3**

In Chapter 3, we showed that you need to burn 8.34 cubic feet of natural gas to heat up 20 gallons of water from room temperature at 70°F to 120°F to produce enough hot water to take a long shower. Let's now calculate the amount (in pounds) of $CO_2$ released into the atmosphere annually due to taking showers, knowing that every 1,000 cubic feet of natural gas when burned produces about 120 pounds of $CO_2$.

Di Studio / Shutterstock.com

amount of $CO_2$ released into atmosphere

$$= \left( 8.34 \; \frac{\text{ft}^3 \text{ of natural gas}}{\text{day}} \right) \left( \frac{365 \; \text{days}}{1 \; \text{year}} \right) \left( \frac{120 \; \text{pounds of } CO_2}{1,000 \; \text{ft}^3 \text{ of natural gas}} \right)$$

$$= 365 \; \text{pounds of } CO_2 \text{ per year}$$

Note that in this analysis we assume that the hot water heater is 100 percent efficient; otherwise, more energy is required and consequently more $CO_2$ is produced.

**EXAMPLE 8.4**

In the United States, every kilowatt-hour of electricity generated in a power plant produces approximately 1.7 pounds of $CO_2$ on average. In Chapter 4, we showed that, when you watch a 46-inch-LCD TV that consumes 250 watts (W) for 4 hours, you consume 1,000 watt-hours (Wh) or 1 kilowatt-hour (kWh) of energy. Let's now calculate the amount (in pounds) of $CO_2$ released into the atmosphere annually due to watching TV. Let us assume you leave your TV on (whether you watch or not) for 4 hours every night for 200 nights. The other 165 nights you do other things.

Pakhnyushchy / Shutterstock.com

amount of $CO_2$ released into atmosphere

$$= \left( 1 \; \frac{\text{kWh}}{\text{night}} \right) \left( \frac{200 \; \text{nights}}{1 \; \text{year}} \right) \left( \frac{1.7 \; \text{pounds of } CO_2}{1 \; \text{kWh}} \right)$$

$$= 340 \; \text{pounds of } CO_2 \text{ per year}$$

Did you ever think that leaving a TV on or surfing the net aimlessly indirectly releases $CO_2$ into the atmosphere?

A typical home in the United States could consume as much as 10,000 kWh of electricity per year. How many pounds of $CO_2$ are released into the atmosphere annually?

In the next section, we discuss the outdoor air quality standards in the United States.

## Before You Go On

Answer the following questions to test your understanding of the preceding section:

1. What do we mean by the word atmosphere?

2. What are the major gases that make up the atmosphere?

3. Which gases in the atmosphere contribute to the greenhouse effect?

4. What is the difference between weather and climate?

5. Which sector of our economy contributes the most to the greenhouse effect?

6. How many pounds of carbon dioxide ($CO_2$) are produced when a gallon of gasoline is burned?

7. How many pounds of carbon dioxide ($CO_2$) are produced when a gallon of diesel fuel is burned?

8. How many pounds of carbon dioxide ($CO_2$) are produced when one kilowatt-hour of electricity is generated?

*Vocabulary—State the meaning of the following terms:*

Atmosphere

Water vapor

Climate

Weather

Greenhouse gases

Troposphere

Stratosphere

Mesosphere

Thermosphere

# LO² 8.2 Outdoor Air Quality Standards in the United States

The United States has one of the most stringent outdoor air quality standards in the world. In the United States, the source of outdoor air pollution is classified into three broad categories: *stationary, mobile, and natural sources.*

Examples of stationary sources include power plants, factories, and dry cleaners. The mobile sources of air pollution consist of cars, buses, trucks, planes, and trains. As the name implies, the sources of natural air pollution include windblown dust, volcanic eruptions, and forest fires.

**The source of outdoor air pollution may be classified into: stationary (e.g., power plants), mobile (e.g., cars), and natural (e.g., windblown dust) sources.**

## Measurement of Pollutants

The Clean Air Act, which sets the standard for six major air pollutants, was signed into law in 1970. The Environmental Protection Agency (EPA) is responsible for setting standards for these six major air pollutants: carbon monoxide (CO), lead (Pb), nitrogen dioxide ($NO_2$), ozone ($O_3$), sulfur dioxide ($SO_2$), and particulate matter (PM).

*Particulate matter* (PM) is defined as a mixture of organic and inorganic solid and liquid particles that are suspended in the air. The components of PM may include sulfate, nitrates, ammonia, sodium chloride, carbon, and dust. Based on their size, the particles of PM are classified into the following types.

1. $PM_{10}$—particles with diameters smaller than 10 microns (1 micron = 0.001 mm)
2. $PM_{2.5}$—particles with diameters smaller than 2.5 microns

Those classified as $PM_{2.5}$ create a greater health risk (in the lungs) because of their size.

*Sulfur dioxide* ($SO_2$) is a colorless gas, and as you may know, it has a strong odor. Sulfur dioxide is released into the atmosphere when we burn fossil fuels in our cars and power plants.

The EPA measures the concentration levels of these pollutants in many urban areas and collects air-quality information by an actual measurement of pollutants from thousands of monitoring sites located throughout the country. The United States national trends from 1980 to 2014 for *carbon monoxide* (CO), lead (Pb), *nitrogen dioxide* ($NO_2$), ozone ($O_3$), sulfur dioxide ($SO_2$), and particulate matter (PM) are shown in Figures 8.7 through 8.12, respectively. Note that close examination of these figures shows (on average) a downward trend.

## Pollution Reduction

According to a study performed by the EPA between 1970 and 1997, the United States population increased by 31 percent and the vehicle miles traveled increased by 127 percent. During this period, the total emission of air pollutants from stationary and mobile sources decreased by 31 percent because of improvements made in the efficiency of cars and in industrial practices, along with the enforcement of the Clean Air Act regulations. The next time you contemplate buying a car, think about this important fact.

Despite this, there are still millions of people who live in areas with unhealthy air quality. The EPA is continuously working to set standards and monitor the emission of pollutants that cause acid rain, damage to bodies of

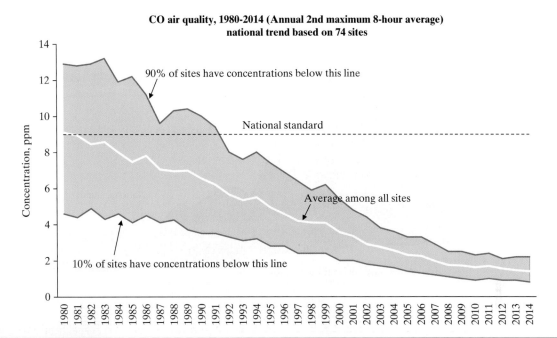

**CO air quality, 1980-2014 (Annual 2nd maximum 8-hour average) national trend based on 74 sites**

90% of sites have concentrations below this line

National standard

Average among all sites

10% of sites have concentrations below this line

Concentration, ppm

**FIGURE 8.7**    Carbon monoxide emission trends
*Source of data:* Data from EPA

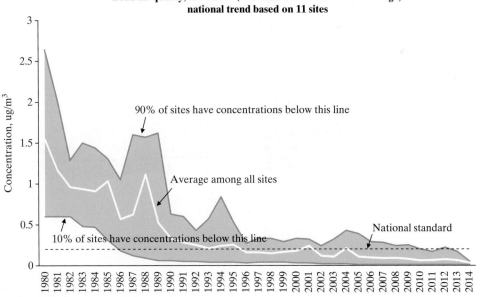

**Lead air quality, 1980-2014 (Annual maximum 3-month average) national trend based on 11 sites**

90% of sites have concentrations below this line

Average among all sites

10% of sites have concentrations below this line

National standard

Concentration, ug/m³

**FIGURE 8.8**    Lead emission trends
*Source of data:* Data from EPA

water and fish (there are currently over a thousand bodies of water in the United States that are under fish consumption advisories), damage to the stratospheric ozone layer, and damage to our buildings and to our national parks. The unhealthy air has more pronounced adverse health effects on children and elderly people. Human health problems associated with poor air quality include various respiratory illnesses and heart or lung diseases.

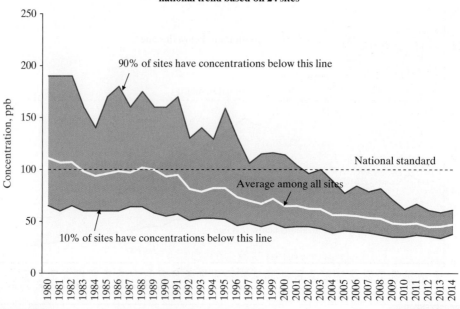

NO₂ air quality, 1980-2014 (Annual 98th percentile of daily max 1-hour average)
national trend based on 24 sites

Nitrogen dioxide emission trends
*Source of data:* Data from EPA

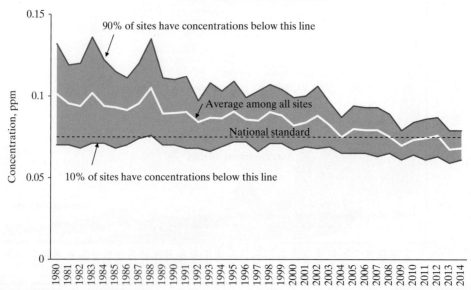

Ozone air quality, 1980-2014 (Annual 4th maximum of daily max 8-hour average)
national trend based on 218 sites

Ozone trends
*Source of data:* Data from EPA

Congress passed amendments to the Clean Air Act in 1990 that required the
EPA to address the effects of many toxic air pollutants by setting new standards. Since 1997, the EPA has issued over twenty air standards to be fully
implemented. The EPA continues to work with the individual states in the
United States to reduce the amount of sulfur in fuels and to set more stringent
emission standards for cars, buses, trucks, and power plants.

**SO₂ air quality, 1980-2014 (annual 99th percentile of daily max 1-hour average)**
**national trend based on 45 sites**

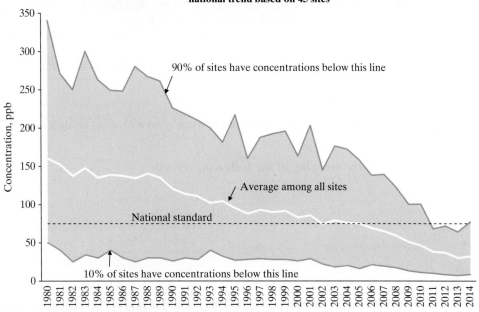

Sulfur dioxide emission trends
*Source of data:* Data from EPA

**PM10 air quality, 1990-2014 (Annual 2nd maximum 24-hour average)**
**national trend based on 193 sites**

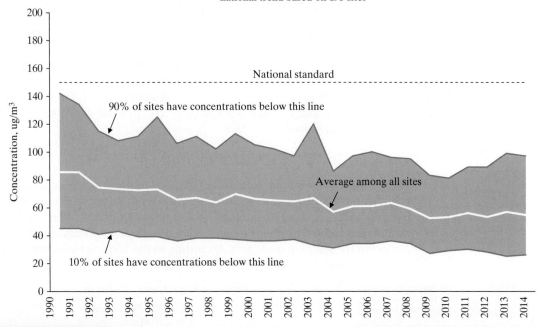

Particulate matter trends
*Source of data:* Data from EPA

We all need to understand that air pollution is a global concern that can affect not only our health but also our climate. It is triggering the onset of global warming, and this is leading to serious natural events. Because we all contribute to this problem, we need to be aware of the consequences of our lifestyles and find ways to reduce pollution.

## Before You Go On

Answer the following questions to test your understanding of the preceding section:

1.  What are the major sources of outdoor air pollution?

2.  What are some of the human health problems associated with poor air quality?

*Vocabulary—State the meaning of the following terms:*

PM

$SO_2$

## LO³  8.3  Indoor Air Quality Standards in the United States

In the previous section, we discussed outdoor air pollution and related health effects. Indoor air pollution can also create health risks. According to EPA studies of human exposure to air pollutants, the indoor levels of pollutants may be two to five times higher than outdoor levels. ***Indoor air quality*** (IAQ) is important in homes, schools, and workplaces. Because most of us spend approximately 90 percent of our time indoors, the indoor air quality is very important to our short-term and long-term health. Moreover, the lack of good indoor air quality can cause sickness or discomfort to building occupants, reducing productivity at the workplace or creating an unfavorable learning environment at school.

### Air Quality and Health Issues

Failure to monitor indoor air quality (IAQ) or to prevent indoor air pollution can also have adverse effects on equipment and the physical appearance of buildings. In recent years, liability issues related to people who suffer dizziness or headaches or other illnesses related to "sick buildings" are becoming a concern for building managers. According to the EPA, some common health symptoms caused by poor indoor air quality are

The factors that influence indoor air quality include ventilation, household cleaners, synthetic building materials, and underground sources such as radon.

*   Headache, fatigue, and shortness of breath
*   Sinus congestion, coughing, and sneezing
*   Eye, nose, throat, and skin irritation
*   Dizziness and nausea

As you know, some of these health symptoms may be caused by other factors and are not necessarily caused by poor air quality. Stress at school, work, or at home could also create health problems with symptoms similar to the ones mentioned. Moreover, individuals react differently to similar problems in their surroundings.

The factors that influence air quality can be classified into several categories: the heating, ventilation, and air-conditioning (HVAC) system; sources of indoor air pollutants; and occupants. In recent years, we have been exposed to more indoor air pollutants for the following reasons.

1. In order to save energy, we are building air-tight houses that have lower air infiltration or exfiltration compared to older structures. In addition, the ventilation rates have also been reduced to save more energy.

2. We are using more synthetic building materials in newly built homes that could give off harmful vapors.

3. We are using more chemical pollutants, such as pesticides and household cleaners, indoors.

As shown in Table 8.2, indoor pollutants can be created by sources within the building or they can be brought in from the outdoors. It is important to keep in mind that the level of contaminants within a building can vary with time. For example, in order to protect floor surfaces from wear and tear in schools and commercial buildings, it is customary to wax them. During the period when waxing is taking place, it is possible, based on the type of chemical used, that anyone near the area might be exposed to harmful vapors. Of course, one simple remedy to this indoor air problem is to wax the floor late on Friday afternoons to avoid exposing too many occupants to harmful vapors. Moreover, this approach will provide some time for the vapor to be exhausted out of the building by the ventilation system over the weekend when the building is not occupied.

**TABLE 8.2**  Typical Sources of Indoor Air Pollutants

| Outside Sources | Building Equipment | Components/Furnishings | Other Indoor Sources |
|---|---|---|---|
| **Polluted outdoor air** Pollen, dust, fungal spores, industrial emissions and vehicle emissions | **HVAC equipment** Microbiological growth in drip pans, ductwork, coils, and humidifiers; improper venting of combustion products; and dust or debris in ductwork | **Components** Microbiological growth on soiled or water-damaged materials; dry traps that allow the passage of sewer gas; materials containing volatile organic compounds, inorganic compounds, and damaged asbestos; and materials that produce particles (dust) | Science laboratories, copy and print areas, food preparation areas, smoking lounges, cleaning materials, emission from trash, pesticides, odors and volatile organic compounds from paint, chalk, and adhesives, occupants with communicable diseases, dryerase markers and similar pens, insects and other pests, personal hygiene products |
| **Nearby sources** Loading docks, odors from dumpsters, unsanitary debris or building exhausts near outdoor air intakes | **Non-HVAC equipment** Emissions from office equipment and emissions from shops, labs, and cleaning processes | | |
| **Underground sources** Radon, pesticides, and leakage from underground storage tanks | | **Furnishings** Emissions from new furnishings and floorings and microbiological growth on or in soiled or water-damaged furnishings | |

*Source:* Data from EPA

# Heating, Ventilation, and Air-Conditioning (HVAC) Systems

The primary purpose of a well-designed heating, ventilation, and air-conditioning (HVAC) system is to provide thermal comfort to its occupants. Based on the building's heating or cooling load, the air that is circulated through the building is conditioned by heating, cooling, humidifying, or dehumidifying. The other important role of a well-designed HVAC system is to filter out the contaminants or to provide adequate ventilation to dilute air-contaminant levels.

The air flow patterns in and around a building also affect the indoor air quality. The air flow pattern inside the building is normally created by the HVAC system. However, the outside air flow around a building envelope that is dictated by wind patterns could also affect the air flow pattern within the building as well. When looking at air flow patterns, the important concept to keep in mind is that air always moves from a high-pressure region to a low-pressure region.

## Methods to Manage Contaminants

There are several ways to control the level of contaminants:

1. Source elimination or removal

2. Source substitution

3. Proper ventilation

4. Exposure control

5. Air cleaning

A good example of *source elimination* is not allowing people to smoke inside a building or not allowing a car engine to run idle near a building's outdoor air intake. In other words, eliminate the source before it spreads out! It is important for designers to keep that idea in mind when designing the HVAC systems for a building—they can avoid placing the outdoor air intakes near loading docks or dumpsters, for example. A good example of *source substitution* is to use a gentle cleaning product rather than a product that gives off harmful vapors when cleaning bathrooms, kitchens, or floors. Local exhaust *control* means removing the sources of pollutants before they can be spread through the air distribution system into other areas of a building. Everyday examples include the use of an exhaust fan in restrooms to force out harmful contaminants. Fume hoods are another example of local exhaust removal in many laboratories.

Clean outdoor air also can be mixed with the inside air to dilute the contaminated air. The American Society of Heating, Refrigerating and Air Conditioning Engineers (ASHRAE) has established a set of codes and standards for how much fresh outside air must be introduced for various applications. *Air cleaning* means removing harmful particulate and gases from the air as it passes through a cleaning system. There are various methods that deal with air contaminant removal, including the use of air filters.

Finally, you can bring the indoor air quality issues to the attention of your friends, classmates, coworkers, and family. We all need to be aware of our environment and try to do our part to create and maintain a healthy indoor air quality.

**EXAMPLE 8.5**

When we breathe in the air inside a building, we breathe in pollutants such as dust, carbon monoxide, radon, mold spores, and chemical fumes that also might be present in the air. Studies have shown that under sedentary conditions, adults breathe in about 7 liters of surrounding air per minute. Let's now calculate the volume of air that we breathe in a day.

$$\text{volume of air in liters} = \left(7 \ \frac{\text{liters}}{\text{minute}}\right)\left(\frac{60 \ \cancel{\text{minutes}}}{1 \ \cancel{\text{hour}}}\right)\left(\frac{24 \ \cancel{\text{hours}}}{1 \ \text{day}}\right)$$

$$= 10{,}080 \ \text{liters/day}$$

$$\text{volume of air in cubic meter} = \left(7 \ \frac{\cancel{\text{liters}}}{\cancel{\text{minute}}}\right)\left(\frac{60 \ \cancel{\text{minutes}}}{1 \ \cancel{\text{hour}}}\right)\left(\frac{24 \ \cancel{\text{hours}}}{1 \ \text{day}}\right)\left(\frac{1 \ \text{m}^3}{1{,}000 \ \cancel{\text{liters}}}\right)$$

$$= 10.1 \ \text{m}^3/\text{day}$$

$$\text{volume of air in cubic feet} = \left(10.1 \ \frac{\cancel{\text{m}^3}}{\text{day}}\right)\left(\frac{3.28 \ \text{ft}}{1 \ \cancel{\text{m}}}\right)^3 = 356 \ \text{ft}^3/\text{day}$$

$$\text{volume of air in gallons} = \left(356 \ \frac{\cancel{\text{ft}^3}}{\text{day}}\right)\left(\frac{7.48 \ \text{gallons}}{1 \ \cancel{\text{ft}^3}}\right)$$

$$= 2{,}660 \ \text{gallons/day}$$

As you can see, we breathe in a large volume of air each day. So it is imperative that the air that we breathe does not contain pollutants that could be harmful to our health.

# Before You Go On

Answer the following questions to test your understanding of the preceding section:

1. What are some common health symptoms that are caused by poor indoor air quality?

2. Give examples of outside sources of indoor air pollutants.

3. Give examples of building equipment that contribute to indoor air pollutants.

4. Explain at least two ways you can control the level of indoor contaminants.

5. Give an example of exposure control.

*Vocabulary—State the meaning of the following terms:*

IAQ

HVAC

## LO⁴  8.4  Global Air Quality Issues

The *World Health Organization (WHO)* is the authority on global health matters, including air-quality-related health issues. It is responsible for setting standards, monitoring these standards, and for providing technical support.

According to the WHO, air pollution is a major global environmental risk to health that causes respiratory infections, heart disease, and lung cancer. Each year, air pollution causes over one million deaths.

The WHO establishes air quality guidelines, assesses health effects of air pollution, and recommends actions to reduce health risks. The latest WHO air quality guidelines recommend limits for the concentration of selected air pollutants such as particulate matter (PM), ozone ($O_3$), nitrogen dioxide ($NO_2$), and sulfur dioxide ($SO_2$). In Section 8.2, we discussed particulate matter and its role in outdoor air pollution. Let's now say a few words

Hung Chung Chih / Shutterstock.com

about ozone. The ozone at ground level is different from the ozone layer that exists in the upper atmosphere that protects us from the sun's harmful radiation. The *ozone* at ground level is created by the reaction of sunlight with industry and vehicle emissions of pollutants such as nitrogen oxides. You may have noticed that this effect is more pronounced during sunny days. Table 8.3 shows a list of these major air pollutants, the WHO's recommended limits, and their adverse health effects. In developing countries, nearly three billion people still burn wood, crop waste, coal, and animal dung in open fires in their homes to cook and to keep themselves warm. The open fires, along with poor ventilation, produce pollutants that are often many times higher than the recommended values shown in Table 8.3. According to the WHO, nearly seven million people a year die prematurely from illnesses attributed to air pollution (2014 data). The percentage of deaths from disease by outdoor and indoor air pollution are shown here.

> According to the World Health Organization, in developing countries, nearly 2 million premature deaths are attributed to indoor air pollution.

**Death caused by outdoor air pollution:**

- 40% — ischemic heart disease;
- 40% — stroke;
- 11% — chronic obstructive pulmonary disease (COPD);
- 6 % — lung cancer; and
- 3 % — acute lower respiratory infections in children.

**Death caused by indoor air pollution:**

- 34% — stroke;
- 26% — ischemic heart disease;
- 22% — COPD;
- 12% — acute lower respiratory infections in children; and
- 6 % — lung cancer.

| TABLE 8.3 | The outdoor and indoor pollutants, the WHO recommended limits, and their health effects ($\mu g$ = microgram = $10^{-6}$ gram) | |
|---|---|---|
| **Pollutant** | **Limits** | **Causes/Health Effects** |
| PM$_{2.5}$ | 10 $\mu g/m^3$ annual mean; 25 $\mu g/m^3$ 24-hour mean | Indoor open fire and leaky stoves in developing countries; Industrial activities/vehicles/power production |
| PM$_{10}$ | 20 $\mu g/m^3$ annual mean; 50 $\mu g/m^3$ 24-hour mean | Cardiovascular and respiratory diseases, lung cancer. |
| O$_3$ | 100 $\mu g/m^3$ 8-hour mean | Vehicles/industrial activities<br><br>Breathing problems, asthma, reduced lung function, lung diseases, and heart disease. |
| NO$_2$ | 40 $\mu g/m^3$ annual mean; 200 $\mu g/m^3$ 1-hour mean | Industrial activities/vehicles/power production/human activities<br><br>Reduced lung function |
| SO$_2$ | 20 $\mu g/m^3$ 24-hour mean; 500 $\mu g/m^3$ 10-minute mean | Industrial/power production/human activities<br>When combined with water, it produces acid rain<br><br>Respiratory problems, reduced lung function, infections of the respiratory tracts. |

## Before You Go On

Answer the following questions to test your understanding of the preceding section:

1. What does WHO stand for?

2. What are the causes of excessive ozone concentration at ground level?

3. Give examples of health problems that are caused by excessive concentration of ozone at ground level.

4. Give examples of the causes of PM in developing countries.

## SUMMARY

### LO¹ Atmosphere, Weather, and Climate

You should recall the characteristics of the atmosphere, know the difference between climate and weather, and understand the greenhouse effect. Air is a mixture of mostly nitrogen and oxygen, as well as small amounts of other gases such as argon, carbon dioxide, sulfur dioxide, and nitrogen oxide. The air surrounding the Earth, depending on its temperature, can be divided into four regions: troposphere, stratosphere, mesosphere, and thermosphere. *Weather* represents atmospheric conditions, such as a temperature that occurs during a period of hours or days. *Climate*, on the other hand, represents the average weather conditions over a long period of time (i.e., decades or centuries). Carbon dioxide plays an important role in sustaining plant life; however, if the atmosphere contains too much carbon dioxide, it does not allow the Earth to cool down effectively by radiation, which results in the greenhouse effect.

## LO² Outdoor Air Quality Standards

The EPA measures the concentration level of pollutants in many urban areas and collects air quality information. The source of outdoor air pollution may be classified into three broad categories: *stationary, mobile, and natural sources*. The Environmental Protection Agency (EPA) is responsible for setting standards for six major air pollutants: carbon monoxide (CO), lead (Pb), nitrogen dioxide ($NO_2$), ozone ($O_3$), sulfur dioxide ($SO_2$), and particulate matter (PM). The EPA is also continuously working to set standards and monitor the emission of pollutants that cause acid rain, which damages bodies of water and fish, buildings, and our national parks. The EPA works with the individual states to reduce the amount of sulfur in fuels and set more stringent emission standards for cars, buses, trucks, and power plants. Because we all contribute to air pollution, we need to be aware of the consequences of our lifestyles and find ways to reduce pollution.

## LO³ Indoor Air Quality Standards

According to EPA studies of human exposure to air pollutants, the indoor levels of pollutants may be many times higher than outdoor levels. Some common health symptoms caused by poor indoor air quality are headaches, fatigue, and shortness of breath. The factors that influence indoor air quality are classified into different categories: the heating, ventilation, and air-conditioning (HVAC) system; sources of indoor air pollutants; and occupants. In order to save energy, we are building air-tight houses with lower ventilation rates and lower air infiltration compared to older structures. We also are using more synthetic building materials in newly built homes that sometimes give off harmful vapors. Moreover, we are using more chemical pollutants, such as pesticides and household cleaners, indoors. As we discussed in this chapter, there are several ways to control the level of contaminants: (1) source elimination or removal, (2) source substitution, (3) proper ventilation, (4) exposure control, and (5) air cleaning. It is very important that you bring indoor air quality issues to the attention of your friends, classmates, and family. We all need to be well-educated in this topic and try to do our part to create and maintain a healthy indoor air quality.

## LO⁴ Global Air Quality Issues

The World Health Organization (WHO) is the authority on global health matters including air-quality-related health issues. It is responsible for setting air quality standards, monitoring these standards, and providing technical support. According to the WHO, air pollution is a major global environmental risk to health that causes respiratory infections, heart disease, and lung cancer. Each year, nearly 2 million premature deaths are attributed to indoor air pollution in developing countries. The latest WHO air-quality guidelines recommend limits for the concentration of selected air pollutants such as particulate matter (PM), ozone ($O_3$), nitrogen dioxide ($NO_2$), and sulfur dioxide ($SO_2$).

## KEY TERMS

Atmosphere 243
Carbon Dioxide 244
Carbon Monoxide 254
Climate 245
Environmental Protection Agency (EPA) 254
Greenhouse Gases 246
Indoor Air Quality 258

Lead 254
Major Air Pollutants 254
Mesosphere 245
Nitrogen Dioxide 254
Outdoor Air Quality 254
Ozone 262
Particulate Matter (PM) 254

Stratosphere 245
Sulfur Dioxide 254
Thermosphere 245
Troposphere 244
Weather 245
World Health Organization (WHO) 262

# Apply What You Have Learned

Each of you is to estimate how much carbon dioxide you produce annually due to driving your cars. If you do not have a car, make estimates for your family vehicles. Write a brief report and state all your assumptions. Discuss your possible lifetime production.

Maksim Toome / Shutterstock.com

## PROBLEMS

*Problems that promote life-long learning are denoted by* 🔑

**8.1** We can reduce the amount of $CO_2$ released into the atmosphere by designing cars with improved fuel economy ratings. What would be the reduction in pounds of $CO_2$ released into the atmosphere by a car with an improved fuel economy rating of 10 mpg? Assume the car is driven 12,000 miles annually.

**8.2** We can reduce the amount of $CO_2$ released into the atmosphere by driving our cars less. What would be the reduction in pounds of $CO_2$ released into the atmosphere by a car that is driven 10,000 miles per year instead of 12,000 miles per year?

**8.3** For Example 8.3, what would be the addition in pounds of $CO_2$ released into the atmosphere if the hot water heater had an efficiency of 85%?

**8.4** For Example 8.4, what would be the reduction in pounds of $CO_2$ released into the atmosphere if the TV consumed 150 W?

**8.5** Estimate the annual reduction in pounds (or kilograms) of $CO_2$ released into the atmosphere if a home that uses 1,500 kWh for lighting replaces the current system with one that is 20% more efficient.

**8.6** Homes in the United States consume on average anywhere between 4,000 kWh and 10,000 kWh annually. Calculate the amount of $CO_2$ released into the atmosphere for a million homes with an annual consumption of 7,000 kWh.

**8.7** What would be the reduction in pounds of $CO_2$ released into the atmosphere if 100 million people walked 3 miles a day instead of driving their cars? Present your results on a daily, weekly, monthly, and yearly basis.

**8.8** The U.S. Federal Highway Administration's recent data regarding the average annual miles driven per driver by age group is shown in the accompanying table.

| Age | Male | Female | Average |
|---|---|---|---|
| 16–19 | 8,206 | 6,873 | 7,624 |
| 20–34 | 17,976 | 12,004 | 15,098 |
| 35–54 | 18,858 | 11,464 | 15,291 |
| 55–64 | 15,859 | 7,780 | 11,972 |
| 65+ | 10,304 | 4,785 | 7,646 |
| Average | 16,550 | 10,142 | 13,476 |

Assume an average fuel economy of 25 miles per gallon, and calculate the amount of $CO_2$ released into the atmosphere by gender and age group. Present your results in a similar tabular form.

8.9    In the northeastern section of the United States, many homes during the winter months are heated by furnaces that burn fuel oil. Calculate the amount of $CO_2$ released into the atmosphere by 100,000 homes if each house consumes on average 800 gallons of fuel oil during this period.

8.10    In many areas where natural gas lines are not available, homes are heated during the winter months by furnaces that burn propane. Calculate the amount of $CO_2$ released into the atmosphere by 100,000 homes if each house consumes on average 600 gallons of propane during this period.

8.11    Estimate your annual electrical energy consumption in kWh and the corresponding $CO_2$ emissions. State all your assumptions.

8.12    Estimate your annual electrical energy consumption in kWh due to the use of your electronic devices (e.g., laptop, tablet, TV) and the corresponding $CO_2$ emissions. State all your assumptions.

8.13    Estimate how much water you consume when taking a shower, and calculate the amount of $CO_2$ released into the atmosphere due to this activity annually.

8.14    Suggest practical ways by which you can reduce the amount of carbon dioxide that is released into the atmosphere due to your activities.

8.15    In order to reduce the level of carbon oxides pollution, a community has banned the recreational use of a wood-burning fireplace. State arguments in favor and against the proposal.

8.16    Investigate the quality of air in Beijing, China. How does it compare to a city like Los Angeles, California? Discuss your findings in a brief report.

8.17    Suggest practical ways by which you can reduce indoor air pollution caused by your activities.

8.18    Investigate the health effects of radon gas emissions in buildings. Present your findings in a brief report.

8.19    Investigate the health effects of emissions from office equipment (e.g., printers, copiers, computers). Present your findings in a brief report.

8.20    Investigate the health effects of emissions from new furnishings and floorings. Present your findings in a brief report.

*"Science may have found a cure
for most evils; but it has found no
remedy for the worst of them all —
the apathy of human beings."*

—HELEN KELLER (1880–1968)

# Water Resources, Consumption Rates, and Quality Standards

## LEARNING OBJECTIVES

**LO¹** Water—Basic Concepts: know about basic water concepts and become familiar with water resources terminology

**LO²** Personal Water Consumption: describe how much water we consume through our daily activities

**LO³** Water Consumption in Agriculture, Commercial, and Industrial Sectors: know how much water is consumed in agriculture, commercial, and industrial sectors

**LO⁴** Drinking Water Standards in the United States: describe the basic drinking water standards in the United States

**LO⁵** Global Water Quality Issues: understand the lack of drinking water in developing countries and the sanitation-related health issues

# *Discussion Starter*

## United Nations Water Statistics

The United Nations (UN) suggests that each person needs 20 to 50 liters of safe, fresh water a day to ensure their basic needs for drinking, cooking, and cleaning.

Source: World Water Assessment Program (WWAP)

More than one in six people worldwide do not have access to this amount of safe, fresh water.

Source: World Health Organization (WHO) and United Nations Children Fund (UNICEF) Joint Monitoring Program on Water Supply and Sanitation (JMP)

Globally, diarrhea is the leading cause of illness and death, and 88 percent of diarrheal deaths are due to a lack of access to sanitation facilities, an inadequate availability of water for hygiene, and unsafe drinking water.

Source: JMP

Today 2.5 billion people, including almost one billion children, live without even basic sanitation. Every 20 seconds, a child dies as a result of poor sanitation. That's 1.5 million preventable deaths each year.

Source: Water Supply and Sanitation Collaborative Council (WSSCC)

In sub-Saharan Africa, treating diarrhea consumes 12 percent of the health budget. On a typical day, more than half the hospital beds are occupied by patients suffering from fecal-related disease.

Source: WSSCC

While the percent of the population with access to improved facilities has increased since 1990 in all regions, the number of people living without access has increased due to slow progress and population growth. In 2008, 2.6 billion people still had no access to improved sanitation facilities.

**To the Students:** Have you ever thought about what your life would be like if you suddenly didn't have access to water and adequate sanitation for a day or two? Could you manage? How much water do you think you use every day through your different activities? How much water do you waste?

## LO[1]  9.1  Water—Basic Concepts

You already know that every living thing needs water to sustain life, and water plays a significant role in our everyday life. It accounts for nearly 60 percent of our body weight, and the amount of water that we need to drink depends on factors such as gender, our activity level, and where we live and work. The Institute of Medicine recommends that under moderate conditions and activity we should drink about two to three liters of water a day to regulate our body temperature, flush toxins out of our kidneys, carry nutrients to our body cells, and moisten tissues in our ears, nose, and throat.

In addition to drinking water, we also need water for other activities such as grooming, laundry, cooking, farming, industrial applications, and fire protection. According to the American Water Works Association (AWWA), in the United States we use about 70 gallons of water on average within our homes per capita per day. The activities that use lots of water include taking showers (~11.5 gallons), washing clothes (~15 gallons), toilets (~18.5 gallons), faucets (~11 gallons), leaks (~9.5 gallons), and dishwashers (~1 gallon). Moreover,

> The total global amount of water available is constant—we don't lose or gain water on the Earth.

according to the AWWA, the daily indoor per capita water use could be reduced to 45 gallons if we were to install more efficient water fixtures and check for leaks.

You also know that nearly two-thirds of the Earth's surface is covered with water, but most of this water cannot be consumed directly; it contains salt and other minerals that must be removed first. To better understand the water cycle, Figure 9.1 is provided. Radiation from the sun evaporates water; water vapors form into clouds; and (under favorable conditions), water vapor eventually turns into rain, sleet, hail, or snow and falls back on the land and into the ocean. On land, depending on the amount of precipitation, part of the water infiltrates the soil, part of it may be absorbed by vegetation, and part of it runs as streams or rivers and collects into natural *reservoirs* called lakes. *Surface water* refers to water in reservoirs, lakes, rivers, and streams. *Groundwater*, on the other hand, refers to water that has infiltrated the ground; surface and groundwater eventually return to the ocean, and the water cycle is completed. In addition to understanding the water cycle, it is also important to realize that the amount of water that is available to us is constant. Even though water can change phase from liquid to vapor or from liquid to ice or snow, the total amount remains constant—we don't lose or gain water on the Earth. For example, when you take a shower,

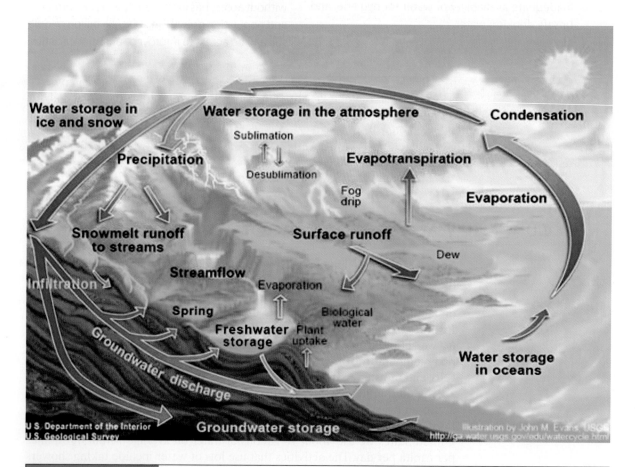

| FIGURE 9.1 | The water cycle |
| --- | --- |
| | (*Source:* USGS) |

the water that you use could end up elsewhere and be used for an entirely different purpose—after it has been filtered and treated, of course.

As we said earlier, everyone knows that we need water to sustain life, but what you may not realize is that water plays a significant role in many commercial, industrial, and agricultural applications. Water is used in all steam power-generating plants to produce electricity. Fuel is burned in a boiler to generate heat, which in turn is added to water to change its phase to steam; steam passes through turbine blades, turning the blades, which in effect runs the generator connected to the turbine, creating electricity. The low-pressure steam liquefies in a condenser and is pumped through the boiler again, closing a cycle. It is important to understand that in order to cool down the steam (i.e., to change its phase from steam to liquid water so that it can be pumped back to the boiler), large quantities of water are drawn into the condenser from nearby rivers or lakes. The water used by the steam power-generation plants to produce electricity is commonly classified as a ***thermoelectric power water supply***.

As we discussed in Chapter 7, electricity is also generated by water stored behind dams. The water is guided into water turbines located in hydroelectric power plants housed within the dam to generate electricity. Recall that the potential energy due to height of water stored behind the dam is converted to kinetic energy (moving energy) as the water flows through and consequently spins the turbine, which turns the electricity generator.

We also need water to grow fruits, vegetables, nuts, cotton, trees, and so on. Wells and irrigation channels provide water for farms and agricultural fields. Large quantities of water are also consumed in mining and industrial activities. For example, in hydraulic mining, water at high pressure is used to remove sediment and extract rocks and minerals. Water is also commonly used as a cooling or cleaning agent in a number of food-processing plants and other industrial applications. Water is used as a cutting tool as well. High-pressure water containing abrasive particles is used to cut marble or metals. Thus, water is not only transported to our homes for our domestic use, but it is also used in many industrial, agricultural, and mining applications. So you see,

Arina P Habich / Shutterstock.com

yexelA / Shutterstock.com

Filip Fuxa / Shutterstock.com

Four Oaks / Shutterstock.com

Joseph Sohm / Shutterstock.com

understanding how we consume water is very important. As is the case with any new areas you explore, the water resources field has its own terminology. Therefore, make sure you spend a little time now to familiarize yourself with the following terms so you can follow the discussions later.

**Storm Water** When it rains or when ice or snow melts over a surface, depending on the properties of the surface, the resulting *storm water* either is absorbed by the surface or flows over it. For example, when it rains in a garden, the water is absorbed by the ground until the soil becomes fully saturated. After the soil is fully saturated, it can absorb no more water; the water either accumulates or flows over the surface, depending on its slope. As you know, water also flows over impervious surfaces such as asphalt, concrete parking lots, sidewalks, paved roads, or building rooftops.

**Surface Water** Water that is available on the surface of the Earth is called *surface water*. For example, creeks, rivers, reservoirs, and lakes are all classified as surface water. A *tributary* is a smaller stream or river that flows into a larger stream or river to eventually create a big river or body of water. A *reservoir* represents man-made or natural bodies of water (such as a lake) that are used to store and control the use and flow of water. In the United States, reservoirs and lakes cover approximately 40 million acres (1 acre = 43,560 square feet).

An *estuary* represents an area where fresh water and salt water merge—in other words, places where rivers meet the ocean or sea. *Wetlands* refer to areas that are covered by water during all or part of a year. The water in wetlands could be present near or at the surface of the soil.

**Aqueduct** A conduit that transports water from one location to another location is called an *aqueduct*. An example of this is a conduit that delivers water from a lake at a high elevation (e.g., in a mountainous area) to a dry region at a lower altitude. Typically, the water movement is gravity driven from a high elevation to a low elevation. Be careful not to confuse an aqueduct with a *levee* or a *dike* which represent man-made or natural ridges or embankments that run parallel to the edge of a river, stream, or lake to prevent flooding. *Flooding* occurs when a stream, river, lake, sea, or ocean overflows its banks and covers areas that are inhabited by people. The *flood stage* denotes the elevation at which a body of water overflows its banks. Moreover, *a flood plain* represents the flat land near a body of water that is affected during a flood.

**Groundwater** Water (from rain, melting ice, and snow) that seeps into the ground and creates wells, springs, or aquifers is *groundwater*. An *aquifer* is basically a geological storage tank that contains large quantities of water which we can tap.

Geological formations act as the surfaces and walls of the storage tank. On the other hand, a water *well* represents a hole that is excavated to reach the underground water.

Poul Riishede / Shutterstock.com

**Water Anomalies** In recent years, much has been said about the melting of glaciers due to *global warming*. A **glacier** represents a relatively large mass of ice that is created by the compaction of snow due to its own weight. An *ice cap* refers to a mass of ice that covers a land area of less than 50,000 square kilometers, whereas the term *ice sheet* is used to refer to areas of ice-covered land greater than 50,000 square kilometers. Another term that you may be familiar with is permafrost. *Permafrost* refers to the soil that has had a temperature below the freezing point of water (0°C) for many years. Finally, **geysers** represent superheated water eruptions due to the high pressure and temperature of the water that lies underneath the Earth.

**Properties of Water** As we mentioned earlier, most of the global water available cannot be consumed directly because it contains salt and other minerals that must be removed first. **Fresh water** refers to water that contains less than one thousand milligrams of dissolved solids per liter (1000 mg/l). Water also can contain substances that would raise its pH value above 7.0. The pH value of water shows the relative acidity of water, with a pH value of 7.0 indicating neutral; below 7.0 indicating an acidity level (the lower the value, the more acidity); and the higher pH value indicating a basic solution (see Figure 9.2). Water with pH values greater than 7.0 is usually harmful for agricultural use.

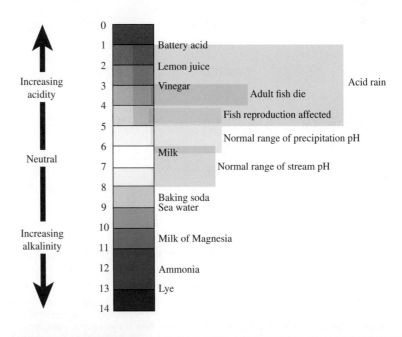

| **FIGURE 9.2** | Acidity and alkalinity |

*Source:* Based on Environment Canada, https://www.ec.gc.ca/eau-water/default.asp?lang=En&n=FDF30C16-1

Another property is **hardness**, which refers to a measure of concentration of calcium and magnesium in water. For example, from your everyday experience, you know that when you shower using hard water, you use more soap; soap does not foam as easily in hard water. **Leaching** describes a process where certain materials—nutrients, salts, pesticides, or contaminants—present in the soil are dissolved in and carried away by water.

## Global Water Distribution

Now that you have some basic understanding of water resource terms, let us consider the distribution of the total available global water. An estimate of the amount of water available in the world is given in Table 9.1. After you examine Table 9.1, it should become clear that most of the water resides in the oceans. Data from Shiklomanov (1993) indicates that 96.5 percent of water on the Earth resides in oceans, seas, and bays. By some estimates, this is equivalent to 321,000,000 cubic miles of water (1,338,000,000 cubic kilometers). The total volume of water available on the Earth is estimated at about 1.4 billion kilometers

> Groundwater is the primary source of our water supply.

| TABLE 9.1 | The Water Budget on the Earth. | | | |
|---|---|---|---|---|
| **Water Source** | **Water Volume(m³)** | **Water Volume(km³)** | **Percent of Fresh Water** | **Percent of Total Water** |
| Oceans, seas, & bays | 321,000,000 | 1,338,000,000 | – | 96.5 |
| Ice caps, glaciers, and permanent snow | 5,773,000 | 24,064,000 | 68.6 | 1.74 |
| Groundwater | 5,614,000 | 23,400,000 | – | 1.7 |
|   Fresh water | 2,526,000 | 10,530,000 | 30.1 | 0.76 |
|   Saline | 3,088,000 | 12,870,000 | – | 0.93 |
| Soil moisture | 3,959 | 16,500 | 0.05 | 0.001 |
| Ground ice and permafrost | 71,970 | 300,000 | 0.86 | 0.022 |
| Lakes | 42,320 | 176,400 | – | 0.013 |
|   Fresh water | 21,830 | 91,000 | 0.26 | 0.007 |
|   Saline | 20,490 | 85,400 | – | 0.007 |
| Atmosphere | 3,095 | 12,900 | 0.04 | 0.001 |
| Swamp water | 2,752 | 11,470 | 0.03 | 0.0008 |
| Rivers | 509 | 2,120 | 0.006 | 0.0002 |
| Biological water | 269 | 1,120 | 0.003 | 0.0001 |

*Source:* Based on Igor Shiklomanov's chapter "World fresh water resources" in Peter H. Gleick (editor), 1993, *Water in Crisis: A Guide to the World's Fresh Water Resources* (Oxford University Press, New York).

cubed ($km^3$). The volume of fresh water is 35 million $km^3$—about 2.5 percent of the total volume—with 24 million $km^3$ or 68.6 percent in the form of ice and permanent snow in the Antarctic and Arctic regions. Ice caps, glaciers, and permanent snow represent 1.74 percent of the total global water budget, which is equivalent to about 5,773,000 cubic miles of water (24,064,000 cubic kilometers). Groundwater represents 1.7 percent of the total, which is equivalent to 5,614,000 cubic miles of water (23,400,000 cubic kilometers).

It is also important to know that even though glaciers and ice caps represent nearly 70 percent of fresh water, because of their location in Greenland and Antarctica, the water is not readily available for human consumption. Because of the lack of easy access to water in glaciers and ice caps, we rely more on the groundwater and surface water near our cities and towns. Groundwater represents over 90 percent of the world's readily available fresh water source, freshwater lakes contain an additional estimated 91,000 $km^3$, and rivers contain 2,120 $km^3$. To get a feel for the order of magnitude of these numbers where the quantity of water is represented by one kilometer cubed, imagine a volume that is defined as 1,000 meters in length (equal to the total length of about 550 people of average height lying down next to each other), 1,000 meters in width, and 1,000 meters in depth.

---

**EXAMPLE 9.1**

irisphoto1 / Shutterstock.com

As mentioned in Chapter 7, the Hoover Dam is one of the Bureau of Reclamations' multipurpose projects on the Colorado River. These projects control floods; they store water for irrigation, municipal, and industrial use; and they provide hydroelectric power, recreation, and fish and wildlife habitat. Lake Mead lies behind the Hoover Dam and contains 28,537,000 acre-feet of water. Let us now express this water volume in gallons and meters cubed ($m^3$).

When following this example, note that an acre-foot is the amount of water required to cover 1 acre to a depth of 1 ft, one acre is equal to 43,560 $ft^2$, and 1 $ft^3$ = 7.48 gallons.

28,537,000 acre-foot of water

$$= (28,537,000 \text{ acre})(1 \text{ ft of water})\left(\frac{43,560 \text{ ft}^2}{1 \text{ acre}}\right)\left(\frac{7.48 \text{ gallons}}{1 \text{ ft}^3}\right)$$

$$= 9.3 \times 10^{12} \text{ gallons of water}$$

28,537,000 acre-foot of water

$$= (28,537,000 \text{ acre})(1 \text{ ft of water})\left(\frac{43,560 \text{ ft}^2}{1 \text{ acre}}\right)\left(\frac{1 \text{ m}}{3.28 \text{ ft}}\right)^3$$

$$= 35.227 \times 10^9 \text{ m}^3 \text{ of water}$$

Think about the magnitude of these numbers: 35 billion cubic meters or 9,300 billion gallons!

# Before You Go On

Answer the following questions to test your understanding of the preceding section:

1. In your own words, explain the water cycle.

2. What do we mean by fresh water?

3. What percentage of the total global water budget is considered fresh water?

4. What is the difference between an aqueduct and a levee or a dike?

5. What do we mean by the term *estuary*?

*Vocabulary—State the meaning of the following terms:*

Surface water

Tributary

Groundwater

Aquifer

Storm water

Glacier

Permafrost

# LO² 9.2 Personal Water Consumption

We now turn our attention to water consumption in our homes due to our daily activities. In the United States, most of the drinking water comes from surface water or groundwater. The Safe Drinking Water Act (SDWA) defines a *public water system (PWS)* as "one that serves piped water to at least 25 persons or to 15 service connections for at least 60 days each year." According to the EPA, there are over 160,000 public water systems in the United States.

> On average, inside our homes we use about 70 gallons of water per capita per day in the United States.

*Community water systems (CWS)* are public water systems that serve people year-round in their homes. Most people in the United States get their water from a community water system. The EPA also regulates other kinds of public water systems, such as those at schools, campgrounds, factories, and restaurants. It is important to note that private water supplies, such as wells that serve one or a few homes, are not regulated by the EPA.

Our daily indoor water consumption per capita is shown in Figure 9.3. As shown, we use the largest amount of water in flushing toilets, followed by washing clothes and taking showers. Figure 9.4 shows our average daily usage

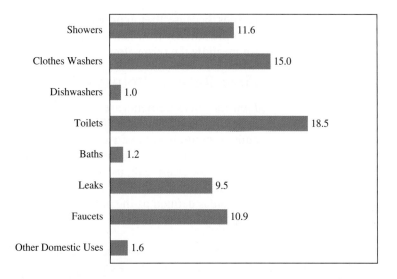

**FIGURE 9.3**      Daily indoor gallons per capita water use in the United States
*Source:* Based on American Water Works Association

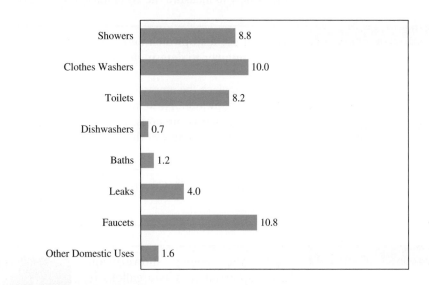

**FIGURE 9.4**      Daily indoor gallons per capita water use for households taking advantage of conservation measures
*Source:* Based on American Water Works Association

if we were to take conservative measures, such as installing low-flow showers and water-efficient toilets in our homes. Before 1992, typical showerheads had flow rates of 5.5 gallons per minute (gpm). Today by government mandate, the new showerhead flow rates cannot exceed more than 2.5 gpm. Today's low-flow showerheads are classified as either *aerating* or *laminar flow*. To reduce the water flow rate, an aerating showerhead mixes air with water and creates a mist, whereas a laminar-flow showerhead creates laminar (no turbulence) streams of water. The aerating showerheads are not commonly used in humid regions, because they tend to increase the humidity level at homes more so than the laminar-flow showerheads. Next, it is important to understand what we mean by the terms flow rate and volume flow rate.

## Flow Rate or Volume Flow Rate

*Flow rate* measurements are necessary to determine the amount of water used in our homes during a specific period of time. City engineers need to know the daily or monthly volumetric water consumption rates to provide an adequate supply of water to our homes. Water suppliers also need to know how much water—how many gallons or cubic meters of water—are used every month by each home so that they can correctly charge their customers. The *volume flow rate* is defined as the volume of the fluid (e.g., water) that flows through a pipe per unit time:

$$\text{volume flow rate} = \frac{\text{volume of fluid}}{\text{time}}$$

Some of the more common units for the volume flow rate include liters per second (l/s), cubic meters per second ($m^3$/s), cubic meters per day ($m^3$/day), gallons per minute (gpm), or gallons per day (gpd). Let us now look at an example that shows how to measure the volumetric flow rate of water coming out of a pipe.

**EXAMPLE 9.2**

You can use a stopwatch and a one-gallon (or a one-liter) empty milk or water container to determine the volumetric flow rate of water coming out of a faucet.

As you place the empty container under the running water, immediately start the stopwatch. Now imagine that it took 25 seconds to fill the one-gallon container completely. To calculate the flow rate of the water coming out of the faucet, you will then perform the following calculations:

$$\overbrace{\frac{1 \text{ gallon}}{25 \text{ s}} = 0.04 \frac{\text{gallons}}{\text{s}}}^{\text{step 1}} = \overbrace{\left(0.04 \frac{\text{gallons}}{\text{s}}\right)\left(\frac{60 \text{ s}}{1 \text{ minute}}\right)}^{\text{step 2}} = 2.4 \frac{\text{gallons}}{\text{minute}} = 2.4 \text{ gpm}$$

**EXAMPLE 9.3**

New water-saving showerheads deliver approximately two gallons of water per minute. Assuming that you shower once a day for five minutes, what is the total amount of water you consume for this activity annually?

Sutichak / Shutterstock.com

$$\text{amount of water consumed showering} = 2\left(\frac{\text{gallons}}{\text{minute}}\right)\left(\frac{5 \text{ minutes}}{\text{day}}\right)\left(\frac{365 \text{ days}}{\text{year}}\right)$$

$$= 3,650\left(\frac{\text{gallons}}{\text{year}}\right)$$

The flow rate of water coming out of faucets is influenced by the water pressure inside the pipe. Therefore, as a well-educated citizen, you also need to understand what we mean by a water pressure of 60 or 80 psi, which are commonly experienced in water lines at homes. *Pressure* provides a measure of intensity of a force acting over an area. It is defined as the ratio of a force over a contact surface area as

$$\text{pressure} = \frac{\text{force}}{\text{area}} \qquad \textbf{9.1}$$

In U.S. Customary units, pressure is usually expressed in **p**ounds per **s**quare **i**nch (psi), which represents the pressure created by a one-pound force over an area of one inch squared (in.$^2$). Therefore, 60 psi represents the pressure for a situation where 60 pounds is acting over an area of one square inch. So, if you had water with a pressure of 60 psi coming out of a pipe with a cross-sectional area of one square inch and you wanted to block the water flow with a piece of wood, it would take a force of 60 pounds to prevent the water from coming out.

In the International System (SI) of units, pressure units are expressed in pascal, where one pascal is the pressure created by one newton force acting over a surface area of 1 meter squared (m$^2$).

---

**EXAMPLE 9.4**

To better understand what the magnitude of a pressure represents, consider the situations shown in Figure 9.5.

Let us first look at the situation depicted in Figure 9.5(a), in which we lay a solid brick that weighs 6.4 pounds (lbf) or 28 newtons (N) and is shaped as a rectangular prism with dimensions of 8½ × 4 × 2½ inches (in.) or 21.6 × 10.2 × 6.4 centimeters (cm) flat on its face. Using Equation (9.1) for this orientation, the pressure at the contact surface is

$$\text{pressure} = \frac{\text{force}}{\text{area}} = \frac{6.4 \text{ lbf}}{(8.5 \text{ in.})(4 \text{ in.})} = 0.19 \, \frac{\text{lbf}}{\text{in.}^2} = 0.19 \text{ psi}$$

In SI units, this is

$$\text{pressure} = \frac{\text{force}}{\text{area}} = \frac{28 \text{ N}}{(0.216 \text{ m})(0.102 \text{ m})} = 1{,}271 \, \frac{\text{N}}{\text{m}^2} = 1{,}271 \text{ Pa}$$

Note again that one pound per square inch is called one psi, and in SI units, one newton (N) per square meter is called one pascal (1 N/1 m$^2$ = 1 Pa).

Now if we were to lay the brick on its end as depicted in Figure 9.5(b), the pressure due to the weight of the brick becomes

$$\text{pressure} = \frac{\text{force}}{\text{area}} = \frac{6.4 \text{ lbf}}{(4 \text{ in.})(2.5 \text{ in.})} = 0.64 \, \frac{\text{lbf}}{\text{in.}^2} = 0.64 \text{ psi}$$

In SI units, this is

$$\text{pressure} = \frac{\text{force}}{\text{area}} = \frac{28 \text{ N}}{(0.102 \text{ m})(0.064 \text{ m})} = 4{,}289 \ \frac{\text{N}}{\text{m}^2} = 4{,}289 \text{ Pa}$$

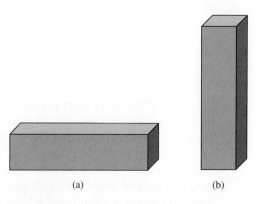

(a)                                    (b)

**FIGURE 9.5**    An experiment demonstrating the concept of pressure: (a) a sold brick resting on its face and (b) a solid brick resting on its end. In position (b), the block creates a higher pressure on the surface

It is important to note here that the weight of the brick is 6.4 pound force (lbf) or 28 newtons (N) regardless of how it is laid. But the pressure that is created at the contact surface depends on the magnitude of the contact surface area. The smaller the contact area, the larger the pressure created by the same force. You already know this from your everyday experiences. Which situation would create more pain, pushing (with the same force) on someone's arm with a finger or a thumbtack?

**EXAMPLE 9.5**

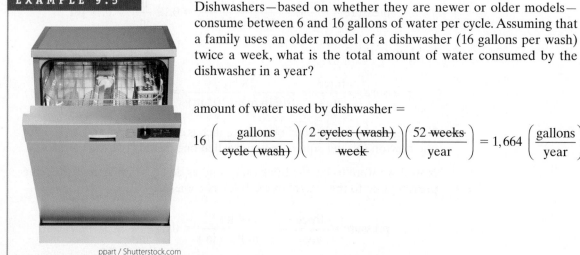

Dishwashers—based on whether they are newer or older models— consume between 6 and 16 gallons of water per cycle. Assuming that a family uses an older model of a dishwasher (16 gallons per wash) twice a week, what is the total amount of water consumed by the dishwasher in a year?

amount of water used by dishwasher =

$$16 \left( \frac{\text{gallons}}{\text{cycle (wash)}} \right) \left( \frac{2 \text{ cycles (wash)}}{\text{week}} \right) \left( \frac{52 \text{ weeks}}{\text{year}} \right) = 1{,}664 \left( \frac{\text{gallons}}{\text{year}} \right)$$

**EXAMPLE 9.6**

Newer clothes washers consume approximately 25 gallons of water per average load of laundry. Assuming that a family does 6 loads of laundry a week, what is the total amount of water consumed annually by this activity?

Shell114 / Shutterstock.com

amount of water consumed by clothes washer =

$$25 \left( \frac{\text{gallons}}{\text{load}} \right) \left( \frac{6 \ \text{loads}}{\text{week}} \right) \left( \frac{52 \ \text{weeks}}{\text{year}} \right) = 7{,}800 \left( \frac{\text{gallons}}{\text{year}} \right)$$

**EXAMPLE 9.7**

An in-bay car-wash system consumes as much as 72 gallons of water per wash. Assuming that you wash your car once every two weeks, what is the total amount of water consumed annually by this activity?

Sigur / Shutterstock.com

amount of water consumed by car wash = $72 \left( \frac{\text{gallons}}{\text{wash}} \right) \left( \frac{1 \ \text{wash}}{2 \ \text{weeks}} \right) \left( \frac{52 \ \text{weeks}}{\text{year}} \right)$

$$= 1{,}872 \left( \frac{\text{gallons}}{\text{year}} \right)$$

## Before You Go On

Answer the following questions to test your understanding of the preceding section:

1.   What is a public water system?

2.   Which one of your daily activities consumes the largest amount of water?

3.   In your own words, explain what is meant by a water pressure of 60 psi.

4.   How would you measure the volume flow rate of water coming out of a showerhead?

*Vocabulary—State the meaning of the following terms:*

Flow rate

gpm

psi

pascal

## LO³   9.3   Water Consumption in Agriculture, Commercial, and Industrial Sectors

Let us begin this section by considering the United Nations reported data as it pertains to water, agriculture, and food security as shown in Figure 9.6. If you study the figure carefully, you should realize that it takes a lot of water to produce the food that we consume on a daily basis. For example, note that it takes 1,000 to 3,000 liters (~265 to 800 gallons) of water to produce just one kilogram (2.2 pounds) of rice or that it takes 13,000 to 15,000 liters (~3,400 to 4,000 gallons) to produce only one kilogram (2.2 pounds) of grain-fed beef.

In addition to food, we have many other needs that require plenty of water. The U.S. Geological Survey (USGS) has been estimating total water consumption in the United States since 1950. They report their findings every five years for both groundwater and surface water sources. In addition, for bookkeeping purposes, the USGS groups major water-consuming activities into broad categories such as *public, domestic, irrigation, livestock, aquaculture, industrial, mining,* and *thermoelectric* power generation, and reports the water-consumption data for each category.

The total water withdrawal levels and population trends for the United States during the 1950 to 2010 period are shown in Figure 9.7. Note that the units of water consumption are expressed in billions of gallons of water per day. During this period, as you can see, the water consumption rate does not increase proportionally as does the population trend. In other words, the rate of population increase is not followed by an increase in water consumption rates. In fact, there is a 13 percent *decline* in the water consumption in 2010

**Water, Agriculture and Food Security**

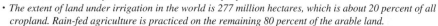

- *The daily drinking water requirement per person is 2 to 4 liters, but it takes 2,000 to 5,000 liters of water to produce one person's daily food.*
- *It takes 1,000 to 3,000 liters of water to produce just one kilogram of rice and 13,000 to 15,000 liters to produce one kilogram of grain-fed beef.*
- *In 2007, the estimated number of undernourished people worldwide was 923 million.*
- *Over the period to 2050, the world's water will have to support the agricultural systems that will feed and create livelihoods for an additional 2.7 billion people.*
- *The extent of land under irrigation in the world is 277 million hectares, which is about 20 percent of all cropland. Rain-fed agriculture is practiced on the remaining 80 percent of the arable land.*
- *The Intergovernmental Panel on Climate Change predicts yields from rain-dependent agriculture could be down by 50 percent by 2020.*
- *Due to climate change, Himalayan snow and ice, which provide vast amounts of water for agriculture in Asia, are expected to decline by 20 percent by 2030.*
- *Irrigation increases the yields of most crops by 100 to 400 percent, and irrigated agriculture currently contributes to 40 percent of the world's food production.*

**FIGURE 9.6**   The U.N. data as it pertains to water, agriculture, and food security
*Source:* United Nations, Based on UN Water Statistics

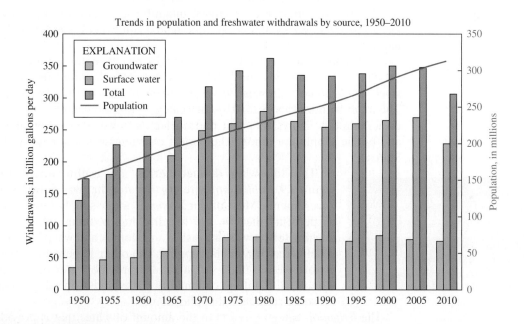

**FIGURE 9.7**   The water withdrawal trends in the United States
*Source:* Based on Maupin, M.A., Kenny, J.F., Hutson, S.S., Lovelace, J.K., Barber, N.L., and Linsey, K.S., 2014 , *Estimated use of water in the United States in 2010: U.S. Geological Survey Circular*

Total Water Use, 2010

**EXPLANATION**

Water withdrawals,
in million gallons
per day

0 to 2,000
2,001 to 5,000
5,001 to 10,000
10,001 to 20,000
20,001 to 38,000

USGS

---

**FIGURE 9.8**    The water withdrawal in each state in 2010
*Source:* Maupin, M.A., Kenny, J.F., Hutson, S.S., Lovelace, J.K., Barber, N.L.,
and Linsey, K.S., 2014, *Estimated use of water in the United States in 2010:
U.S. Geological Survey Circular*

when compared to 2005. This decline in consumption in the United States may be attributed to conservation measures that use water more effectively for different activities.

The water consumption by each state is shown in Figure 9.8. As you can see, California withdraws the largest quantity of water. The states with the largest percentage of water consumption are California, Texas, Idaho, Florida, and Illinois.

As we mentioned previously, the USGS places major water-consuming activities into categories and labels them as *public, domestic, irrigation, livestock, aquaculture, industrial, mining,* and *thermoelectric* power generation. The water withdrawals in 2010 by each category are shown in Figure 9.9. When examining this figure, note that thermoelectric power generation, irrigation, and public supply are among the largest consumers of water in the United States.

> It is customary to group major activities that consume water into public, domestic, irrigation, livestock, aquaculture, industrial, mining, and thermoelectric power.

The following defined terms are useful when studying Figure 9.9. *Public supply* refers to water that was drawn by public governments. Most of our *domestic water* supply (approximately 85 percent) that we use for drinking, cooking, taking showers, flushing toilets, washing dishes and clothes, and watering our lawns is delivered by a public supplier. The other 15 percent comes from private wells. Public suppliers also provide water for businesses, schools, firefighting, community parks and swimming pools, and for some commercial applications.

The *irrigation* category refers to the amount of water that is provided by man-made systems for agricultural purposes. Areas that lack sufficient rainfall, such as California, require man-made irrigation systems to grow crops and fruits. In Figure 9.9, the amount of water used for agricultural activities such

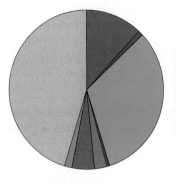

2010 withdrawals by category,
in million gallons per day

| | |
|---|---:|
| Public supply | 42,000 |
| Self-supplied domestic | 3,600 |
| Irrigation | 115,000 |
| Livestock | 2,000 |
| Aquaculture | 9,420 |
| Self-supplied industrial | 15,900 |
| Mining | 5,320 |
| Thermoelectric power | 161,000 |

Values do not sum to 355,000
Mgal/d because of independent
rounding

| | |
|---|---|
| **FIGURE 9.9** | Total water withdrawals in the United States by category (2010 data)<br>***Source:*** Maupin, M.A., Kenny, J.F., Hutson, S.S., Lovelace, J.K., Barber, N.L., and Linsey, K.S., 2014, *Estimated use of water in the United States in 2010: U.S. Geological Survey Circular* |

as dairy operations, feed lots, and providing drinking water for livestock is represented by the *livestock* category. ***Aquaculture*** refers to the farming of fish, shrimp, and other animals that live in water. The farming of plants and algae that live in water is also grouped into this category.

The water that is used for industrial purposes such as making paper, chemicals, or steel is classified as *industrial*. Approximately 80 percent of the water used in this category comes from wells and rivers, and the remaining portion is publically supplied. The water used for the excavation of rocks and minerals is classified as *mining water*. Earlier, we explained that the water used in power plants to generate electricity is classified as a *thermoelectric power* supply.

| | |
|---|---|
| **EXAMPLE 9.8** | According to the USGS, in 2010, about 355 billion gallons of water per day were consumed to address various needs in the United States. Let us now visualize how much water this volume represents in terms of how many square miles of land can be covered to a depth of 5 feet each year.<br><br>When following this example, again note that $1 \text{ ft}^3 = 7.48$ gallons and $1$ mile $= 5,280$ ft.<br><br>$$\text{volume} = \left(\frac{355 \times 10^9 \text{ gallons}}{\text{day}}\right)\left(\frac{1 \text{ ft}^3}{7.48 \text{ gallons}}\right)\left(\frac{365 \text{ days}}{\text{year}}\right)$$<br>$$= (x \text{ miles})\left(\frac{5,280 \text{ ft}}{1 \text{ mile}}\right)(x \text{ miles})\left(\frac{5,280 \text{ ft}}{1 \text{ mile}}\right)(5 \text{ ft})$$<br><br>And solving for $x$ we get $x = 352.5$ miles.<br>Think about the amount of water consumed in the United States in 2010 for all activities. The water will cover a square area defined by 352.5 miles by 352.5 miles to a depth of 5 feet! |

**EXAMPLE 9.9**

Referring to Example 9.8 to better visualize how much water we consume in the United States each year, how many hours do you have to drive your car at the speed of 60 miles per hour (mph) to go around the area that is covered by water to a depth of 5 feet? Can you do it without refilling the gas tank?

Speed is defined as distance travelled over time. Then, the time that it takes to cover only one side of the area is given by

$$\text{speed} = \frac{\text{distance}}{\text{time}}$$

$$\frac{60 \text{ miles}}{\text{hour}} = \frac{352.5 \text{ miles}}{\text{time (in hours)}}$$

Solving for variable *time*, we get *time* = 5.875 hours. To go around the area, we need to multiply this number by 4 since there are four sides to a square, which results in 23.5 hours. Therefore, it would take nearly a complete day to go around the given area that is filled with water to a depth of 5 feet.

As you can tell by now, you would need to refill the gas tank many times and take some rest stops along the way!

## Before You Go On

Answer the following questions to test your understanding of the preceding section:

1.  What are the major categories of water consumption? In your own words, explain at least three categories.

2.  Which of the categories discussed in this section consumes the largest amount of water?

3.  Which states are among the top water-consuming states? Name at least three states.

4.  In your own words, explain what is meant by the aquaculture category.

5.  In your own words, explain what is meant by the public supply category.

*Vocabulary—State the meaning of the following terms:*

Thermoelectric power

Aquaculture

# LO⁴  9.4  Drinking Water Standards in the United States

Ilya Andriyanov / Shutterstock.com

In Section 9.2, we explained where our tap water comes from. In this section, we discuss the drinking water standards. The U.S. *Environmental Protection Agency* (EPA) sets the standards for the maximum level of contaminants that can be in our drinking water and still be considered safe to drink. The EPA, along with state and local departments of health and environment, monitors and enforces drinking water standards. The *Safe Drinking Water Act* (SDWA) was passed by the United States Congress in 1974; the highlights of the SDWA are shown in Figure 9.10.

As you would expect, human activities and naturally occurring microorganisms contribute to the contaminant level in our water supply. For example, in agriculture, fertilizers and pesticides or animal waste in large cattle, pig, or poultry farms contribute to pollution. Other human activities such as mining, construction, manufacturing goods, dry cleaning, landfills, or waste-water-treatment plants also contribute to pollution. Basically, the EPA sets two standards for the level of water contaminants:

> The EPA sets two standards for the level of water contaminants: the maximum contaminant level goal and the maximum contaminant level. It monitors the level of contaminants such as asbestos, cyanide, mercury, lead, and nitrate in our water supply.

1. *Maximum contaminant level goal (MCLG)*
2. *Maximum contaminant level (MCL)*

The MCLG represents the maximum level of a given contaminant in the water that causes no known harmful health effects. On the other hand, the MCL, which may represent slightly higher

---

**Highlights of the Safe Drinking Water Act https://www.epa.gov/sdwa**

- Authorizes EPA to set enforceable health standards for contaminants in drinking water
- Requires public notification of water systems violations and annual reports (Consumer Confidence Reports) to customers on contaminants found in their drinking water—https://www.epa.gov/ccr
- Establishes a federal-state partnership for regulation enforcement
- Includes provisions specifically designed to protect underground sources of drinking water—https://www.epa.gov/sourcewaterprotection
- Requires disinfection of surface water supplies, except those with pristine, protected sources
- Establishes a multi-billion-dollar state revolving loan fund for water system upgrades—https://www.epa.gov/drinkingwatersrf
- Requires an assessment of the vulnerability of all drinking water sources to contamination—www.epa.gov/safewater/protect

---

**FIGURE 9.10**   The highlights of the Safe Drinking Water Act

*Source:* Environmental Protection Agency, Drinking Water: Past, Present and Future, EPA-816-F-00-002

levels of contaminants in the water, is the level of contaminants that are legally enforceable. The EPA attempts to set the MCL close to the MCLG, but this goal may not be attainable because of economic or technical reasons. Examples of drinking water standards are shown in Table 9.2.

**Viral and Bacterial Contamination** Human and animal wastes can contribute to microbial contamination of water supplies that can lead to outbreaks of waterborne diseases. Sometimes at water treatment facilities, some

**TABLE 9.2**    **Examples of Drinking Water Standards**

| Contaminant | MCGL | MCL | Source of Contaminant by Industries |
|---|---|---|---|
| Antimony | 6 ppb | 6 ppb | copper smelting, refining, porcelain plumbing fixtures, petroleum refining plastics, resins, storage batteries |
| Asbestos | 7 M.L. (million fibers per liter) | 7 M.L. | asbestos products, chlorine, asphalt felts and coating, auto parts, petroleum refining, plastic pipes |
| Barium | 2 ppm | 2 ppm | copper smelting, car parts, inorganic pigments, gray ductile iron, steel works, furnaces, paper mills |
| Beryllium | 4 ppb | 4 ppb | copper rolling and drawing, nonferrous metal smelting, aluminum foundries, blast furnaces, petroleum refining |
| Cadmium | 5 ppb | 5 ppb | zinc and lead smelting, copper smelting, inorganic pigments |
| Chromium | 0.1 ppm | 0.1 ppm | pulp mills, inorganic pigments, copper smelting, steel works |
| Copper | 1.3 ppm | 1.3 ppm | primary copper smelting, plastic materials, poultry slaughtering, prepared feeds |
| Cyanide | 0.2 ppm | 0.2 ppm | metal heat treating, plating, and polishing |
| Lead | zero | 15 ppb | lead smelting, steel works and blast furnaces, storage batteries, china plumbing fixtures |
| Mercury | 2 ppb | 2 ppb | electric lamps, paper mills |
| Nickel | 0.1 ppm | 0.1 ppm | petroleum refining, gray iron foundries, primary copper, blast furnaces, steel |
| Nitrate | 10 ppm | 10 ppm | nitrogenous fertilizer, fertilizing mixing, paper mills, canned foods, phosphate fertilizers |
| Nitrite | 1 ppm | 1 ppm | fertilizers |
| Selenium | 0.05 ppm | 0.05 ppm | metal coatings, petroleum refining |
| Thallium | 0.5 ppb | 2 ppb | primary copper smelting, petroleum refining, steel works, blast furnaces |

*Source:* Environmental Protection Agency, Table of Regulated Drinking Water Contaminants

Boil Water Notices for
Microbial Contaminants

When microorganisms such as those that indicate fecal contamination are found in drinking water, water suppliers are required to issue "Boil Water Notices." Boiling water for one minute kills the microorganisms that cause disease. Therefore, these notices serve as a precaution to the public. www.epa.gov/safewater/fag/emerg.html

**FIGURE 9.11**    The EPA's Boil Water Notices
*Source:* Environmental Protection Agency, https://www.epa.gov/ground-water-and-drinking-water/emergency-disinfection-drinking-water

disease-causing microorganisms—commonly known as ***pathogens***—are not captured by filtration or killed by disinfection means. As a result, when pathogens find their way into our drinking water supply, they can cause serious illnesses among the elderly, infants, and those with compromised immune systems. When detected in water supplies, local public health departments issue boiling water advisories, as shown in Figure 9.11.

**Contamination from Fertilizers** Excessive levels of nitrate (found in fertilizers) in drinking water can also cause illness in the elderly and infants. In fact, the *"blue baby syndrome"* is attributed to high levels of nitrate in tap water that is used in making baby formula and foods. Water that is contaminated with nitrate should not be boiled, because when the water boils, some of it evaporates, resulting in less water with the same amount of nitrate. This results in a more harmful increase in the concentration level.

**Metal Contamination** Traces of metals can also be found in our drinking water supply. For example, lead, which has no nutritional value, can find its way into our tap water from lead pipes in older buildings or soldering used to connect pipes. Older homes with lead piping systems can have lead concentration levels above normal. Lead dissolves into the water running through old pipes; its concentration depends on factors such as pH level, water temperature, and water hardness. According to the EPA " . . . new homes are also at risk: pipes legally considered to be 'lead-free' may contain up to eight percent lead. These pipes can leach significant amounts of lead into the water for the first several months after their installation." As is the case with water with abnormal levels of nitrate, water that is contaminated with lead also should not be boiled, because as explained previously, boiling leads to an increase in the concentration level. Lead can build up in bones, and concentration levels above 0.015 milligrams per liter could damage nervous and reproductive systems, resulting in a setback in mental development and behavioral problems in children. In adults, long-term exposure could result in kidney problems or high blood pressure. Often, lead poisoning manifests itself as a blue line in the gums.

## LO⁵   9.5   Global Water Quality Issues

John Wollwerth / Shutterstock.com

The United Nations (UN) estimates nearly 1.5 billion people depend on groundwater for their drinking water supply. The United Nations also estimates that approximately 700 km³ of groundwater is withdrawn annually around the world. Rivers with an estimated 263 international river basins constitute approximately 45 percent of the Earth's surface water. Canada has the most freshwater lakes in the world (about 50 percent). As a good global citizen, there are additional important facts that you need to know about the global water supply. According to the UN Vital Water Graphics Report (*excerpts from this report are given here*):

1. Fresh water resources are unevenly distributed, with much of the water located far from human populations. Many of the world's largest river basins run through thinly populated regions. There are an estimated 263 international rivers, covering 45.3 percent of the land surface of the Earth (excluding Antarctica).

2. Groundwater represents about 90 percent of the world's readily available fresh water resources, and some 1.5 billion people depend upon groundwater for their drinking water.

3. Agricultural water use accounts for about 75 percent of the total global consumption—mainly through crop irrigation—while industrial use accounts for about 20 percent. The remaining 5 percent is used for domestic purposes.

4. It is estimated that two out of every three people will live in water-stressed areas by the year 2025. In Africa alone, it is estimated that 25 countries will be experiencing water stress (below 1,700 m³ per capita per year) by 2025. Today, 450 million people in 29 countries suffer from water shortages.

5. Clean water supplies and sanitation remain major problems in many parts of the world, with 20 percent of the global population lacking access to safe drinking water. Around 1.1 billion people globally do not have access to improved water supply sources, while 2.4 billion people do not have access to any type of improved sanitation facilities. About 2 million people die every year due to waterborne diseases; most of them are children less than five years old. A wide variety of human activities also affect the coastal and marine environment. Population pressures, increasing demand for space and resources, and poor economic performance all undermine the sustainable use of our oceans and coastal areas.

6. Serious problems affecting the quality and use of these ecosystems include the alteration and destruction of habitats and ecosystems. For example, estimates show that almost 50 percent of the world's coasts are threatened by development-related activities. In marine fisheries, most areas are producing significantly lower yields than in the past. Substantial

According to the World Health Organization, each year two million deaths are attributed to unsafe water, poor sanitation, and poor hygiene.

increases are never again likely to be recorded for global fish catches. In contrast, inland and marine aquaculture production is increasing and now contributes 30 percent of the total global fish yield. The impact of climate change is projected to include a significant rise in the level of the world's oceans. This will cause some low-lying coastal areas to become completely submerged and will increase human vulnerability in other areas. Small Island Developing States (SIDS), which are highly dependent upon marine resources, are especially vulnerable, due to both the effects of the sea-level rising and to changes in marine ecosystems.

The World Health Organization (WHO) is the authority on global health matters, including water-quality-related health issues. It is responsible for setting standards and for monitoring and providing technical support. According to the WHO, nearly one billion people lack access to clean drinking water; as a result, each year millions of deaths are attributed to unsafe water. The mission of the water sanitation and health division of the World Health Organization (WHO) is

*"the reduction of water and waste related disease and the optimization of the health benefits of sustainable water and waste management."*

Here are some significant facts about water-related diseases as reported recently by the WHO.

- Two million annual deaths are attributed to unsafe water, poor sanitation, and lack of hygiene.
- More than 50 countries still report cholera to the WHO. The *cholera* bacterium, which is commonly found in water, causes an illness that infects the intestine with the bacterium. An estimated five million cases and over 100,000 deaths are reported annually. Symptoms include diarrhea and vomiting.
- Millions of people are exposed to unsafe levels of naturally occurring arsenic that causes cancer.
- Malaria causes over one million deaths annually. Malaria is a parasitic infectious disease that is transmitted by mosquitoes that breed in water.
- An estimated 260 million people are infected by parasitic worms that cause schistosomiasis. This illness causes abdominal pain, blood in stool, enlarged liver, and an increased risk of bladder cancer.
- Diarrhea killed approximately 2.2 million people in 1998.
- Wastewater in agriculture is associated with serious public health risks.
- Four percent of the global disease burden could be prevented by improving water supply, sanitation, and hygiene.

Source: WHO, http://www.who.int/water_sanitation_health/facts_figures/en/

Here are some additional troubling facts on sanitation as reported by the WHO.

- The regions with the lowest access to adequate sanitation are sub-Saharan Africa (31%), southern Asia (36%), and Oceania (53%). Underlying issues that add to the challenge in many countries include a weak infrastructure, an inadequate human resource base, and scarce resources to improve the situation.

John Wollwerth / Shutterstock.com

- The lack of sanitation facilities forces people to defecate in the open, in rivers, or near areas where children play or food is prepared, which increases the risk of transmitting disease. The Ganges River in India has 1.1 million liters of raw sewage dumped into it every minute. This is a startling figure, considering that one gram of feces may contain 10 million viruses, one million bacteria, 1,000 parasite cysts, and 100 worm eggs.
- Examples of diseases transmitted through water contaminated by human waste include diarrhea, cholera, dysentery, typhoid, and hepatitis A. In Africa, 115 people die every hour from diseases linked to poor sanitation, poor hygiene, and contaminated water.
- Health-care facilities need proper sanitation and must practice good hygiene to control infection. Worldwide, between 5 and 30 percent of patients develop one or more avoidable infections during stays in health-care facilities.
- Each year more than 200 million people are affected by droughts, floods, tropical storms, earthquakes, forest fires, and other hazards. Sanitation is an essential component in emergency response and rehabilitation efforts to stem the spread of diseases, rebuild basic services in communities, and help people return to normal daily activities.
- Studies show that improved sanitation reduces diarrhea death rates by a third. Diarrhea, which is largely preventable, is a major killer; it is responsible for 1.5 million deaths every year, mostly among children less than five years old living in developing countries.
- Adequate sanitation encourages children to be at school, particularly girls. Access to latrines raises school attendance rates for children; an increase in girls' enrollment can be attributed to the provision of separate, sanitary facilities in underdeveloped countries.

- Hygiene education and the promotion of hand washing are simple, cost-effective measures that can reduce diarrhea cases by up to 45 percent. Even when ideal sanitation is not available, instituting good hygiene practices in communities leads to better health. Proper hygiene goes hand-in-hand with the use of improved facilities to prevent disease.
- The economic benefits of sanitation are persuasive. Every U.S. dollar invested in improved sanitation translates into an average return of 9 dollars. Those benefits are experienced specifically by poor children and disadvantaged communities that need them most.

Source: WHO

In the discussion opener of this chapter, we asked if you have ever thought about what your life would be like if you suddenly didn't have access to water and adequate sanitation for a day or two. Imagine what it would be like to wake up tomorrow morning to discover you had no safe water source—nor the ability to acquire any.

## *Before You Go On*

Answer the following questions to test your understanding of the preceding section:

1. What are the major sources of water contaminants?

2. Explain the difference between the maximum contaminant level goal and maximum contaminant level.

3. Give examples of industries that contribute to nitrate contamination.

4. Give examples of industries that contribute to asbestos contamination.

5. How do pathogens find their way into our tap water?

6. Would boiling water with excess nitrate levels reduce its concentration?

7. Give examples of diseases that are transmitted through water contaminated by human waste.

### *Vocabulary—State the meaning of the following terms:*

EPA

Pathogen

MCLG

MCL

# SUMMARY

## LO¹ Water—Basic Concepts

You should be familiar with the water cycle and realize that the total amount of water available on the Earth remains constant. Even though water can change its phase from liquid to solid (ice) or from liquid to vapor, we don't lose or gain water on earth. You also should be familiar with water resource terminology, such as surface water, groundwater, aquifer, tributary, estuary, aqueduct, and so on. You also should recognize the importance of the distribution of the total available global water.

## LO² Personal Water Consumption

You should be familiar with how much water you consume to address your personal needs. It is important to realize that we use the most amount of water in flushing toilets, followed by washing clothes and taking showers. Therefore, we should consider taking conservative measures such as installing low-flow showers and water-efficient toilets in our homes. You should be able to estimate your annual water consumption at home.

## LO³ Water Consumption in Agriculture, Commercial, and Industrial Sectors

The U.S. Geological Survey (USGS) has been estimating the total water consumption in the United States since 1950. They report their findings every five years for both groundwater and surface water sources. For bookkeeping purposes, the USGS groups major water-consuming activities into broad categories such as public, domestic, irrigation, livestock, aquaculture, industrial, mining, and thermoelectric power generation, and reports the data for each category. The public supply refers to water that was drawn by the government. Most of our domestic water supply is delivered by a public supplier. Public suppliers also provide water for businesses, schools, firefighting, community parks and swimming pools, and (at times) for commercial applications. The irrigation category refers to the amount of water that was provided by man-made systems for agricultural purposes. The amount of water that is used for agricultural activities such as dairy operations, feed lots, and providing drinking water for livestock is represented by the livestock category. Aquaculture refers to the farming of species that live in the water, such as fish and shrimp. The farming of plants and algae that live in water is

also grouped into this category. Water that is used for industrial purposes, such as making paper, chemicals, or steel, is classified as industrial. The water used for the excavation of rocks and minerals is classified as mining water. Finally, the water used in power plants to generate electricity is classified as thermoelectric power. Thermoelectric power generation, irrigation, and the public supply sectors are among the largest consumers of water in the United States.

## LO⁴ Drinking Water Standards in the United States

The U.S. Environmental Protection Agency (EPA) sets the standards for the maximum level of contaminants that can be in our drinking water and still be considered safe to drink. Human activities and naturally occurring microorganisms contribute to the level of contaminants in our water supply. The EPA sets two standards for the level of water contaminants: (1) the maximum contaminant level goal (MCLG) and (2) the maximum contaminant level (MCL). The MCLG represents the maximum level of a given contaminant in the water that causes no known harmful health effects. On the other hand, the MCL, which may represent slightly higher levels of contaminants in the water, is the level of contaminants that are legally enforceable.

## LO⁵ Global Water Quality Issues

The World Health Organization (WHO) is responsible for setting water standards, monitoring the standards, and providing technical support. According to the WHO, nearly one billion people lack access to clean drinking water; as a result, millions of deaths are attributed to unsafe water each year. Examples of diseases transmitted through water contaminated by human waste include diarrhea, cholera, dysentery, typhoid, and hepatitis. Diarrhea, which is largely preventable, is responsible for 1.5 million deaths every year, mostly among children younger than five years old living in developing countries. Hygiene education and the promotion of hand washing are simple, cost-effective measures that can reduce diarrhea cases by up to 45 percent. More than 50 countries also report cholera to the WHO. An estimated 2.6 billion people lack access to adequate sanitation globally. According to the WHO, wastewater used in agricultural processes is also associated with serious public health risks.

## KEY TERMS

Aquaculture 285
Aqueduct 272
Aquifer 272
Cholera 291
Community Water System
   (CWS) 276
Diarrhea 292
Dike 272
Estuary 272
Flood 272
Fresh Water 273

Geyser 273
Glacier 273
Groundwater 270
Hardness 274
Leaching 274
Levee 272
Maximum Contaminant Level
   (MCL) 287
Maximum Contaminant Level
   Goal (MCLG) 287
Pathogens 289

Public Water System (PWS) 276
Reservoir 270
Storm Water 272
Surface Water 270
Thermoelectric Power Water
   Supply 271
Tributary 272
Wetlands 272

## *Apply What You Have Learned*

Each of you is to estimate how much water you consume each year for taking showers, flushing toilets, doing laundry, and washing dishes. To determine your shower water consumption:

■ Obtain a container of a known volume and time how long it takes to fill the container.

■ Calculate the volumetric flow rate in gallons per minute (or liters per minute). Then, measure the time that you spend on average when taking showers. Calculate the volume of the water you consume on a daily basis. Multiply this daily value by 365 to get the yearly value.

For the other activities, look up the size of your toilet water tank, clothes washing machine,

Shell114 / Shutterstock.com

Sutichak / Shutterstock.com

ppart / Shutterstock.com

Sutichak / Shutterstock.com

and dishwashing machine. Estimate on average how many times per day, week, or month you use each of them. Calculate the volume of the water you consume and determine the yearly value.

Compile your findings into a single, brief report, and present it to the class.

# PROBLEMS

*Problems that promote life-long learning are denoted by* ⊶.

**9.1**  Assuming that a large family uses a dishwasher that uses 10 gallons of water per wash three times a week, what is the total amount of water consumed by the dishwasher in a year?

**9.2**  An old showerhead delivers approximately 3 gallons of water per minute. Assuming that a person showers twice a day for 5 minutes each time, what is the total amount of water that is consumed for this activity annually?

**9.3** ⊶  Investigate how low-flow toilets work. Estimate how many gallons of water could be saved per year by using high-efficiency toilets for a family of four.

**9.4**  A car-wash system consumes 60 gallons of water per wash. Assuming that a person washes her car once every week, what is the total amount of water consumed annually by this activity?

**9.5** ⊶  In the United States, your water consumption charges are typically based on

    (a)  a monthly administrative fee

    (b)  meter charge — depending on the size of the line connected to the meter — for example, 5/8 inch for residential or 1 to 4 inches for commercial use; the larger the line size, the more expensive the meter charge

    (c)  usage charge — based on gallons used per month

Moreover, the sewer charge is proportional to the amount of water used in a month. Look up the water and sewer charges for your city. Investigate the water and sewer charges in your town. Express your findings in a brief report.

**9.6**  Assuming a household water consumption of 70 gallons per day per capita, what was the total amount of water that was consumed during 2010 by all of the people in the United States? (*Note:* The population of the United States in 2010 was about 309 million people.) How much water would have been saved if the per capita consumption was reduced to 60 gallons per day?

**9.7** ⊶  Investigate how much water a leaky faucet wastes in one week, one month, and one year. Perform an experiment by placing a container under a leaky faucet and actually measure the amount of water accumulated in an hour. You can simulate a leaky faucet by not completely closing the faucet and letting it drip.

You are to design the experiment. Think about the parameters that you need to measure. Express your findings in gallons/day, gallons/week, gallons/month, and gallons/year. At this rate, how much water is wasted by 10 million households with leaky faucets? Write a brief report to discuss your findings.

**9.8** ⊶  Using the concepts discussed in this chapter, measure the volumetric flow rate of water out of a drinking fountain.

**9.9** ⊶  Visit a home appliance store or go online to look up the water consumption of at least three different brands and sizes of dishwashing machines. Create a table that shows an estimate of annual water consumption for each machine for a family of four. State all your assumptions.

**9.10** ⊶  Visit a home appliance store or go online to look up the water consumption of at least three clothes-washing machines. Create a table that shows an estimate of annual water consumption for each machine for a family of four.

**9.11** ⊶  Investigate household water consumption rates in Europe and compare them to averages for the United States. Discuss your findings in a brief report.

**9.12** ⊶  Investigate how much water is consumed for irrigation in your state (or country if you live outside the United States).

**9.13** ⊶  Investigate how much water is consumed for thermoelectric power generation in your state (or country if you live outside the United States).

**9.14** There are over 200,000 water main breaks each year in the United States because the water infrastructure is aging. Investigate what is being done to address this problem and to create a sustainable national water infrastructure. Discuss your findings in a brief report.

**9.15** Investigate how much water is consumed annually on your campus. Suggest at least three ways to reduce consumption by individuals. Present your results in a brief report.

**9.16** Estimate how much water is consumed during sporting events (basketball and football games) on your campus. Suggest ways to reduce consumption. State your assumptions and present your results in a brief report.

**9.17** It is recommended that we drink at least 2 liters of water per day. Estimate how much drinking water you will consume in your lifetime. State your assumptions, and present your lifetime drinking-water consumption in both liters and meters cubed ($m^3$).

**9.18** Imagine 100 million adults who will live another 65 years. Also assume that each person would use $0.27 \ m^3$ of water every day on average. How much water would be consumed by this population over their expected remaining lives?

**9.19** According to the USGS in 2010, about 161 billion gallons of water per day were consumed in the thermoelectric power category in the United States. How much water does this volume represent in terms of square miles of land covered to a depth of 4 ft each year?

**9.20** According to the USGS in 2010, about 355 billion gallons of water per day were consumed to address various needs in the United States. Show that this volume is equal to 397,000 thousand acre-feet per year. (*Note*: An acre-foot is the amount of water required to cover 1 acre to a depth of 1 ft, 1 acre is equal to $43,560 \ ft^2$, and $1 \ ft^3 = 7.48$ gallons.)

Everett–Art / Shutterstock.com

*"It is much better to know something about everything than to know everything about one thing."* — Blaise Pascal (1623–1662)

# Understanding the Materials We Use in our Daily Lives

## LEARNING OBJECTIVES

**LO¹** Earth—Our Home: describe different layers of the Earth, its structure, and its properties

**LO²** The Phases of Matter and Properties of Materials: explain the phases of matter and the important properties of materials

**LO³** Metals: describe different metals and their applications

**LO⁴** Plastics, Glass, Composites, and Wood: describe the compositions and applications of these materials

**LO⁵** Concrete: describe the basic ingredients of and construction practices using concrete

# Discussion Starter

## WHAT IS YOUR SMART PHONE MADE OF?

To make smart phones, manufacturers use precious **raw materials that must be extracted and processed,** and consume **natural resources** and energy that can affect our air, land, and water as well as plants and animals.

**A smart phone is made up of many parts which use these materials, such as:**

**LCD DISPLAYS**: Glass, plastic (made from crude oil), and liquid crystalline

**CIRCUIT BOARDS**: Copper, gold, lead, silver, and palladium

**RECHARGEABLE BATTERIES**: Lithium metallic oxide

*Source:* Based on EPA.

According to the Environmental Protection Agency (EPA), for every one million smart phones recycled, 35,274 pounds (16,000 kg) of copper, 772 pounds (350 kg) of silver, 75 pounds (34 kg) of gold, and 33 pounds (15 kg) of palladium are recovered.

To the Students: Take a few minutes and think about where you live: furnishings, appliances, electronic devices, and other products you use every day—including food and drink containers—and then answer the following questions. Justify your answers with assumptions and simple calculations. How much material—such as metal (e.g., aluminum, copper, steel), plastics, glass, wood, and concrete—do you think you will consume in your lifetime to maintain a good standard of living? How much of these materials do you think will be recycled?

# LO¹ 10.1 Earth—Our Home

In this chapter, we look more closely at materials that make up the products that we use every day. We will examine solid materials, such as metals and their alloys, plastics, glass, and wood, along with those that solidify over time, such as concrete. But first, we need to take a closer look at our home—the Earth—to better understand where the raw materials used to make products come from.

The Earth is the third planet from the sun. It has a spherical shape with a diameter of 7,926.4 miles or 12,756.3 kilometers (km) and an approximate mass of $13.17 \times 10^{24}$ pounds or $5.98 \times 10^{24}$ kilograms (kg). To better represent the Earth's structure, it is divided into major layers that are located above and below its surface. For example, the *atmosphere* represents the air layer that covers the surface of the Earth. The air extends approximately 90 miles or 140 kilometers (km) from the surface of the Earth to a point called the "edge of space." We discussed the atmosphere in detail in Chapter 8. We also explained the bodies of water that cover the Earth's surface in Chapter 9.

Our knowledge of what is inside the Earth and its composition continues to improve with research. Each day we learn from studies of the Earth's surface and near-surface rocks, interior heat-transfer rates, gravity and magnetic fields, and earthquakes. The results of these studies suggest that the Earth's interior is made up of different layers with different characteristics, and its mass is composed mostly of iron, oxygen, and silicon (approximately 32 percent iron, 30 percent oxygen, and 15 percent silicon). Earth also contains other elements, such as sulfur, nickel, magnesium, and aluminum. The structure below the Earth's surface is generally grouped into four layers: the *crust*, *mantle*, *outer core*, and *inner core* (see Table 10.1). This classification is based on the properties of materials and the manner by which the materials move or flow.

> The structure below the Earth's surface is commonly grouped into the crust, mantle, outer core, and inner core.

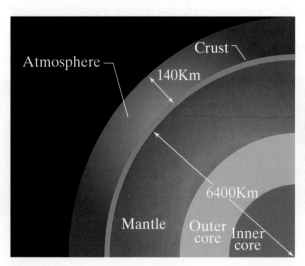

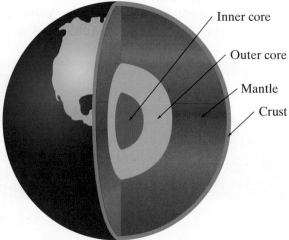

| TABLE 10.1 | The Approximate Mass for Each Layer of the Earth | | |
|---|---|---|---|
| | Approximate Mass (lbm) | Approximate Mass (kg) | Percentage of the Earth's Total Mass |
| Atmosphere | $1.12 \times 10^{19}$ | $5.1 \times 10^{18}$ | 0.000086 |
| Oceans | $3.08 \times 10^{21}$ | $1.4 \times 10^{21}$ | 0.024 |
| Crust | $5.73 \times 10^{22}$ | $2.6 \times 10^{22}$ | 0.44 |
| Mantle | $8.9 \times 10^{24}$ | $4.04 \times 10^{24}$ | 68.47 |
| Core | | | |
| Outer core | $4.03 \times 10^{24}$ | $1.83 \times 10^{24}$ | 31.01 |
| Inner core | $2.12 \times 10^{23}$ | $9.65 \times 10^{22}$ | 1.63 |

The ***crust*** makes up about 0.5 percent of the Earth's total mass and 1 percent of its volume; because of the ease of access to materials near the crust's surface, its composition and structure has been studied extensively. It has a maximum thickness of approximately 25 miles or 40 kilometers (km). The crust is thicker under the continents and thinner under the oceanic floors. Scientists have been able to collect samples of the crust up to a depth of 12 kilometers; however, because the drilling expenses increase with depth, the progress to deeper locations has been slow. The Earth's crust—the oceanic floors and the continents—is made up of about twelve plates, which continuously move at slow rates (a few centimeters per year). Moreover, the boundaries of these plates (where they come together) mark regions of earthquake and volcanic activities. As you may also know, over time the collisions of these plates have created mountain ranges around the world.

As shown in Table 10.1, most of the mass of the Earth comes from the mantle. The ***mantle*** is made up of molten rock that lies underneath the crust and makes up nearly 84 percent of the Earth's volume. Unlike the crust, what we know of the mantle composition is based on our studies of the propagation of sound waves, heat flow, earthquakes, and magnetic and gravity fields. Rooted in these studies and additional investigations is the suggestion that the lower part of the mantle is made up of iron and magnesium silicate minerals. The mantle starts approximately 25 miles or 40 kilometers (km) below the Earth's surface and extends to a depth of 1,800 miles or 2,900 kilometers.

The *inner* and *outer cores* make up about 33 percent of the Earth's mass and 15 percent of its volume. Our knowledge of the structures of the inner and outer core comes primarily from the study of the behavior and speed of shear and compression waves in the core. Based on these studies, the ***inner core*** is considered to be solid, while the ***outer core*** is thought to be fluid and composed mainly of iron. The outer core starts at a depth of 1,800 miles or 2,900 kilometers and extends to a depth of 3,200 miles or 5,200 kilometers. The inner core is located between 3,200 and 4,000 miles or 5,200 and 6,400 kilometers below the Earth's surface. Let us now consider an example pertaining to this subject.

**EXAMPLE 10.1**

The volume of a sphere is given by

$$\text{volume} = \frac{4}{3}\pi R^3$$

where $\pi \approx 3.14$ and R represents the radius of the sphere. In this section, we said that the *inner* and *outer cores* make up about 15 percent of the Earth's volume. Using the volume formula for the sphere, let's verify this number.

$$\text{volume of entire Earth} = \frac{4}{3}\pi R^3 = \frac{4}{3}(3.14)(6,400 \times 10^3)^3 = 1.1 \times 10^{21}\,\text{m}^3$$

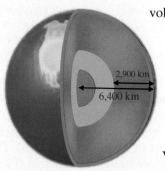

$$\begin{array}{l}\text{volume of inner and} \\ \text{outer cores}\end{array} = \frac{4}{3}\pi R^3 = \frac{4}{3}(3.14)[6,400 \times 10^3 - 2,900 \times 10^3]^3$$

$$= 1.8 \times 10^{20}\,\text{m}^3$$

2,900 km

6,400 km

$$\frac{\text{volume of inner and outer cores}}{\text{volume of entire Earth}} = \frac{1.8 \times 10^{20}\,\text{m}^3}{1.1 \times 10^{21}\,\text{m}^3} \approx 0.16 \text{ or } 16\%$$

Very close to 15 percent!

# Before You Go On

Answer the following questions to test your understanding of the preceding section:

1.  Name the different layers that make up the Earth.

2.  Which layer of the Earth contains the largest amount of mass?

3.  Which layer of the Earth is the thickest?

4.  What are the major chemical components of the Earth?

*Vocabulary—State the meaning of the following terms:*

Mantle

Oceanic crust

Continental crust

Inner core

Outer core

## LO² 10.2 The Phases of Matter and Properties of Materials

When you look around, you find that matter exists in various forms and shapes. You also notice that matter changes shape when its conditions or its surrounding conditions are changed. All solid objects, liquids, gases, and living things are made of matter, and matter itself is made up of atoms or chemical elements. As you may know, there are 106 known chemical elements to date. Atoms of similar characteristics are grouped together and shown in the periodic table of chemical elements (Figure 10.1). Atoms are made up of even smaller particles we call *electrons*, *protons*, and *neutrons*. In your chemistry class, you have studied these ideas in more detail. Atoms are the basic building blocks of all matter. They are combined naturally or in a laboratory setting to create molecules. For example, as you already know, water molecules are made of two atoms of hydrogen and one atom of oxygen. A glass of water is made of billions and billions of homogeneous water molecules. A molecule is the smallest portion of a given matter that still possesses its microscopic characteristic properties.

### Phases of Matter

Matter can exist in four states: *solid*, *liquid*, *gaseous*, and *plasma*, depending on its own and its surrounding conditions. Consider the water that we

| IA | IIA | IIIB | IVB | VB | VIB | VIIB | VIIIB | | | IB | IIB | IIIA | IVA | VA | VIA | VIIA | VIIIA |
|---|---|---|---|---|---|---|---|---|---|---|---|---|---|---|---|---|---|
| 1 **H** 1.0079 | | | | | | | | | | | | | | | | | 2 **He** 4.003 |
| 3 **Li** 6.941 | 4 **Be** 9.012 | | | | | | | | | | | 5 **B** 10.811 | 6 **C** 12.011 | 7 **N** 14.007 | 8 **O** 15.999 | 9 **F** 18.998 | 10 **Ne** 20.180 |
| 11 **Na** 22.990 | 12 **Mg** 24.305 | | | | | | | | | | | 13 **Al** 26.982 | 14 **Si** 28.086 | 15 **P** 30.974 | 16 **S** 32.066 | 17 **Cl** 35.453 | 18 **Ar** 39.948 |
| 19 **K** 39.098 | 20 **Ca** 40.078 | 21 **Sc** 44.956 | 22 **Ti** 47.88 | 23 **V** 50.942 | 24 **Cr** 51.996 | 25 **Mn** 54.938 | 26 **Fe** 55.845 | 27 **Co** 58.933 | 28 **Ni** 58.69 | 29 **Cu** 63.546 | 30 **Zn** 65.39 | 31 **Ga** 69.723 | 32 **Ge** 72.61 | 33 **As** 74.922 | 34 **Se** 78.96 | 35 **Br** 79.904 | 36 **Kr** 83.8 |
| 37 **Rb** 85.468 | 38 **Sr** 87.62 | 39 **Y** 88.906 | 40 **Zr** 91.224 | 41 **Nb** 92.906 | 42 **Mo** 95.94 | 43 **Tc** 98 | 44 **Ru** 101.07 | 45 **Rh** 102.906 | 46 **Pd** 106.42 | 47 **Ag** 107.868 | 48 **Cd** 112.411 | 49 **In** 114.82 | 50 **Sn** 118.71 | 51 **Sb** 121.76 | 52 **Te** 127.60 | 53 **I** 126.905 | 54 **Xe** 131.29 |
| 55 **Cs** 132.905 | 56 **Ba** 137.327 | 57 **La** 138.906 | 72 **Hf** 178.49 | 73 **Ta** 180.948 | 74 **W** 183.84 | 75 **Re** 186.207 | 76 **Os** 190.23 | 77 **Ir** 192.22 | 78 **Pt** 195.08 | 79 **Au** 196.967 | 80 **Hg** 200.59 | 81 **Tl** 204.383 | 82 **Pb** 207.2 | 83 **Bi** 208.980 | 84 **Po** 209 | 85 **At** 210 | 86 **Rn** 222 |
| 87 **Fr** 223 | 88 **Ra** 226.025 | 89 **Ac** 227.028 | 104 **Rf** 261 | 105 **Db** 262 | 106 **Sg** 263 | 107 **Bh** 262 | 108 **Hs** 265 | 109 **Mt** 266 | 110 **Uun** 269 | 111 **Uuu** 272 | 112 **Uub** 277 | 114 | | 116 | | 118 | |

| Lanthanide series | 58 **Ce** 140.115 | 59 **Pr** 140.908 | 60 **Nd** 144.24 | 61 **Pm** 145 | 62 **Sm** 150.36 | 63 **Eu** 151.964 | 64 **Gd** 157.25 | 65 **Tb** 158.925 | 66 **Dy** 162.5 | 67 **Ho** 164.93 | 68 **Er** 167.26 | 69 **Tm** 168.934 | 70 **Yb** 173.04 | 71 **Lu** 174.967 |
|---|---|---|---|---|---|---|---|---|---|---|---|---|---|---|
| Actinide series | 90 **Th** 232.038 | 91 **Pa** 231.036 | 92 **U** 238.029 | 93 **Np** 237.048 | 94 **Pu** 244 | 95 **Am** 243 | 96 **Cm** 247 | 97 **Bk** 247 | 98 **Cf** 251 | 99 **Es** 252 | 100 **Fm** 257 | 101 **Md** 258 | 102 **No** 259 | 103 **Lr** 262 |

**FIGURE 10.1**  The chemical elements to date (2016)

drink every day. As you already know, under certain conditions, water exists in a solid form that we call *ice*. At a standard atmospheric pressure, water exists in a solid form as long as its temperature is kept under 32°F (0°C). Under standard pressure, if you were to heat the ice and consequently change its temperature, the ice would melt and change into a liquid form. Under standard atmospheric pressure, the water remains liquid up to a temperature of 212°F (100°C) as you continue heating it. If you were to carry out this experiment further by adding more heat, eventually the water changes its phase from liquid into a gas. This phase of water is commonly refer red to as *steam*. If you had the means to heat the water to even higher temperatures, ones exceeding 3,600°F (2,000°C), you would find that you can break up the water molecules into their atoms, and eventually the atoms break up into free electrons and nuclei that we call *plasma*.

> Matter can exist in four states: *solid, liquid, gaseous,* or *plasma,* depending on its own and the surrounding conditions.

In general, the properties of a material depend on its phase. For example, as you know from your everyday experience, the density of ice is different from liquid water (ice cubes float in liquid water), and the density of liquid water is different from that of steam. Next, we will consider some of the important properties of materials.

kubais / Shutterstock.com

Valentyn Volkov / Shutterstock.com

focal point / Shutterstock.com

## Properties of Materials

Think about all of the products we use in our everyday lives. These products include TVs, furniture, cars, aircraft, computers, clothing, toys, home appliances, heating and cooling equipment, health-care devices, and tools and machines that make various products. Designers of products consider important factors such as cost, efficiency, reliability, and safety when designing products, and they perform tests to make certain that the products they design

arka38 / Shutterstock.com

vovan / Shutterstock.com

Alexandru Nika / Shutterstock.com

You can more / Shutterstock.com

Oleksiy Mark / Shutterstock.com

risteski goce / Shutterstock.com

Yulia Nikulyasha Nikitina / Shutterstock.com

BassKwong / Shutterstock.com

DM7 / Shutterstock.com

withstand various conditions. Designers also are continuously searching for ways to improve already existing products such as cars, cell phones, appliances, and electronics.

Take a moment to think about our infrastructure, including the buildings, highways, mass transit systems, airports, communication systems, water distribution systems, and power plants that supply power to our homes, manufacturing companies, and offices. All of the raw materials that are used to make various things come from the Earth. As a result, there are many people behind the scenes who are responsible for both finding suitable ways, and designing the necessary equipment, to extract raw materials, petroleum, and natural gas from the Earth.

When we use a product such as a smart phone, an electronic tablet, a car, a clothes washing machine, an oven, or a refrigerator, we need to be mindful of *what* type of materials went into making the product, *where* the materials came from, *how* much energy it took to extract the raw materials and to produce the product, and eventually, what it takes to dispose of it. As we mentioned previously, there are a number of factors that designers consider when selecting a material for a specific application. For example, they consider the properties of material such as density, strength, flexibility, machinability, durability, thermal expansion, electrical and thermal conductivity, and resistance to corrosion. They also consider the cost of the material.

> The properties of a material may be divided into three groups: *electrical*, *mechanical*, and *thermal*.

In general, the properties of a material may be divided into three groups: *electrical*, *mechanical*, and *thermal*. In electrical and electronic applications, for example, the electrical resistivity of materials such as a wire is important. How much resistance to flow of electricity does the material offer? In many structural and aerospace applications, the mechanical properties of materials are important. These properties include the strength of the material and ***strength-to-weight ratio***. For example, we want our cars and planes to be light and yet strong enough to withstand forces. In applications dealing with fluids (liquids and gases) such as delivering natural gas to our homes and gasoline to storage facilities, properties such as density, viscosity, and vapor pressure are important. For example, did you know that in order to reduce smog during summer months the petroleum refineries are required to produce gasoline with a lower vapor pressure? That is because the summer gasoline with the lower vapor pressure will not evaporate as quickly as the winter gasoline with a higher vapor pressure. The thermal expansion of materials that make up our roads and bridges is also important because of the temperature fluctuations that occur from winter to summer.

After having read the preceding paragraphs, some of you may not be certain why you need to know about the properties of materials and may have questions similar to the following: "Do I really need to know about the basic properties of materials?", "Why is it important for me to learn about basic properties of materials?", and "I am not studying to be a designer or an engineer, so why do I need to know these things?"

You may recall from Chapter 1 that, as good global citizens, we need to realize that the choices we make every day affect all of us. We need to change our behaviors especially with respect to the way we consume the finite resources available on Earth. We need to become lifelong learners so that we can make informed decisions and anticipate and react to the global changes caused by technological innovations. Therefore, as good global citizens, it is important to have a basic understanding of material properties so that you can be familiar with the specifications of a new product and make wise choices when purchasing it. The meaning of some material properties is summarized next.

demarcomedia / Shutterstock.com

**Electrical Resistivity** The value of ***electrical resistivity*** is a measure of the resistance of material to the flow of electricity. For example, plastics and ceramics typically have high resistivity, whereas metals typically have low resistivity. Among the best conductors of electricity are silver and copper. You already know that most of the electrical wires are made of copper. The materials with less electrical resistivity dissipate less heat and have less loss associated with them (more energy efficient).

**Density** Density is defined as mass per unit volume; ***density*** is a measure of how compact the material is for a given volume. For example, the average density of aluminum alloys is 2,700 kg/m$^3$ and steel has a density of 7,850 kg/m$^3$. By comparison, aluminum has a density that is approximately one-third the density of steel. It is important to realize that it takes less energy to transport light products to marketplace.

Roman Samokhin / Shutterstock.com

© age fotostock / Alamy

©IM_photo / Shutterstock.com

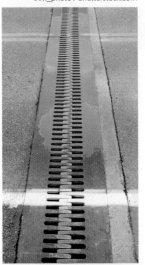

Evoken / Shutterstock.com

El Nariz / Shutterstock.com

Alberto Loyo / Shutterstock.com

**Tensile Strength** The *tensile strength* of a piece of material is determined (when pulled) by measuring the maximum tensile load a material specimen in the shape of a rectangular bar or cylinder can carry without failure. The tensile strength or ultimate strength of a material is expressed as the maximum tensile force per unit cross-sectional area of the specimen. For example, steel has a higher tensile strength than aluminum.

**Compression Strength** Some materials are stronger in compression (when pushed) than they are in tension (pulled); concrete is a good example. The *compression strength* of a piece of material is determined by measuring the maximum compressive load a material specimen in the shape of cylinder or cube can carry without failure. The ultimate compressive strength of a material is expressed as the maximum compressive force per unit cross-sectional area of the specimen.

**Strength-to-Weight Ratio** As the term implies, this is the ratio of the strength of the material to its specific weight (weight of the material per unit volume). For example, materials such as aluminum alloys with high strength-to-weight ratios are used in planes.

**Thermal Expansion** The coefficient of linear expansion can be used to determine the change in the length (per original length) of a material that occurs if the temperature of the material is changed. *Thermal expansion* is an important material property to consider when designing products and structures (e.g., roads, bridges) that are expected to experience a relatively large temperature swing during their service lives.

**Thermal Conductivity** *Thermal conductivity* is a property that shows how good the material is in transferring thermal energy (heat) from a high-temperature region to a low-temperature region within the material. When selecting a material for insulation purposes, we consider materials with very low thermal conductivity.

LianeM / Shutterstock.com

**Heat Capacity** The value of *heat capacity* represents the amount of thermal energy required to raise the temperature of 1 kilogram mass of a material by 1°C, or using U.S. Customary Units, this is the amount of thermal energy required to raise one pound mass of a material by 1°F.

**Viscosity** The value of *viscosity* of a fluid represents a measure of how easily the given fluid can flow. The higher the viscosity value is, the more resistance the fluid offers to flow. For example, it requires less energy to transport water in a pipe than it does to move oil.

In the next section, we examine the application and properties of common solid materials—such as metals and their alloys, plastics, glass, and wood—and those that solidify over time—such as concrete. We also examine the application and chemical composition of some common materials that are found in the products that we use every day. First let's look at an example.

**EXAMPLE 10.2**

NPeter / Shutterstock.com

In this example, we calculate the average density of the Earth using

$$\text{Earth's average density} = \frac{\text{Earth's mass}}{\text{Earth's volume}}$$

We computed the volume of the Earth in Example 10.1, and the mass of the Earth is $5.98 \times 10^{24}$ kg. Now, if we divide the mass of the Earth by its volume, we get

$$\text{Earth's average density} = \frac{5.98 \times 10^{24}\, \text{kg}}{1.1 \times 10^{21}\, \text{m}^3} \approx 5{,}400\ \frac{\text{kg}}{\text{m}^3}$$

As you can see, the average density of the Earth is approximately 5.4 times the density of water ($1{,}000$ kg/m$^3$). How does the average density of Earth compare to the densities of metals? As you read the following sections, pay close attention to the densities of metals.

## Before You Go On

Answer the following questions to test your understanding of the preceding section:

1.  What are the phases of matter?

2.  Give three examples of material properties.

3.  Explain what is meant by the density of a material.

**Vocabulary—State the meaning of the following terms:**

Density

Viscosity

Heat capacity

Thermal conductivity

Thermal expansion

## LO³ 10.3  Metals

In this section, we briefly examine the chemical composition and common application of metals. We discuss light metals, copper and its alloys, iron, and steel.

BACHTUB DMITRII / Shutterstock.com        aodaodaodaod / Shutterstock.com

Jirat Teparaksa / Shutterstock.com

## Lightweight Metals

Aluminum, titanium, and magnesium, because of their small densities (relative to steel), are commonly referred to as *lightweight metals*. Because of their relatively high strength-to-weight ratios, lightweight metals are used in many structural and aerospace applications.

**Aluminum** *Aluminum* and its alloys have densities that are approximately one-third the density of steel. Pure aluminum is very soft; thus it is generally used in electronic applications and in making reflectors and foils. Because pure aluminum is soft and has a relatively small tensile strength, it is alloyed with other metals to make it stronger, easier to weld, and to increase its resistance to corrosive environments. Aluminum is commonly alloyed with copper (Cu), zinc (Zn), magnesium (Mg), manganese (Mn), silicon (Si), and lithium (Li). Generally speaking, aluminum and its alloys resist corrosion; they are easy to mill and cut; and they can be brazed or welded. Aluminum parts also can be joined using adhesives. They are good conductors of electricity and heat and thus have relatively high thermal conductivity and low electrical resistance values. Aluminum is fabricated in sheets, plates, foil, rods, and wire and is extruded to make window frames or automotive parts. You are already familiar with everyday examples of common aluminum products, including beverage cans, household aluminum foil, non-rust staples in tea bags, and so on. The use of aluminum in various sectors of our economy is shown in Figure 10.2.

As shown in Figure 10.2, most of the aluminum produced is consumed by the packaging, transportation, building, and electrical sectors. Because of their light weight and strength, composite materials (we explain composite materials later in this chapter) are being substituted for aluminum in aerospace applications, for example, in military and commercial planes, helicopters, and satellites. As you already know from your everyday experience

> Aluminum, titanium, and magnesium, because of their small densities (relative to steel), are commonly referred to as *lightweight metals*.

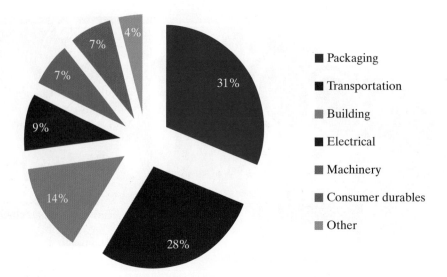

31%

28%

14%

9%

7%

7%

4%

■ Packaging
■ Transportation
■ Building
■ Electrical
■ Machinery
■ Consumer durables
■ Other

**FIGURE 10.2**   Aluminum use by various sectors

with packing materials, plastics, paper, and glass are also used as alternatives to aluminum; in construction, steel and wood serve as substitutes for aluminum.

**Titanium** *Titanium* has an excellent strength-to-weight ratio. Titanium is used in applications where relatively high temperatures from 750°F up to 1,100°F (400°C up to 600°C) are expected. Titanium alloys are used in the fan and the compressor blades of engines of commercial and military airplanes. In fact, without the use of titanium alloys, the engines on commercial airplanes would not have been possible. Like aluminum, titanium is alloyed with other metals to improve its properties.

Titanium alloys show excellent resistance to corrosion. Titanium is quite expensive compared to aluminum, and it is heavier than aluminum, having a density which is roughly one-half that of steel. Because of their relatively high strength-to-weight ratios, titanium alloys also are used in both commercial and military airplane airframes (fuselage and wings) and landing gear components. Titanium alloys are a metal of choice in many products; you can find them in golf clubs, bicycle frames, tennis racquets, and spectacle frames. Because of their excellent corrosion resistance, titanium alloys have been used in the tubing in desalination plants as well. Replacement hips and other joints are examples of other applications where titanium is currently being used.

dvande / Shutterstock.com    Rudy Umans / Shutterstock.com    Chayatorn Laorattanavech / Shutterstock.com

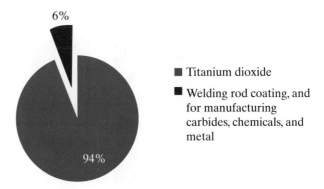

6%

94%

■ Titanium dioxide
■ Welding rod coating, and for manufacturing carbides, chemicals, and metal

**FIGURE 10.3** The percentage of titanium mineral concentrate use by sector

As shown in Figure 10.3, about 94 percent of the titanium mineral concentrate is consumed as titanium dioxide ($TiO_2$), which is commonly used as pigments in paint, plastics, and paper. The remaining 6 percent is used in the manufacturing of welding rods, chemicals, and metals.

**Magnesium** With its silvery white appearance, *magnesium* is another lightweight metal that looks like aluminum. However, it is lighter, having a density of approximately 1,700 kg/m³. Pure magnesium does not provide good strength for structural applications, and because of this fact it is alloyed with other elements such as aluminum, manganese, and zinc to improve its properties. Magnesium and its alloys are used in nuclear applications, in dry cell batteries, in aerospace applications, and in some automobile parts as sacrificial anodes to protect other metals from corrosion. In the United States, magnesium oxide and other compounds are recovered from seawater and lake brines. Magnesium compounds are used in agricultural, chemical, and industrial applications.

Mariusz Szczygiel / Shutterstock.com

## Copper and Its Alloys

*Copper* is a good conductor of electricity. Because of this property, copper is commonly used in many electrical applications, including home wiring. Copper and many of its alloys are also good conductors of heat; this thermal property makes copper a good choice for heat exchanger applications in air conditioning and refrigeration systems. Copper alloys are also used as tubes, pipes, and fittings in plumbing and heating applications.

Copper is alloyed with zinc, tin, aluminum, nickel, and other elements to modify its properties. When copper is alloyed with zinc, it is commonly called **brass**. The mechanical properties of brass depend on the exact composition of percent copper and percent zinc. **Bronze** is an alloy of copper and tin. When copper is alloyed with aluminum, it is

Constantine Pankin / Shutterstock.com

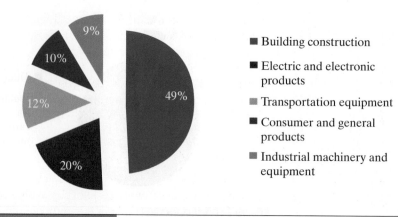

| | |
|---|---|
| ■ | Building construction |
| ■ | Electric and electronic products |
| ■ | Transportation equipment |
| ■ | Consumer and general products |
| ■ | Industrial machinery and equipment |

**FIGURE 10.4**    Copper use by various sectors

---

**EXAMPLE 10.3**

Blackspring / Shutterstock.com

According to the Aluminum Association, 60.2 billion aluminum cans were recycled in a recent year. Given that a 12-ounce empty aluminum soda can has an approximate mass of 0.029 pound-mass (lbm) or 13 grams, let us calculate how many tons of aluminum were recycled.

$$\text{amount of aluminum recycled} = (60.2 \times 10^9 \text{ cans})\left(\frac{0.029 \text{ lbm}}{1 \text{ can}}\right)\left(\frac{1 \text{ ton}}{2,000 \text{ lbm}}\right)$$

$$= 872,900 \text{ tons} \approx 8.73 \times 10^5 \text{ tons}$$

It is important for us to know that it takes less energy to recycle aluminum than to produce new aluminum. The recycling takes approximately 8 percent of the energy required to produce new aluminum.

---

referred to as *aluminum bronze*. Copper and its alloys are also used in water tubes, heat exchangers, hydraulic brake lines, pumps, and screws.

The percentage of copper consumption in various sectors of our economy is shown in Figure 10.4. As shown in this figure, the majority of the copper extracted is consumed in building construction and electrical and electronic products. Other materials are used in the place of copper in our homes. For example, today we use plastic water or drain pipes, optical fibers instead of copper wires, or aluminum alloys in heating and cooling devices such as home or automobile radiators.

## Iron and Steel

Joe Gough / Shutterstock.com

Steel is a common material that is used in the framework of buildings, bridges, in the bodies of appliances such as refrigerators, ovens, dishwashers, washers and dryers, and in cooking utensils. Steel is an alloy of iron with approximately 2 percent or less carbon. Pure iron

OZaiachin / Shutterstock.com

is soft and thus not good for structural applications, but the addition of even a small amount of carbon to iron hardens it and gives steel better mechanical properties such as greater strength.

The properties of steel can be modified by adding other elements, such as chromium, nickel, manganese, silicon, and tungsten. For example, chromium is used to increase the resistance of steel to corrosion. In general, steel can be classified into three broad groups:

1. Carbon steels containing approximately 0.015 to 2 percent carbon
2. Low-alloy steels having a maximum of 8 percent alloying elements
3. High-alloy steels containing more than 8 percent of alloying elements
   Carbon steels constitute most of the world's steel consumption; thus you will commonly find them in the body of appliances and cars. Low-alloy steels have good strength and are commonly used as machine or tool parts and as structural members. High-alloy steels, such as *stainless steel*, contain approximately 10 to 30 percent chromium and up to 35 percent nickel. The 18/8 stainless steels, which contain 18 percent chromium and 8 percent nickel, are commonly used for tableware and kitchenware products. Finally, *cast iron* is also an alloy of iron that has 2 to 4 percent carbon. Note

Svetlana Lukienko / Shutterstock.com

that the addition of extra carbon to the iron changes its properties completely. In fact, cast iron is a brittle material, whereas most iron alloys containing less than 2 percent carbon are ductile. As shown in Figure 10.5, most of the steel consumption goes to steel service centers, construction, and automotive industries.

> Steel is an alloy of iron with approximately 2% or less carbon. The addition of carbon to iron gives steel greater strength.

## Nickel

The percentage of the United States nickel consumption by sector and end-uses is shown in Figures 10.6 and 10.7. As shown in Figure 10.6, over 40 percent of the nickel consumed in the United States goes into stainless and alloy steel production. High-alloy steels, such as stainless steels, contain approximately 10 to 30 percent chromium and up to 35 percent nickel. Figure 10.7 shows the percentage of end-uses of nickel, with transportation leading at 32 percent. Stainless steel accounts for over 60 percent of nickel use worldwide.

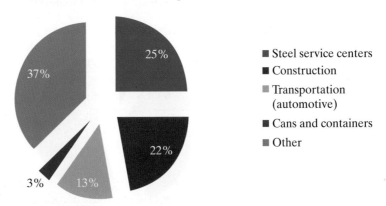

- Steel service centers
- Construction
- Transportation (automotive)
- Cans and containers
- Other

**FIGURE 10.5**    The percentage of steel consumption by sector

5%
9%
44%
42%
Stainless and alloy steel production
Nonferrous alloys and superalloys
Electroplating
Other

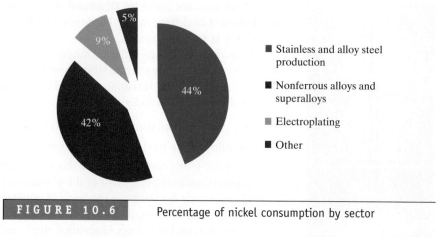

FIGURE 10.6 Percentage of nickel consumption by sector

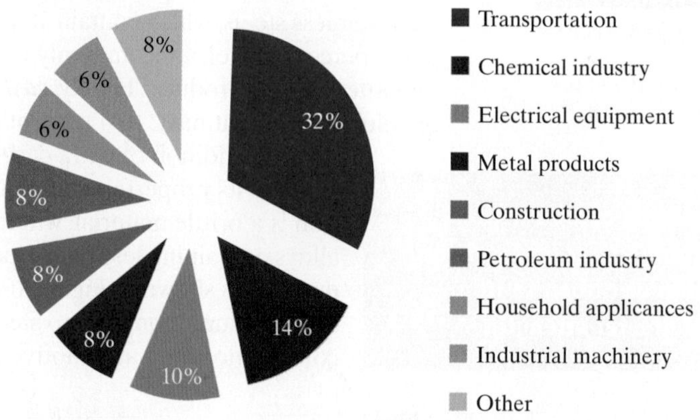

8%
6%
6%
32%
8%
8%
8%
14%
10%
Transportation
Chemical industry
Electrical equipment
Metal products
Construction
Petroleum industry
Household applicances
Industrial machinery
Other

FIGURE 10.7 Percentage of nickel consumption by end-use

## Zinc

As mentioned previously, copper is alloyed with zinc, aluminum, nickel, and other elements to modify its properties. Zinc also is alloyed with other materials to increase the resistance of that material to corrosion. As shown in Figure 10.8,

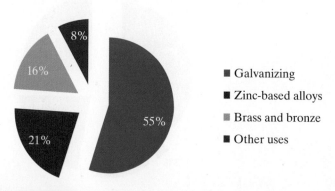

8%
16%
55%
21%
Galvanizing
Zinc-based alloys
Brass and bronze
Other uses

FIGURE 10.8 The percentage of zinc consumption by end-use

55 percent of the zinc consumed is for galvanizing, and 16 percent is used for making brass and bronze. Zinc is also consumed by the rubber, chemical, and paint industries.

---

**EXAMPLE 10.4**

Ivonne Wierink / Shutterstock.com

The body of a typical clothes dryer is constructed from approximately 100 pounds (45 kg) of steel. In a recent year, 6.5 million clothes dryers were sold in the United States. Let us estimate how many tons (or kilograms) of steel went into making the dryers.

amount of steel consumed making the dryers

$$= (6.5 \times 10^6 \ \text{dryer}) \left( \frac{100 \ \text{lbm}}{1 \ \text{dryer}} \right) \left( \frac{1 \ \text{ton}}{2,000 \ \text{lbm}} \right) = 325,000 \ \text{tons}$$

$$= 295 \times 10^6 \ \text{kg}$$

Now think about the amount of steel that goes into making other appliances, such as dishwashers, refrigerators, clothes washing machines, and ovens each year!

---

# Before You Go On

Answer the following questions to test your understanding of the preceding section:

1.  What is a lightweight metal?

2.  What is the difference between steel and iron?

3.  Give examples of applications in which titanium is used.

4.  Give examples of aluminum use.

5.  Give examples of copper use.

*Vocabulary—State the meaning of the following terms:*

Steel

Bronze

Brass

18/8 stainless steel

## LO⁴  10.4  Plastics, Glass, Composites, and Wood

### Plastics

In the latter part of the 20th century, plastics increasingly became the material of choice for many applications. They are lightweight, strong, inexpensive, and easily made into various shapes. Over 100 million metric tons of plastic are produced annually worldwide. Of course, this number increases as the demand for inexpensive, durable, and disposable material grows. Most of you are already familiar with examples of plastic products, including grocery and trash bags, home-cleaning containers, vinyl siding, polyvinyl chloride (PVC) piping, valves, and fittings that are readily available in home improvement centers. Styrofoam™ plates and cups, soft drink containers, plastic forks, knives, spoons, and sandwich bags are other examples of plastic products that are consumed every day.

Africa Studio / Shutterstock.com

*Polymers* are the backbone of what we call plastics. They are chemical compounds that have large, molecular, chainlike structures. Plastics are often classified into two categories: *thermoplastics* and *thermosets*. When heated to certain temperatures, the **thermoplastics** can be molded and remolded. For example, when you recycle Styrofoam dishes, they can be heated and reshaped into cups, bowls, or other shapes. By contrast, **thermosets** cannot be remolded into other shapes by heating. The application of heat to thermosets does not soften the material for remolding; instead, the material simply breaks down. There are many other ways of classifying plastics; for instance, they may be classified on the basis of their chemical composition, molecular structure, the way the molecules are arranged, or their densities. For example, based on their chemical composition, polyethylene, polypropylene, polyvinyl chloride, and polystyrene are the most commonly produced plastics. A grocery bag is an example of a product made from high-density polyethylene (HDPE). However, note that in a broader sense both polyethylene and polystyrene are thermoplastics. In general, the way molecules of a plastic are arranged influences its mechanical and thermal properties.

Plastics have relatively small thermal and electrical conductivity values. Some plastic materials—such as Styrofoam cups—are designed to have air trapped in them to reduce the heat conduction even more. Plastics are easily colored by using various metal oxides. For example, titanium oxide and zinc oxide are used to give a plastic sheet its white color. Carbon is used to give plastic sheets their black color—as in black trash bags. Depending on the application, other additives are added to the polymers to obtain specific characteristics, such as rigidity, flexibility, enhanced strength, or a longer life span, which prevent any change in the appearance or mechanical properties of the plastic over time. As with other materials, research is being performed every day to make plastics stronger and more durable, to control the aging process, to make them less susceptible to sun damage, and to control water and gas diffusion through them. The latter is especially important when the goal is to add shelf-life to food that is wrapped in plastic.

Silicone is a synthetic compound that consists of silicon, oxygen, carbon, and hydrogen. Be sure not to confuse it with silicon, which is a nonmetallic chemical.

## Silicon

***Silicon*** is a nonmetallic chemical element that is used quite extensively in the manufacturing of transistors and various electronic and computer chips. Pure silicon is not found in nature; it is found in the form of silicon dioxide in sands and rocks or found combined with other elements such as aluminum, calcium, sodium, or magnesium in the form that is commonly referred to as *silicates*. Silicon, because of its atomic structure, is an excellent semiconductor—a material whose electrical conductivity properties can be changed to act either as a conductor of electricity or as an insulator (preventer of electricity flow). Silicon is also used as an alloying element with other elements such as iron and copper to give steel and brass certain desired characteristics.

Be sure not to confuse silicon with ***silicones***, which are synthetic compounds consisting of silicon, oxygen, carbon, and hydrogen. You find silicones in lubricants, varnishes, and water-proofing products.

F. ENOT / Shutterstock.com

## Glass

*Glass* is commonly used in products such as windows, light bulbs, housewares (such as drinking glasses), chemical containers, beverage and beer containers, and decorative items. The composition of the glass depends on its application.

**Silica Glass** The most widely used form of glass is soda-lime-silica glass. The materials used in making soda-lime-silica glass include sand (silicon dioxide), limestone (calcium carbonate), and soda ash (sodium carbonate). Other materials are added to create desired characteristics for specific applications. For example, bottle glass contains approximately 2 percent aluminum oxide, and glass sheets contain about 4 percent magnesium oxide. Metallic oxides are also added to give glass various colors. For example, silver oxide gives glass a yellowish stain, and copper oxide gives glass its bluish, greenish color, with the degree of color depending on the amount added to the composition of the glass. Optical glasses have specific chemical compositions and are quite expensive. The composition of optical glass influences its refractive index and its light-dispersion properties.

Glass that is made completely from silica (silicon dioxide) has properties that are sought after by many industries such as fiber optics, but it is expensive to manufacture because the sand has to be heated to temperatures exceeding $1,700°C$. Silica glass has a low coefficient of thermal expansion, high electrical resistivity, and high transparency to ultraviolet light. Because silica glass has a low coefficient of thermal expansion, it is used in high-temperature applications. Ordinary glass has a relatively high coefficient of thermal expansion; therefore, when its temperature is changed suddenly, it breaks easily due to thermal stresses

Dmitry Kalinovsky / Shutterstock.com

zentilia / Shutterstock.com

developed by the temperature rise. Glass cookware contains boric oxide and aluminum oxide to reduce its coefficient of thermal expansion.

**Glass Fibers** *Glass fibers* are commonly used today in fiber optics, which is the branch of science dealing with the transmission of data, voice, and images through thin glass or plastic fibers. Every day, copper wires are replaced by transparent glass fibers in telecommunications to connect computers together in networks. The glass fibers typically have an outer diameter of 0.0125 millimeters (mm) or 12 microns with an inner transmitting core diameter of 0.01 millimeters (mm) or 10 microns. Infrared light signals in the wavelength ranges of 0.8 to 0.9 meters (m) or 1.3 to 1.6 meters (m) are generated by light-emitting diodes or semiconductor lasers and travel through the inner core of glass fiber.

The optical signals generated in this manner can travel to distances as far as 100 kilometers (km) without any need to amplify them again. Plastic fibers made of polymethylmethacrylate, polystyrene, or polycarbonate are also used in fiber optics. These plastic fibers are, in general, cheaper and more flexible than glass fibers. But when compared to glass fibers, plastic fibers require more amplification of signals due to their greater optical losses. They are generally used in networking computers in a building.

## Composites

Because of their light weight and good strength, composite materials are becoming increasingly the materials of choice for a number of products and aerospace applications. Today you find composite materials in artificial teeth, military and commercial planes, helicopters, satellites, fast-food restaurant tables and chairs, and many sporting goods. They are also commonly used to repair the bodies of automobiles. In comparison to conventional materials (such as metals), composite materials can be lighter and stronger. For this reason, composite materials are used extensively in aerospace applications.

*Composites* are created by combining two or more solid materials to make a new material that has properties that are superior to those of the individual components. Composite materials consist of two main ingredients: *matrix materials* and *fibers*. Fibers are embedded in matrix materials, such as ceramics, plastics, aluminum, or other metals.

> Composites are created by combining two or more solid materials to make a new material that has properties that are superior to those of the individual components.

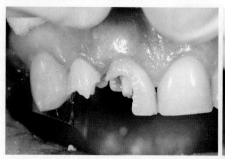
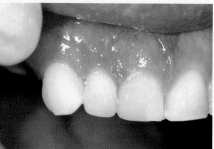

Lighthunter / Shutterstock.com

fabiodevilla / Shutterstock.com

Depending upon what type of host matrix material is used to create the composite material, the composites may be classified into three classes:

1. polymer-matrix composites
2. metal-matrix composites
3. ceramic-matrix composites

Glass, graphite, and silicon-carbide fibers are among those used in the construction of composite materials.

The strength of the fibers is increased when embedded in the matrix material, and the composite material created in this manner is lighter and stronger. Moreover, once a crack starts in a single material due to either excessive loading or imperfections in the material, the crack propagates to the point of failure. In a composite material, on the other hand, if one or a few fibers fail, it does not necessarily lead to failure of other fibers or the material as a whole. Furthermore, the fibers in a composite material can be oriented either in a certain direction or many directions to offer more strength in the direction of expected loads. Therefore, composite materials are designed for specific load applications. For instance, if the expected load is uniaxial— meaning that it is applied in a single direction—then all the fibers are aligned in the direction of the expected load. For applications expecting multi-direction loads, the fibers are aligned in different directions to make the material equally strong in various directions.

## Wood

In Chapter 6, we discussed wood as a fuel and that, because of its abundance in many parts of the world, it has been a material of choice for many applications throughout history. Wood is a *renewable* resource, and because of its ease of workability and strength it has been used to make many products. Wood

Iriana Shiyan / Shutterstock.com

also has been used as fuel in stoves and fireplaces. Today, wood is still used in a variety of products ranging from telephone poles to toothpicks. Common examples of wood products include hardwood flooring, roof trusses, furniture frames, kitchen cabinets, wall supports, doors, decorative items, window frames, trimming in luxury cars, tongue depressors, clothespins, baseball bats, bowling pins, fishing rods, and wine barrels. Wood is also the main ingredient that is used to make various paper products. Whereas a steel structural member is susceptible to rust, wood, on the other hand, is prone to fire, termites, and rotting.

Wood is an anisotropic material, meaning that its properties are direction-dependent. For example, under axial loading (when pulled), wood is stronger in a direction parallel to a grain than it is in a direction across the grain. However, wood is stronger in a direction normal to the grain when it is bent. The properties of wood also depend on its moisture content; the lower the moisture content, the stronger the wood is. The density of wood is generally a good indication of its strength. As a rule of thumb, the higher the density of wood, the higher is its strength. Moreover, any defects (such as knots) affect the load-carrying capacity of wood. Of course, the location of the knot and the extent of the defect also directly affect its strength.

Timber is commonly classified as *softwood* and *hardwood*. **Softwood** timber is made from trees that have cones (coniferous), such as pine, spruce, and Douglas fir. On the other hand, **hardwood** timber is made from trees that have broad leaves or flowers. Examples of hardwoods include walnut, maple, oak, and beech. This classification of wood into softwood and hardwood should be used with caution, because there are some hardwood timbers that are actually softer than softwoods.

symbiot / Shutterstock.com

---

**EXAMPLE 10.5**

A 500-ml plastic PET water bottle has an approximate mass of 10 grams or 0.022 pound mass (lbm). Assuming that we drink at least 2 liters of water every day, how many pounds of plastic went into making enough 500-ml water bottles for 1 billion people?

amount of plastic consumed making water bottles

$$= \left( \frac{4 \times 10^9 \ \text{bottles}}{1 \ \text{day}} \right) \left( \frac{0.022 \ \text{lbm}}{1 \ \text{bottle}} \right) \left( \frac{365 \ \text{days}}{1 \ \text{year}} \right) \left( \frac{1 \ \text{ton}}{2,000 \ \text{lbm}} \right)$$

$$= 16,060,000 \ \text{tons} \approx 16 \ \text{million tons!}$$

Now think about the amount of plastic that goes into making other products each year!

# Before You Go On

Answer the following questions to test your understanding of the preceding section:

1. Give examples of plastics in use.

2. What is the difference between thermoplastics and thermosets?

3. What is the difference between silicon and silicone?

4. What materials are used in making soda-lime-silica glass?

*Vocabulary—State the meaning of the following terms:*

Polymer

Thermoplastics

Silicon

Silicone

Composite material

Softwood

Hardwood

## LO⁴ 10.5 Concrete

Today, concrete is commonly used in the construction of roads, bridges, buildings, tunnels, and dams. What is normally called *concrete* consists of three main ingredients: *aggregate, cement,* and *water.* Aggregate refers to materials such as gravel and sand, and cement refers to the bonding material that holds the aggregate together. The type and size (fine to coarse) of aggregate used in making concrete varies depending on the application. The amount of water used in making concrete (water-to-cement ratio) also influences its strength. Of course, the mixture must have enough water so that the concrete can be poured and have a consistent cement paste that completely wraps around all aggregates. The ratio of amount of cement to aggregate used in making concrete also affects the strength and durability of concrete.

> Concrete is a mixture of cement, aggregate (such as sand and gravel), and water.

Another factor that influences the cured strength of concrete is the temperature of its surroundings when it is poured. Calcium chloride is added to cement when the concrete is poured in cold climates. This addition of calcium chloride accelerates the curing process to counteract the effect of the low temperature of the surroundings. You may have also noticed that, as you walk by newly poured concrete for a driveway

logoboom / Shutterstock.com

or sidewalk, water is sprayed onto the concrete for some time after it is poured. This is to control the rate of contraction of the concrete as it sets.

Concrete is a brittle material that supports compressive loads much better than it does tensile loads. Because of this fact, concrete is commonly ***reinforced*** with steel bars or a steel mesh of thin metal rods to increase its load-bearing capacity, especially in the sections where tensile stress is expected. Concrete is poured into forms that contain the metal mesh or steel bars. Reinforced concrete is used in foundations, floors, walls, and columns. Another common construction practice is the use of ***precast concrete***. Precast concrete slabs, blocks, and structural members are fabricated in less time with less cost in factory settings where surrounding conditions are controlled. The precast concrete parts are then moved to the construction site where they are erected. This practice saves time and money.

As we mentioned previously, concrete has a higher compressive strength than tensile strength. Because of this fact, concrete is also ***prestressed*** in the following manner. Before concrete is poured into forms that have the steel rods or wires, the steel rods or wires are stretched; after the concrete has been poured and after enough time has elapsed, the tension in the rods or wires is released. This process, in turn, compresses the concrete. The prestressed concrete then acts as a compressed spring, which becomes uncompressed under

wrangler / Shutterstock.com

the action of tensile loading. Therefore, the prestressed concrete section will not experience any tensile stress until the section has been completely uncompressed. It is important to note once again the reason for this practice is that concrete is weak under tension.

The percentage of cement use by various sectors is shown in Figure 10.9. As you might expect, the economy and construction projects in particular define the amount of cement produced.

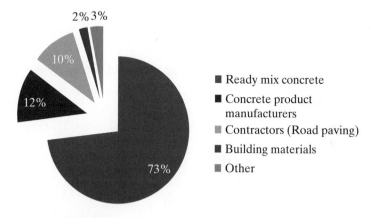

2% 3%
10%
12%
73%

■ Ready mix concrete
■ Concrete product manufacturers
■ Contractors (Road paving)
■ Building materials
■ Other

**FIGURE 10.9**    The percentage of cement use by sectors

**EXAMPLE 10.6**

Typical two-car driveways are 18 feet wide, 20 feet long, and 4 inches thick. According to the U.S. Census Bureau's recent survey, there are approximately 133 million housing units in the United States. Let us now estimate the volume of concrete used in making two-car driveways for 10 million houses. If the average capacity of a concrete delivery truck is approximately 200 ft³, how many truck loads did it take to pour the concrete?

volume of concrete consumed in making 10 million driveways

$$= (10 \times 10^6 \ \cancel{\text{driveways}}) \left( \frac{(18 \text{ ft})(20 \text{ ft})(4 \ \cancel{\text{in}})\left( \dfrac{1 \text{ ft}}{12 \ \cancel{\text{in}}}\right)}{1 \ \cancel{\text{driveway}}} \right) = 12 \times 10^8 \text{ ft}^3$$

number of truck loads

$$= (12 \times 10^8 \ \cancel{\text{ft}^3}) \left( \frac{1 \text{ truck load}}{200 \ \cancel{\text{ft}^3}} \right) = 6,000,000 \text{ truck loads}$$

1.2 billion cubic feet of concrete and 6 million truck loads! Now think about the volume of the concrete poured every year to build sidewalks, roads, buildings, and so on. Also, think about how much energy it takes to extract and deliver the materials that make up the concrete to job sites.

karamysh / Shutterstock.com

## Before You Go On

Answer the following questions to test your understanding of the preceding section:

1. What are the main ingredients of concrete?

2. How is prestressed concrete made?

3. What is precast concrete?

*Vocabulary—State the meaning of the following terms:*

Aggregate

Precast concrete

Prestressed concrete

# SUMMARY

## LO¹  Earth—Our Home

You should have a good understanding of the structure of the Earth: its size, layers, and its chief chemical composition. You also should be familiar with the basic elements such as aluminum, zinc, iron, copper, nickel, and magnesium that we extract from the Earth to make products. The Earth has a spherical shape with a diameter of 7,926.4 miles (12,756.3 kilometers) and an approximate mass of $13.17 \times 10^{24}$ pounds ($5.98 \times 10^{24}$ kilograms). It is divided into major layers that are located above and below its surface. The Earth's mass is composed mostly of iron, oxygen, and silicon (approximately 32 percent iron, 30 percent oxygen, and 15 percent silicon). It also contains other elements such as sulfur, nickel, magnesium, and aluminum. The structure below the Earth's surface is generally grouped into four layers: crust, mantle, outer core, and inner core. The crust makes up about 0.5 percent of the Earth's total mass and 1 percent of its volume. The mantle is made up of molten rock that lies underneath the crust and makes up nearly 84 percent of the Earth's volume. The inner core and outer core make up about 33 percent of the Earth's mass and 15 percent of its volume. The inner core is considered to be solid, while the outer core is thought to be fluid and composed mainly of iron.

## LO²  The Phases of Matter and Properties of Materials

You should understand the phases of matter and the basic properties of materials, such as density, thermal conductivity, and viscosity. Matter may exist as solid, liquid, gaseous, or plasma, and can change phase when its condition or its surroundings are changed. A good example is water: it may exist in a solid *(ice)*, liquid *(water)*, or gaseous *(steam)* form.

Material properties depend on many factors, including the exact chemical composition and how the material was processed. Material properties also change with temperature and time as the material ages. In your own words, you should be able to explain some of the basic material properties. For example, the value of electrical resistivity is a measure of the resistance of a material to the flow of electricity; density is a measure of how compact the material is for a given volume; or thermal conductivity is a property of a material that shows how good the material is in transferring thermal energy (*heat*) from a high-temperature region to a low-temperature region within the material.

## LO³  Metals

You should also be familiar with common applications of basic materials, such as light metals and their alloys, along with steel and its alloys. Aluminum, titanium, and magnesium, because of their small densities (relative to steel), are commonly referred to as *lightweight metals* and are used in many structural and aerospace applications.

Aluminum and its alloys have densities that are approximately one-third the density of steel. Aluminum is commonly alloyed with other metals such as copper, zinc, and magnesium. Everyday examples of common aluminum products include beverage cans, household aluminum foil, staples that won't rust in tea bags, building insulation, and so on.

Titanium has an excellent strength-to-weight-ratio. Titanium is used in applications where relatively high temperatures, from 400 up to 600°C, are expected. Titanium alloys are used in the fan and compressor blades of engines of commercial and military airplanes. Titanium alloys also are used in golf clubs, bicycle frames, and spectacle frames.

Magnesium is another lightweight metal that looks like aluminum but is lighter than it. It is commonly alloyed with other elements such as aluminum, manganese, and zinc to improve its properties. Magnesium and its alloys are used in nuclear applications, in dry cell batteries, and in aerospace applications.

Copper is a good conductor of electricity and heat; because of these properties, it is commonly used in many electrical, heating, and cooling applications. Copper alloys are also used as tubes, pipes, and fittings in plumbing. When copper is alloyed with zinc, it is commonly called *brass*. *Bronze* is an alloy of copper and tin.

Steel is a common material used in the framework of buildings, bridges, the body of appliances such as refrigerators, ovens, dishwashers, clothes-washers and dryers, and in cooking utensils. Steel is an alloy of iron with approximately 2 percent or less carbon. The properties of steel can be modified by adding other elements, such as chromium, nickel, manganese, silicon, and tungsten. The 18/8 stainless steels, which contain 18 percent chromium and 8 percent nickel, are commonly used for tableware and kitchenware products. Cast iron is also an alloy of iron that has 2 to 4 percent carbon.

## LO⁴  Plastics, Glass, Composites, and Wood

Plastic products include grocery and trash bags, soft drink containers, home-cleaning containers, vinyl siding, polyvinyl chloride (PVC) piping, valves, and fittings. Styrofoam™ plates and cups, plastic forks, knives, spoons, and sandwich bags are other examples of plastic products that are consumed every day. *Polymers* are the backbone of what we call plastics. They are chemical compounds that have large, chainlike molecular structures. Plastics are often classified into two categories: *thermoplastics* and *thermosets*. When heated to certain temperatures, thermoplastics can be molded and remolded. By contrast, thermosets cannot be remolded into other shapes by heating.

Silicon is a nonmetallic chemical element that is used quite extensively in the manufacturing of transistors and various electronic and computer chips. It is found in the form of silicon dioxide in sands and rocks or combined with other elements such as aluminum or calcium or sodium or magnesium in the form that is commonly referred to as *silicates*. Silicon, because of its atomic structure, is an excellent semiconductor, which is a material whose electrical conductivity properties can be changed to act either as a conductor of electricity or as an insulator (preventer of electricity flow).

Glass is commonly used in products such as windows, light bulbs, houseware such as drinking glasses, chemical containers, beverage and beer containers, and decorative items. The composition of the glass depends on its application. The most widely used form of glass is soda-lime-silica glass. The materials used in making soda-lime-silica glass include sand (silicon dioxide), limestone (calcium carbonate), and soda ash (sodium carbonate). Other materials are added to create desired characteristics for specific applications. Silica glass fibers are commonly used today in fiber optics, which is a branch of science that deals with transmitting data, voice, and images through thin glass or plastic fibers.

Composite materials are found in military planes, helicopters, satellites, commercial planes, fast-food restaurant tables and chairs, and many sporting goods. In comparison to conventional materials, such as metals, composite materials can be lighter and stronger. Composite materials consist of two main ingredients: matrix materials and fibers. Fibers are embedded in matrix materials, such as aluminum or other metals, plastics, or ceramics. Glass, graphite, and silicon carbide fibers are examples of fibers used in the construction of composite materials. The strength of the fibers is increased when embedded in the matrix material, and the composite material created in this manner is lighter and stronger.

Common examples of wood products include hardwood flooring, roof trusses, furniture frames, wall supports, doors, decorative items, window frames, kitchen cabinets, trimming in luxury cars, tongue depressors, clothespins, baseball bats,

bowling pins, fishing rods, and wine barrels. Timber is commonly classified as either *softwood* or *hardwood*. Softwood timber is made from trees that have cones (coniferous), such as pine, spruce, and Douglas fir. On the other hand, hardwood timber is made from trees that have broad leaves or flowers.

## LO⁵ Concrete

Concrete is used in the construction of roads, bridges, buildings, tunnels, and dams. It consists of three main ingredients: aggregate, cement, and water. Aggregate refers to materials such as gravel and sand, and cement refers to the bonding material that holds the aggregate together. Concrete is usually *reinforced* with steel bars or steel mesh that consists of thin metal rods to increase its load-bearing capacity. Another common construction practice is the use of *precast concrete*. Precast concrete slabs, blocks, and structural members are fabricated in less time with lower costs in factory settings where the surrounding conditions are controlled. Because concrete has a higher compressive strength than tensile strength, it is *prestressed* by pouring it into forms that have steel rods or wires. The steel rods or wires are stretched, so the prestressed concrete then acts as a compressed spring, which becomes uncompressed under the action of tensile loading.

## KEY TERMS

Brass 311
Bronze 311
Cast Iron 313
Composites 318
Compression Strength 307
Concrete 321
Crust 301
Density 306
Electrical Resistivity 306
Glass Fibers 318

Hardwood 320
Heat Capacity 307
Inner Core 301
Lightweight Metals 309
Mantle 301
Outer Core 301
Precast Concrete 322
Prestressed Concrete 322
Reinforced Concrete 322
Silicon 317

Silicone 317
Softwood 320
Stainless Steel 313
Strength-to-Weight Ratio 306
Tensile Strength 307
Thermal Conductivity 307
Thermal Expansion 307
Thermoplastics 316
Thermosets 316
Viscosity 307

# Apply What You Have Learned

1. Every day we use a wide range of paper products at home and school. These paper products are made from different paper grades. Wood pulp is the main ingredient used in making a paper product, and a common practice is to grind the wood first and cook it with some chemicals. Investigate the composition, processing methods, and annual consumption rate of the following grades of paper products in the United States or your country. Write a brief report discussing your findings. The paper products to investigate should include: printing paper, sanitary paper, glassine and waxing paper, paper bags, cardboard boxes, and paper towels.

HomeStudio / Shutterstock.com

2. The 18/8 stainless steels, which contain 18 percent chromium and 8 percent nickel, are commonly used for tableware and kitchenware products. Investigate how much iron, carbon, chromium, and nickel are consumed in a typical home with standard kitchen and silverware sets. Write a brief report to your instructor discussing your assumptions and findings.

AlenKadr / Shutterstock.com

# PROBLEMS

*Problems that promote life-long learning are denoted by* 🔑

**10.1** Identify and list at least five different materials that are used in making a car.

**10.2** Name at least five different materials that are used in making a refrigerator.

**10.3** Identify and list at least five different materials that are used in making your TV set or computer.

**10.4** List at least five different materials that are used in making a building envelope (walls, floors, roofs, windows, and doors).

**10.5** List at least five different materials used to fabricate window and door frames.

**10.6** List some of the materials used in the fabrication of compact fluorescent lights.

**10.7** Identify at least five products around your home that contain plastics.

**10.8** 🔑 In a brief report, discuss the advantages an disadvantages of using Styrofoam or paper for coffee or tea cups.

**10.9** 🔑 As you already know, roofing materials keep water from penetrating into the roof structure. There is a wide range of roofing products available on the market today. For example, asphalt shingles, which are made by impregnating a dry felt with hot asphalt, are used on some houses. Other houses use wood shingles, such as red cedar or redwood. A large number of houses use interlocking clay tiles as roofing materials. Investigate the properties and the characteristics of various roofing materials. Write a brief report discussing your findings.

**10.10** Visit a home improvement center (hardware/lumber store) in your town and gather information about various types of insulating materials that are used in houses. Write a brief report discussing the advantages, disadvantages, and characteristics of various insulating materials, including their thermal characteristics in terms of R-value.

**10.11** Investigate the characteristics of titanium alloys used in sporting equipment such as bicycle frames, tennis racquets, and golf club shafts. Write a brief summary report discussing your findings.

**10.12** Investigate the characteristics of titanium alloys used in medical implants for hips and other joint replacements. Write a brief summary report discussing your findings.

**10.13** Cobalt-chromium alloys, stainless steel, and titanium alloys are three common biomaterials that have been used as surgical implants. Investigate the use of these biomaterials, and write a brief report discussing the advantages and disadvantages of each.

**10.14** Endoscopy refers to a medical examination of the inside of a human body by means of inserting a lighted optical instrument through a body opening. Fiberscopes operate in the visible wavelengths and consist of two major components. One part consists of a bundle of fibers that illuminates the examined area, and the other transmits the images of the examined area to the physician through a display device. Investigate the design of fiberscopes or the fiber-optic endoscope, and discuss your findings in a brief report.

**10.15** Crystal glass tableware that sparkles is sought after by many people as a sign of affluence. This crystal commonly contains lead monoxide. Investigate the properties of crystal glass in detail, and write a brief report discussing your findings.

**10.16** You all have seen grocery bags that have labels and printed information on them. Investigate how information is printed on plastic bags. For example, a common practice includes using a wet-inking process; another process makes use of lasers and heat-transfer decals. Discuss your findings in a brief report.

**10.17** Teflon™ and Nylon™ are trade names of plastics that are used in many products. Look up the actual chemical names of these products, and give at least five examples of where they are used.

**10.18** Investigate how the following basic wood products are made: plywood, particle board, veneer, and fiberboard. Also investigate common methods of wood preservation. Discuss your findings in a brief report. What is the environmental impact of both the production and use of treated wood products in this question?

**10.19** Investigate the common uses of cotton and its typical properties. Discuss your findings in a brief report.

**10.20** Look around your home and estimate how many meters or feet of *visible* copper wire are in use. Consider extension and power cords for common items such as your hairdryer, TV, cell phone charger, computer charger, lamps, printer cable, refrigerator, microwave oven, and so on. Write a brief report and discuss your findings.

**10.21** How many cans or glasses of soda or juice do you drink every day? Estimate your annual aluminum and or glass consumption. Express your results in kilograms or pounds per year.

**10.22** Investigate how much steel was used in making the following appliances: clothes dryer, dishwasher, refrigerator, and oven. Discuss your findings in a brief report.

**10.23** This is a group assignment. Investigate how much concrete is used to make a sidewalk or walkway. Estimate the amount of concrete used to make walkways on your campus. Discuss your findings and assumptions in a brief report.

**10.24** Estimate the amount of paper that you use every year. Consider your printing habits and needs, loose and bonded paper, and book, magazine, and newspaper consumption. How much wood would it take to meet your demand? Discuss your assumptions and findings in a brief report.

**10.25** Investigate where the major nickel mines are located in the world and how much they produce. Discuss your findings in a brief report.

**10.26** Investigate where the major zinc mines are located in the world and how much they produce. Discuss your findings in a brief report.

**10.27** Investigate where the major aluminum mines are located in the world and how much they produce. Discuss your findings in a brief report.

**10.28** By some estimates, we consume twice as many goods as we did fifty years ago. As a result, our appetite for raw materials—from wood to steel—keeps increasing. The rise in world population and our increased standard of living only exacerbates this problem. What can you do to reverse this trend? Discuss your suggestions as backed up by data in a brief report.

Stocksnapper / Shutterstock.com

*"The selfish spirit of commerce knows no country, and feels no passion or principle but that of gain."*

— THOMAS JEFFERSON (1743–1826)

# Municipal and Industrial Waste and Recycling

## LEARNING OBJECTIVES

**LO¹**   Municipal Waste: describe how we generate waste through the activities of our daily lives

**LO²**   Industrial Waste: describe industrial waste and how much of it is generated

**LO³**   Recycling and Composting: understand the importance of recycling and composting

# *Discussion Starter*

## Municipal Solid Waste (MSW) Generation in the United States

Each year, the U.S. Environmental Protection Agency (EPA) collects and reports data on the generation and disposal of waste in the United States. According to the latest EPA data, in 2012,

*"Americans generated about 251 million tons of trash and recycled and composted nearly 86.6 million tons of this material. On average, 1.51 pounds out of 4.38 pounds waste was recycled (or decomposed) per person per day. These values correspond to a 34.5 percent recycling rate."*

**Source**: United States Environmental Protection Agency, Municipal Solid Waste Generation, Recycling, and Disposal in the United States: Facts and Figures for 2012, https://www.epa.gov/sites/production/files/2015-09/documents/2012_msw_fs.pdf

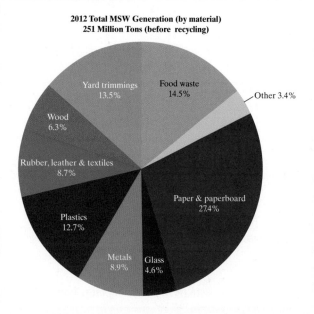

2012 Total MSW Generation (by material)
251 Million Tons (before recycling)

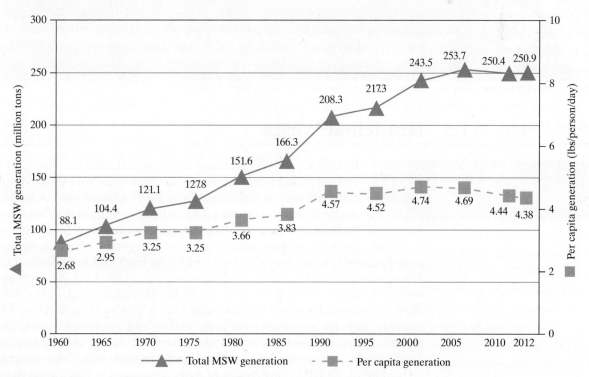

Municipal solid waste (MSW) generation rates from 1960 to 2012

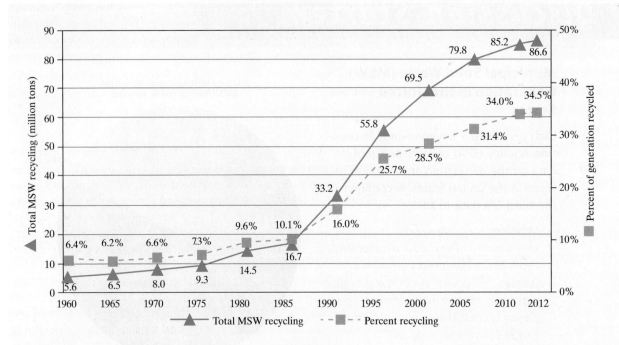

Municipal solid waste (MSW) recycling rates from 1960 to 2012

**To the Students:** What do you think is your MSW footprint every year? How much material do you think you recycle each year? What material do you think constitutes the largest amount of your waste?

## LO¹  11.1  Municipal Waste

Waste can be classified into municipal and industrial waste.

In Chapter 10, we discussed common materials that are used in making products that we use in our daily lives. We examined solid materials such as metals and their alloys, plastics, glass, wood, and materials that solidify over time such as concrete. In this chapter, we look at what happens to these materials when they are discarded.

Waste can be classified into two broad categories of *municipal* and *industrial* waste. **Municipal waste** is basically the trash that we, as individuals, throw away every day. It consists of items such as food scraps, packaging materials, bottles, tissue paper, and so on. On the other hand, as the name implies, **industrial waste** refers to waste that is produced in industry. This category includes materials from construction and renovation, demolition materials, medical discards, and waste generated during the exploration, development, and production of fossil fuels, mining rocks, and minerals. According to the EPA, the amount of waste generated in the United States has been increasing from 2.7 pounds per person per day in 1960 to 4.4 pounds per person per day in 2012. In 2012, we generated about 250 million tons of

The amount of municipal waste generated in the United States has been increasing from 2.7 pounds per person per day in 1960 to 4.4 pounds per person per day in 2010.

municipal waste in the United States. Think about this number! As good global citizens we should be concerned not only about the space that our waste occupies in landfills, but we also need to think carefully about the entire life cycle of a piece of material or a product. For example, when you throw away a piece of wrapping paper, you need to think about *all* of the natural resources used: the trees that were harvested to make the paper and the amount of energy (e.g., fossil fuel) that was consumed to produce, process, transport, and finally dispose of it.

The amount of trash generated in the United States by material type is shown in Figure 11.1. When examining this figure, note that items such as food scraps, plastics, and paper and paperboard make up a large portion of our waste. To better understand the amount and type of products that are discarded, it is customary to group trash into additional sub-categories. For example, the EPA reports the amount of waste generation and recovery using sub-categories such as *durable goods, nondurable goods, containers and packaging materials, and plastic packaging.*

***Durable goods*** cover products such as small and major appliances, furniture, carpets and rugs, rubber tires, and lead-acid batteries. By ***nondurable goods***, the EPA refers to products such as office paper, newspapers, books, magazines, paper plates and cups, tissue paper and paper towels, disposable

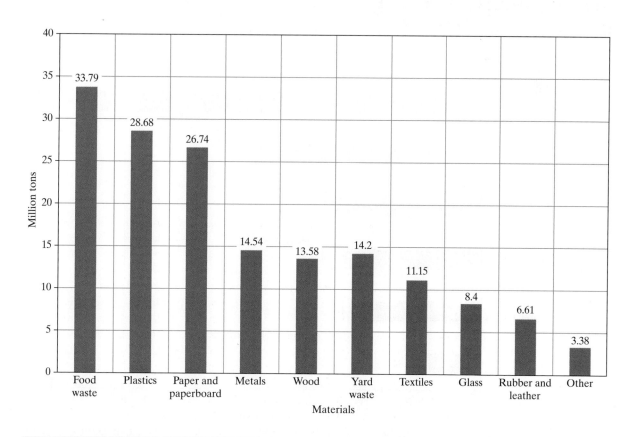

**FIGURE 11.1**  The amount of municipal solid waste discarded (by material) in 2010
*Source:* United States Environmental Protection Agency

diapers, plastic plates and cups, trash bags, clothing and footwear, towels, sheets, and so on. The *containers and packaging* category covers glass packing (beer and soft drink bottles, wine and liquor bottles, and jars), steel packing (cans), aluminum packing (beer and soft drink cans and foil), paper and paper packing (corrugated boxes, bags and sacks, wrapping paper, etc.). *Plastic packaging* includes polyethylene terephthalate (PET) bottles and jars, high-density polyethylene (HDPE) bottles, and bags, sacks, and wraps. Other wastes include food scraps and yard trimmings. Let us now look at each of these categories in more detail.

## Plastics in Products

The amount of plastics in durable goods, in nondurable goods, and in containers and packing that was generated, recovered, and discarded in 2010 is shown in Figure 11.2. The detailed breakdown of plastics found in various products is shown in Table 11.1. When examining Table 11.1, note that we generated about 890,000 tons of plastic plates and cups, 980,000 tons of trash bags, 2,670,000 tons of plastic bottles and jars, and 3,930,000 tons of bags, sacks, and wraps. How much are you contributing to this waste?

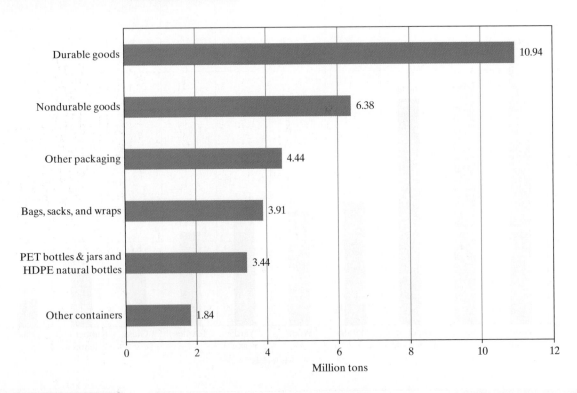

| | |
|---|---|
| **FIGURE 11.2** | The amount of plastic products generated in MSW (2010 data) |
|  | *Source:* Based on data from the Franklin Associates |

| TABLE 11.1 | Plastics in Products in MSW (2010 Data, in Thousands of Tons and Percent of Generation by Resin) |

| Products category | Generation (Thousand tons) | Recovery (Thousand tons) | Recovery (Percent of generation) | Discards (Thousand tons) |
|---|---|---|---|---|
| **Durable Goods** | | | | |
| PET | 160 | | | |
| HDPE | 1,170 | | | |
| PVC | 240 | | | |
| LDPE/LLDPE | 2,000 | | | |
| PP | 4,070 | | | |
| PS | 580 | | | |
| Other resins | 2,740 | | | |
| *Total Plastics in Durable Goods* | **10,960** | **700** | **6.4%** | **10,260** |
| **Nondurable Goods** | | | | |
| Plastic Plates and Cups[§] | | | | |
| LDPE/LLDPE | 20 | | | 20 |
| PLA | 10 | | | 10 |
| PP | 140 | | | 140 |
| PS | 720 | | | 720 |
| *Subtotal Plastic Plates and Cups* | **890** | **Neg.** | **Neg.** | **890** |
| Trash Bags | | | | |
| HDPE | 230 | | | 230 |
| LDPE/LLDPE | 750 | | | 750 |
| *Subtotal Trash Bags* | **980** | | | **980** |
| All Other Nondurables[*] | | | | |
| PET | 440 | | | 440 |
| HDPE | 510 | | | 510 |
| PVC | 270 | | | 270 |
| LDPE/LLDPE | 1,180 | | | 1,180 |
| PLA | 30 | | | 30 |
| PP | 1,290 | | | 1,290 |
| PS | 210 | | | 210 |
| Other resins | 600 | | | 600 |
| *Subtotal All Other Nondurables* | **4,530** | | | **4,530** |
| **Total Plastics in Nondurable Goods, by resin** | | | | |
| PET | 440 | | | 440 |
| HDPE | 740 | | | 740 |
| PVC | 270 | | | 270 |
| LDPE/LLDPE | 1,950 | | | 1,950 |

(continued)

**TABLE 11.1**     Plastics in Products in MSW (2010 Data, in Thousands of Tons and Percent of Generation by Resin) *(continued)*

| Products category | Generation (Thousand tons) | Recovery (Thousand tons) | Recovery (Percent of generation) | Discards (Thousand tons) |
|---|---|---|---|---|
| **Total Plastics in Nondurable Goods, by resin *(continued)*** | | | | |
| PLA | 40 | | | 40 |
| PP | 1,430 | | | 1,430 |
| PS | 930 | | | 930 |
| Other resins | 600 | | | 600 |
| ***Total Plastics in Nondurable Goods*** | **6,400** | **Neg.** | **Neg.** | **6,400** |
| **Plastic Containers & Packaging** | | | | |
| Bottles and Jars* | | | | |
| PET | 2,670 | 560 | 21.0% | 2,110 |
| Natural Bottles† | | | | |
| HDPE | 800 | 220 | 27.5% | 580 |
| Other Plastic Containers | | | | |
| HDPE | 1,450 | 280 | 19.3% | 1,170 |
| PVC | 30 | Neg. | | 30 |
| LDPE/LLDPE | 30 | Neg. | | 30 |
| PP | 240 | 20 | 8.3% | 220 |
| PS | 80 | Neg. | | 80 |
| ***Subtotal Other Containers*** | **1,830** | **300** | **16.4%** | **1,530** |
| Bags, Sacks, and Wraps | | | | |
| HDPE | 690 | 30 | 4.3% | 660 |
| PVC | 50 | | | 50 |
| LDPE/LLDPE | 2,380 | 420 | 17.6% | 1,960 |
| PP | 680 | | | 680 |
| PS | 130 | | | 130 |
| ***Subtotal Bags, Sacks, and Wraps*** | **3,930** | **450** | **11.5%** | **3,480** |
| Other Plastics Packaging‡ | | | | |
| PET | 710 | Neg. | | 710 |
| HDPE | 600 | 40 | 6.7% | 560 |
| PVC | 320 | Neg. | | 320 |
| LDPE/LLDPE | 1,070 | Neg. | | 1,070 |
| PLA | 10 | Neg. | | 10 |
| PP | 1,110 | 40 | 3.6% | 1,070 |
| PS | 340 | 20 | 5.9% | 320 |
| Other resins | 290 | 30 | 10.3% | 260 |
| ***Subtotal Other Packaging*** | **4,450** | **130** | **2.9%** | **4,320** |

| TABLE 11.1 | Plastics in Products in MSW (2010 Data, in Thousands of Tons and Percent of Generation by Resin) *(continued)* |
|---|---|

| Products category | Generation (Thousand tons) | Recovery (Thousand tons) | Recovery (Percent of generation) | Discards (Thousand tons) |
|---|---|---|---|---|
| **Total Plastics in Containers and Packaging, by resin** | | | | |
| PET | 3,380 | 560 | 16.6% | 2,820 |
| HDPE | 3,540 | 570 | 16.1% | 2,970 |
| PVC | 400 | | | 400 |
| LDPE/LLDPE | 3,480 | 420 | 12.1% | 3,060 |
| PLA | 10 | | | 10 |
| PP | 2,030 | 60 | 3.0% | 1,970 |
| PS | 550 | 20 | 3.6% | 530 |
| Other resins | 290 | 30 | 10.3% | 260 |
| ***Total Plastics in Containers and Packaging*** | **13,680** | **1,660** | **12.1%** | **12,020** |
| **Total Plastics in MSW, by resin** | | | | |
| PET | 3,980 | 560 | 14.1% | 3,420 |
| HDPE | 5,450 | 570 | 10.5% | 4,880 |
| PVC | 910 | | | 910 |
| LDPE/LLDPE | 7,430 | 420 | 5.7% | 7,010 |
| PLA | 50 | | | 50 |
| PP | 7,530 | 60 | 0.8% | 7,470 |
| PS | 2,060 | 20 | 1.0% | 2,040 |
| Other resins | 3,630 | 730 | 20.1% | 2,900 |
| ***Total Plastics in MSW*** | **31,040** | **2,360** | **7.6%** | **28,680** |

§ Due to source data aggregation, PET cups are included in "Other Plastic Packaging".

* All other nondurables include plastics in disposable diapers, clothing, footwear, etc.

** Injection stretch blow molded PET containers as defined in the *2008 Report on Postconsumer PET Container Recycling Activity Final Report*. National Association for PET Container Resources.
Prior to 2010, caps and labels recovered with PET bottles and jars were included in the PET recovery estimate. Beginning in 2010, these recovered materials were included with other plastic packaging.

† White translucent homopolymer bottles as defined in the *2007 United States National Postconsumer Plastics Bottles Recycling Report*. American Chemistry Council and the Association of Postconsumer Plastic Recyclers.

Neg. = negligible, less than 5,000 tons

HDPE = High density polyethylene

LDPE = Low density polyethylene

LLDPE = Linear low density polyethylene

PET = Polyethylene terephthalate

PLA = Polylactide

PP = Polypropylene

PS = Polystyrene

PVC = Polyvinyl chloride

‡ Other plastic packaging includes coatings, closures, lids, PET cups, caps, clamshells, egg cartons, produce baskets, trays, shapes, loose fill, etc. Some detail of recovery by resin omitted due to lack of data.

*Source*: Based on Franklin Associates, A Division of ERG

**EXAMPLE 11.1**

According to the 2010 census, the United States total population reached 308,745,538 (nearly 309 million) people. Given the number of people and the 250 million tons of waste that was produced in that year, let us calculate the waste generation per person per day.

We can calculate the average waste generation per person per day (for 2010) by taking the following steps:

waste generation per person per day

$$= \left( \frac{250{,}000{,}000 \; \cancel{\text{tons}} \text{ of trash}}{\cancel{\text{year}} \cdot \cancel{\text{US population}}} \right) \left( \frac{2{,}000 \text{ pounds}}{1 \; \cancel{\text{ton}}} \right) \left( \frac{1 \; \cancel{\text{year}}}{365 \text{ days}} \right) \left( \frac{\cancel{\text{US population}}}{308{,}745{,}538 \text{ persons}} \right)$$

$$= 4.43 \left( \frac{\text{pounds}}{\text{person} \cdot \text{day}} \right)$$

Note: the units read pounds per person per day.

**EXAMPLE 11.2**

In 2010, 2,670,000 tons of plastic bottles and jars were produced. Let us examine what this value represents on average in terms of plastic bottles and jars generated per person per year.

plastic bottle and jar generation per person per year

$$= \left( \frac{2{,}670{,}000 \; \cancel{\text{tons}}}{\cancel{\text{year}} \cdot \cancel{\text{US population}}} \right) \left( \frac{2{,}000 \text{ pounds}}{1 \; \cancel{\text{ton}}} \right) \left( \frac{\cancel{\text{US population}}}{308{,}745{,}538 \text{ persons}} \right)$$

$$= 17.3 \left( \frac{\text{pounds}}{\text{person} \cdot \text{year}} \right)$$

Note: the units read pounds per person per year.

## Paper and Paperboard Products

Each year in the United States, we also generate and discard a relatively large amount of paper and paperboard products. Table 11.2 presents the amount of paper and paperboard products from nondurable goods and in containers

Schlegelfotos / Shutterstock.com                imanhakim / Shutterstock.com

| **TABLE 11.2** | Paper and Paperboard Products in MSW, (2010 in thousands of tons and percent of generation) |
|---|---|

| Product category | Generation (Thousand tons) | Recovery (Thousand tons) | Recovery (Percent of generation) | Discards (Thousand tons) |
|---|---|---|---|---|
| **Nondurable Goods** | | | | |
| Newspapers/Mechanical Papers[†] | 9,880 | 7,070 | 71.6% | 2,810 |
| Books | 990 | | | |
| Magazines | 1,590 | | | |
| Office-type Papers[*] | 5,260 | | | |
| Standard Mail[**] | 4,340 | | | |
| Other Commercial Printing | 2,480 | | | |
| Paper Plates and Cups | 1,350 | | | |
| Other Nonpackaging Paper[*] | 4,190 | | | |
| Subtotal Nondurable Goods | | | | |
|     excluding Newspapers/ Mechanical Papers[§] | 23,690 | 10,650 | 45.0% | 13,040 |
| ***Total Paper and Paperboard Nondurable Goods*** | **33,570** | **17,720** | **52.8%** | **15,850** |
| **Containers and Packaging** | | | | |
| Corrugated Boxes | 29,050 | 24,690 | 85.0% | 4,360 |
| Gable Top/Aseptic Cartons[‡] | 540 | | | |
| Folding Cartons | 5,470 | | | |
| Other Paperboard Packaging | 90 | | | |
| Bags and Sacks | 1,040 | | | |
| Other Paper Packaging | 1,490 | | | |
| Subtotal Containers and Packaging | | | | |
|     excluding Corrugated Boxes[§] | 8,630 | 2,160 | 25.0% | 6,470 |
| ***Total Paper and Paperboard Containers and Packaging*** | **37,680** | **26,850** | **71.3%** | **10,830** |
| ***Total Paper and Paperboard[^]*** | **71,250** | **44,570** | **62.6%** | **26,680** |

[†] Starting in 2010, newsprint and groundwood inserts expanded to include directories and other mechanical papers previously counted as Other Commercial Printing.

[*] High-grade papers such as copy paper and printer paper, both residential and commercial.

[**] Formerly called Third Class Mail by the U.S. Postal Service.

[***] Includes paper in games and novelties, cards, etc.

[§] Valid default values for separating out paper and paperboard sub-categories for recovery and discards were not available.

[‡] Includes milk, juice, and other products packaged in table-top cartons and liquid food aseptic cartons.

[^] Table 4 does not include 10,000 tons of paper used in durable goods and 50,000 tons tissue in disposable diapers (Table 1).

Neg. = Less than 5,000 tons or 0.05 percent.

*Source*: Based on Franklin Associates, A Division of ERG

**EXAMPLE 11.3**

In 2010, 3,490,000 tons of tissue paper and paper towel waste were produced in the United States. Let us examine what this value represents on average in terms of waste per person per year.

tissue paper and paper towel waste per person per year

$$= \left( \frac{3,490,000 \text{ tons}}{\text{year} \cdot \text{US population}} \right) \left( \frac{2,000 \text{ pounds}}{1 \text{ ton}} \right) \left( \frac{\text{US population}}{308,745,538 \text{ persons}} \right)$$

$$= 22.6 \left( \frac{\text{pounds}}{\text{person} \cdot \text{year}} \right)$$

and packaging that was generated, recovered, and discarded in 2010. When studying Table 11.2, note that we generated nearly 1,590,000 tons of magazine waste, 5,260,000 tons of office-type papers, or 1,350,000 tons of paper plate and cup waste. What paper and paperboard products contribute to your waste footprint?

## Metal Products

Think about all of the aluminum soft drink containers that are thrown away every day. In 2010, the United States generated nearly 1,900,000 tons of aluminum waste from soft drink cans, beer cans, and foil. Add to that the appliances, automobiles, and metal cans that are also discarded. Table 11.3 shows the amount of metal products from durable and nondurable goods and in containers and packaging that was generated, recovered, and discarded in the United States.

## Glass Products

Glass products make up another large portion of our waste, as shown in Figure 11.3. The detailed quantities of glass from durable goods and containers and packaging that was generated, recovered, and discarded in 2010 is shown in Table 11.4.

Evan Lorne / Shutterstock.com

vladimir salman / Shutterstock.com

**TABLE 11.3** Metal Products in MSW (2010 in thousands of tons and percent of generation)

| Product category | Generation (Thousand tons) | Recovery (Thousand tons) | Recovery (Percent of generation) | Discards (Thousand tons) |
|---|---|---|---|---|
| **Durable Goods** | | | | |
| Ferrous Metals* | 14,160 | 3,820 | 27.0% | 10,340 |
| Aluminum** | 1,310 | Neg. | Neg. | 1,310 |
| Lead† | 1,540 | 1,480 | 96.1% | 60 |
| Other Nonferrous Metals‡ | 560 | Neg. | Neg. | 560 |
| *Total Metals in Durable Goods* | 17,570 | 5,300 | 30.2% | 12,270 |
| **Nondurable Goods** | | | | |
| Aluminum | 200 | Neg. | Neg. | 200 |
| **Containers and Packaging** | | | | |
| **Steel** | | | | |
| Cans | 2,300 | 1,540 | 67.0% | 760 |
| Other Steel Packaging | 440 | 350 | 79.5% | 90 |
| *Total Steel Packaging* | 2,740 | 1,890 | 69.0% | 850 |
| **Aluminum** | | | | |
| Beer and Soft Drink Cans§ | 1,370 | 680 | 49.6% | 690 |
| Other Cans | 70 | NA | | 70 |
| Foil and Closures | 460 | NA | | 460 |
| *Total Aluminum Packaging* | 1,900 | 680 | 35.8% | 1,220 |
| *Total Metals in Containers and Packaging* | 4,640 | 2,570 | 55.4% | 2,070 |
| | | | | |
| **Total Metals** | 22,410 | 7,870 | 35.1% | 14,540 |
| *Ferrous* | 16,900 | 5,710 | 33.8% | 11,190 |
| *Aluminum* | 3,410 | 680 | 19.9% | 2,730 |
| *Other Nonferrous* | 2,100 | 1,480 | 70.5% | 620 |

* Ferrous metals (iron and steel) in appliances, furniture, tires, and miscellaneous durables.

** Aluminum in appliances, furniture, and miscellaneous durables.

† Lead in lead-acid batteries.

‡ Other nonferrous metals in appliances and miscellaneous durables.

§ Aluminum can recovery does not include used beverage cans imported to produce new beverage cans.

Neg. = Less than 5,000 tons or 0.05 percent.

Details may not add to totals due to rounding.

NA = Not Available

*Source:* Based on Franklin Associates, A Division of ERG

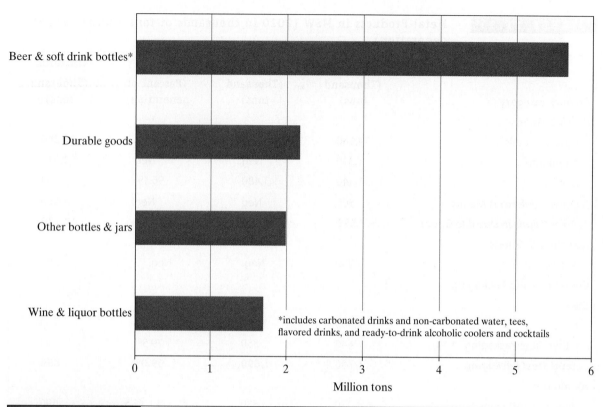

*includes carbonated drinks and non-carbonated water, tees, flavored drinks, and ready-to-drink alcoholic coolers and cocktails

| FIGURE 11.3 | Glass products generated in MSW (2010 data) |
|---|---|
| | *Source:* Based on data from the Franklin Associates |

**TABLE 11.4    Glass Products in MSW, (2010 in thousands of tons and percent of generation)**

| Product category | Generation (Thousand tons) | Recovery (Thousand tons) | Recovery (Percent of generation) | Discards (Thousand tons) |
|---|---|---|---|---|
| **Durable Goods·** | 2,170 | Neg. | Neg. | 2,170 |
| Containers and Packaging | | | | |
|   Beer and Soft Drink Bottles·· | 5,670 | 2,350 | 41.4% | 3,320 |
|   Wine and Liquor Bottles | 1,700 | 420 | 24.7% | 1,280 |
|   Other Bottles and Jars | 1,990 | 360 | 18.1% | 1,630 |
| ***Total Glass Containers*** | **9,360** | **3,130** | **33.4%** | **6,230** |
| ***Total Glass*** | **11,530** | **3,130** | **27.1%** | **8,400** |

· Glass as a component of appliances, furniture, consumer electronics, etc.

·· Includes carbonated drinks and non-carbonated water, teas, flavored drinks, and ready-to-drink alcoholic coolers and cocktails.

 Neg. = Less than 5,000 tons or 0.05 percent.

 Details may not add to totals due to rounding.

*Source:* Based on Franklin Associates, A Division of ERG

## Rubber and Leather

What happens to the tires or shoes that we throw away? How much rubber or leather do we generate each year, and how much of it is discarded or recovered? Table 11.5 shows the amount of rubber and leather products in municipal solid waste, including durable and nondurable goods. How much waste do you contribute in this area during your lifetime?

JCVStock / Shutterstock.com

| **TABLE 11.5** | Rubber and Leather Products in MSW, (2010 in thousands of tons and percent of generation) |

| Product category | Generation (Thousand tons) | Recovery (Thousand tons) | Recovery (Percent of generation) | Discards (Thousand tons) |
|---|---|---|---|---|
| **Durable Goods** | | | | |
| Rubber in Tires[1] | 3,300 | 1,170 | 35.5% | 2,130 |
| Other Durables[2] | 3,440 | Neg. | Neg. | 3,440 |
| *Total Rubber & Leather* | | | | |
| *Durable Goods* | 6,740 | 1,170 | 17.4% | 5,570 |
| **Nondurable Goods** | | | | |
| Clothing and Footwear | 790 | Neg. | Neg. | 790 |
| Other Nondurables | 250 | Neg. | Neg. | 250 |
| *Total Rubber & Leather* | | | | |
| *Nondurable Goods* | **1,040** | **Neg.** | **Neg.** | **1,040** |
| *Total Rubber & Leather* | **7,780** | **1,170** | **15.0%** | **6,610** |

[1] Automobile and truck tires. Does not include other materials in tires.

[2] Includes carpets and rugs and other miscellaneous durables.

Neg. = Less than 5,000 tons or 0.05 percent.

Details may not add to totals due to rounding.

*Source:* Based on Franklin Associates, A Division of ERG

**EXAMPLE 11.4**

Based on how far you drive your car each year (e.g., 10,000 or 12,000 miles) and the quality of your car tires, you may change your tires every 4 to 5 years. Let us now look at how many tires you would throw away from driving your current and future cars for the next 50 years.

number of tires thrown away during next 50 years per cars you own

$$= \left( \frac{4 \text{ tires}}{4 \text{ \sout{years}}} \right) \left( \frac{50 \text{ \sout{years}}}{\text{cars you own}} \right) = 50 \left( \frac{\text{tires}}{\text{cars you own}} \right)$$

Think about it! For 100 million car owners, this value adds up to 5 billion tires!

As you can see from the result of Example 11.4, discarded tires are a major source of waste. Each year, we discard hundreds of millions of tires that find their way into stockpiles and landfills. Fortunately, we are finding new uses for these discarded tires. For example, the United States Army Corps of Engineers—a federal agency with nearly 38,000 military and civilian personnel and the world's largest public engineering design and construction management agency—utilized discarded tires to protect a marshland from wave actions in Alabama. The EPA's Office of Research and Development is conducting research to also use discarded tires as rubberized asphalt, for bridge erosion protection, or as highway sound barriers.

## Before You Go On

Answer the following questions to test your understanding of the preceding section:

1. What are the two broad categories of waste?

2. Give examples of municipal waste.

3. Give examples of industrial waste.

4. Give examples of durable and nondurable goods.

*Vocabulary—State the meaning of the following terms:*

Municipal waste

Durable goods

Nondurable goods

## LO² 11.2 Industrial Waste

Robert Asento / Shutterstock.com

Candus Camera / Shutterstock.com

Now that you have some basic understanding of municipal waste, let us turn our attention to industrial waste. Industrial waste makes up a significant portion of solid waste in the United States. It consists of construction, renovation, and demolition materials. It also represents the waste that is created during the exploration, development, and production of fossil fuels (e.g., coal, natural gas, crude oil). Medical waste, that is waste materials generated at clinics, dental offices, veterinary and human hospitals, and medical research laboratories are also classified as industrial waste. Examples of medical waste include discarded needles, blood-soaked bandages, surgically removed body organs or parts, cultures, surgical gloves, and instruments. According to the EPA, industrial facilities—such as manufacturers of organic and inorganic chemicals, plastics, resins, steel, clay, glass, concrete, paper, and food—dispose over seven billion tons of solid waste every year. During demolition projects dealing with structures such as buildings, roads, or bridges, many materials such as concrete, beams, wood, asphalt, glass, bricks, steel, doors, and windows are thrown away.

### Hazardous Waste

As the name implies, ***hazardous waste*** refers to waste that, if improperly disposed, could be harmful to human health and the environment. Household examples of hazardous waste include most cleansers, paint, pesticides, batteries, and used car oil. These products may contain toxic or corrosive ingredients. For example, putting paint in the trash or pouring it down a drain or a storm sewer would eventually lead to environmental problems.

### Solid Waste Management

As a society, we can manage our waste in a number of ways. To start with, we can consider source reduction at the manufacturing level. We can change the way we design and package to reduce the amount of plastics, paper, metal, or glass that we use for products. Material substitution is another way we can manage our industrial waste. For example, when designing products, we should use materials that are environmentally friendly. We also can purchase products with longer life spans. For example, we can purchase a high-mileage tire instead of a low-mileage tire. When applicable, we can change our buying habits and purchase food and other consumable items in bulk quantities to reduce the amount of packaging materials. We can buy cereal in bags instead of boxes, or purchase a coffee brick instead of a coffee can. We can always reuse some materials or share them with others. For example, we can find secondary uses for glass and plastic containers, waste paper, or clothing. We can also borrow or rent items for temporary use, or buy or sell things at a garage sale. We could consider buying extended warranties or repairing things instead of throwing them away. As you can see, there are a number of ways we can reduce our waste.

As we stated before, the most important thing to remember is that you should not only think about the space that your trash occupies in a landfill, but also consider the entire life cycle of the trash. Remind yourself when you throw something away what natural resources were consumed to make the product. Also think about the amount of energy that was used to produce, process, transport, and finally dispose of it!

## Landfills

Let us now look at where our trash ends up. Today's landfills are designed to receive both municipal and industrial waste and are built in areas far away from flood plains, wetlands, and other sensitive locations to protect the environment. Their designs incorporate composite liners to protect the groundwater and the underlying soil from the solid waste stream. They also have environmental monitoring systems to check for harmful gas emissions and signs of groundwater contamination. They have leachate collection and removal systems. *Leachate* refers to a liquid that acts as a trap to draw harmful

Huguette Roe / Shutterstock.com

substances. Leachate could contain a high concentration of environmentally harmful materials, and if not properly removed, leachate potentially drains from a landfill and harms the environment.

To protect the public health, landfills are regularly covered with layers of soil to control insects, contain litter, and reduce odor. Some hazardous materials cannot be sent to landfills, including paints, pesticides, batteries, and motor oil. Unfortunately, in many developing countries landfills represent a public health hazard because they do not adhere to stringent standards. The regional distribution of landfill facilities in the United States is shown in Figure 11.4.

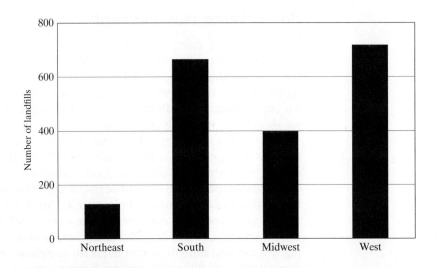

| FIGURE 11.4 | Number of landfills in the United States (2010) |
| --- | --- |

*Source:* Based on *BioCycle*, October 2010

## Before You Go On

Answer the following questions to test your understanding of the preceding section:

1. Give examples of construction, renovation, and demolition waste.

2. Give examples of medical waste.

3. Give examples of hazardous waste.

4. Discuss at least three ways we can manage our waste.

5. Describe some of the basic characteristics of today's landfill design.

*Vocabulary—State the meaning of the following terms:*

Industrial waste

Medical waste

Hazardous waste

Leachate

## LO³  11.3  Recycling and Composting

We can manage waste by *reducing* (the amount that we throw away), *recycling*, and *composting*. As mentioned previously, according to the EPA, we generated about 250 million tons of trash and only recycled and composted 85 million tons of this material in 2010. This ratio represents a recycling and composting rate of 34 percent:

recycling and composting rate

$$= \left( \frac{85 \text{ million tons recycled and composted}}{250 \text{ million tons waste generated}} \right)(100\%)$$

$$= 34\%$$

Think about it! Only one-third of all trash that we generated was recycled or composted. On a per-person per day basis, out of 4.43 pounds of trash, only 1.5 pounds of the trash was recycled and composted. Although we can do better, it is important to realize that as a society—over the years—we have been increasing our recycling and composting activities. Our recycling trend during the past five decades is shown in Figure 11.5. From examining this figure, note that in 1960 the recycling and composting rate was only 6.4 percent.

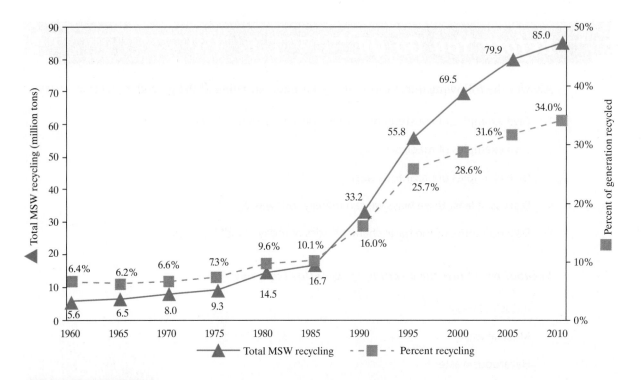

The total and the percent of total municipal solid waste recycled in the United States
from 1960 through 2010
*Source:* United States Environmental Protection Agency

## Recycling

As you already know, recycling has many advantages. When we recycle,
we reduce landfill use, we consume less energy, and we use our natural resources more effectively. To promote recycling, communities take
advantage of different schemes. Some communities offer curbside services,
whereas in others, residents may have to drop off recyclables at designated
centers or return products such as bottles and cans to collect a refund that
was deposited during the purchase of the products. Once recyclables are
collected, cleaned, and sorted, they are offered as a commodity for sale.
Manufacturers that purchase the recycled commodity then use them in
making their products. You are familiar with many products that contain
recycled materials, such as cardboard boxes and aluminum, glass, and
plastic containers.

Let us now examine the available recycling data in more detail. The number of material recovery facilities for each region of the United States is shown
in Table 11.6. As of 2010, there were 633 facilities with a total capacity of
nearly 100,000 tons per day. The percentage of population served by curbside
recycling in 2010 is shown in Figure 11.6. It is clear from examining Figure 11.6
that the northeast has the highest percentage of curbside pick-up, while the
midwest has the lowest.

The percentage of materials recovered by weight of total recovery in
2010 is shown in Figure 11.7. As shown, paper and paperboard has the highest

| Region | Number | Estimated Tons Per Day (tpd) |
|--------|--------|------------------------------|
| NORTHEAST | 153 | 27,186 |
| SOUTH | 195 | 24,754 |
| MIDWEST | 153 | 23,118 |
| WEST | 132 | 23,391 |
| U.S. Total | **633** | **98,449** |

**TABLE 11.6** Material Recovery Facilities (2010)

*Source:* Based on Governmental Advisory Associates, Inc.

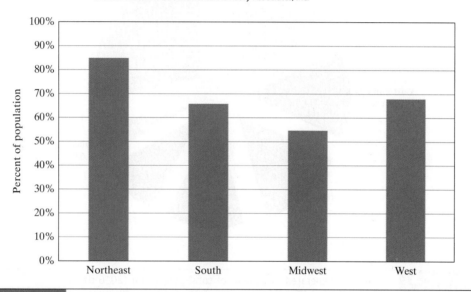

**FIGURE 11.6** The Percent of population served by curbside recycling (2010 data)
*Source:* EPA Report, Office of Resource Conservation and Recovery, November 2011

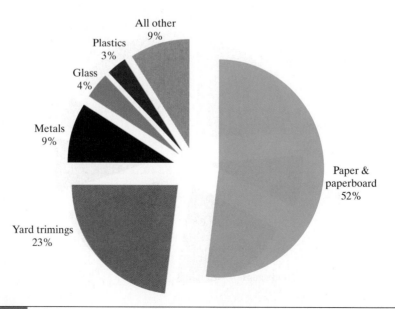

**FIGURE 11.7** Materials recovered in percent by weight of total recovery (2010 data)
*Source:* EPA Report, Office of Resource Conservation and Recovery, November 2011

percentage followed by yard trimmings. There are many other ways that we can study our waste and recycling patterns, as shown in Figures 11.8 through 11.13. The data contained in these figures are self-explanatory. Spend some time and study them carefully.

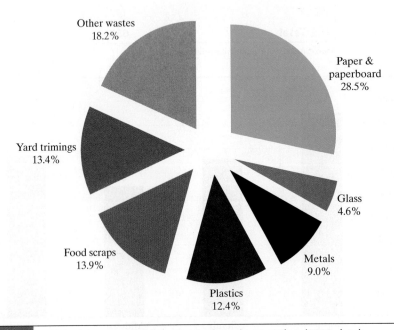

**FIGURE 11.8**   Materials generated in percent of total generation (2010 data)
*Source:* EPA Report, Office of Resource Conservation and Recovery, November 2011

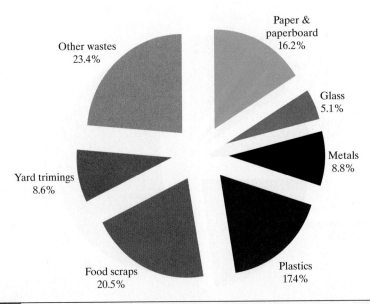

**FIGURE 11.9**   Materials discarded in percent of total discard (2010 data)
*Source:* EPA Report, Office of Resource Conservation and Recovery, November 2011

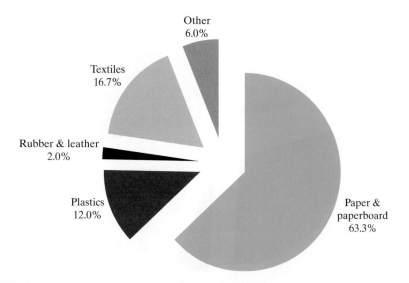

**FIGURE 11.10**    Nondurable goods generated in percent of total generation (2010 data)
*Source:* EPA Report, Office of Resource Conservation and Recovery, November 2011

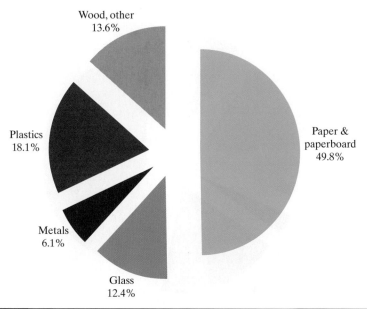

**FIGURE 11.11**    Containers and packing generated in percent of total generation (2010 data)
*Source:* EPA Report, Office of Resource Conservation and Recovery, November 2011

As we discussed in Chapter 10, metals are found in many everyday products. As a result, it is important to carefully study the recycling of metals such as iron, steel, aluminum, copper, zinc, and nickel. The source of data for the following section is the 2010 U.S. Geological Survey Report.

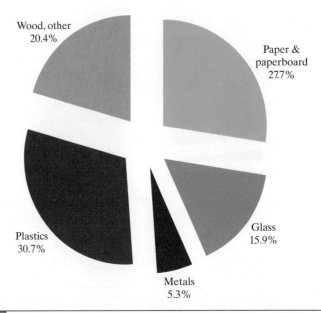

**FIGURE 11.12**    Containers and packing discarded in percent of total generation (2010 data)
*Source:* EPA Report, Office of Resource Conservation and Recovery, November 2011

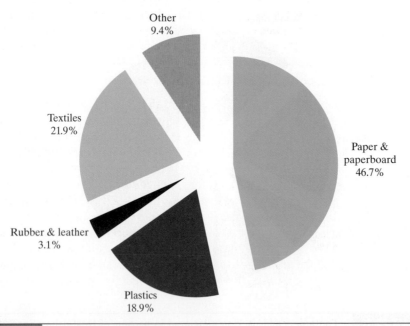

**FIGURE 11.13**    Nondurable goods discarded in percent of total generation (2010 data)
*Source:* EPA Report, Office of Resource Conservation and Recovery, November 2011

**Iron and Steel** Let's begin with steel, since steel consumption is a good indicator of economic growth; an increase in steel consumption usually means an increase in industrial activities. Moreover, most of the foundries in the United States are designed to recycle iron and steel scraps. Recycling or the re-melting of iron and steel scraps also requires less energy (as compared to producing

PAKULA PIOTR / Shutterstock.com

materials from raw minerals), reduces the need for landfill space, and thus has less impact on the environment. Recycled steel from discarded automobiles, appliances, cans, and construction steel (e.g., plates, beams, rebar) make up the major source for new products. In 2010, the bodies of millions of end-of-life vehicles were shredded and recycled in foundries. In fact, the amount of recycled steel produced exceeded the demand for steel for domestic production of new vehicles. The recycling rates for appliances, cans, and construction materials in 2009 were approximately 90, 66, and 70 percent, respectively.

**Aluminum** In 2010, 2.7 million tons of aluminum were recovered from manufacturing scrap (59%) and old discarded aluminum products (41%), which translates to 24 percent of the apparent aluminum consumption. The *apparent consumption* is formally defined as the sum of the domestic production plus recovery from old aluminum scrap plus net import.

**Copper** In 2010, approximately 1,730,000 metric tons of copper were consumed; out of this amount, 160,000 tons were recovered through recycling. The recycled amount accounted for about 9 percent of total consumption in that year.

**Zinc and Nickel** In 2010, approximately 85,000 tons of zinc were recovered from recycling activities. This figure is equivalent to about 41 percent of the total amount of zinc produced in the United States. In the same year, about 100,000 tons of nickel were recovered from recycling activities.

## Composting

Evan Lorne / Shutterstock.com

Each year we throw away large quantities of food, fruit and vegetable skins and cores, grass clippings, yard trimmings, coffee grounds, tea bags, egg shells, and so on. This category makes up nearly 30 percent of our trash. Composting offers an environmentally friendly alternative to throwing food and yard trimmings in landfills. **Composting** refers to the biological decomposition or decay of food wastes, yard trimmings, and other organic materials. Normally, composting leads to a dark, soil-like material that is used as fertilizer to provide nutrients for plants and microorganisms. In addition to reducing the amount of waste that we ship to landfills, composting reduces the need for chemical fertilizers. Materials that are commonly composted include:

- Coffee grounds and filters
- Fruits and vegetables
- Leaves
- Nut shells
- Tea bags
- Yard trimmings
- Hair and fur
- Grass clippings
- Egg shells
- Shredded newspaper
- Sawdust

Yard trimmings can be combined with a variety of organic wastes to create products with certain chemical characteristics. The United States' municipal solid-waste composting capacity in 2010 is shown in Figure 11.14. As shown in this figure, the western states lead with a capacity of nearly 7 tons per million people. The number of facilities in the United States that compost yard trimmings is shown in Figure 11.15. The midwest has over 1,400 facilities to handle yard trimmings.

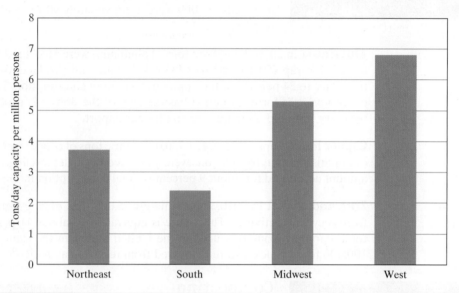

**FIGURE 11.14**  Municipal solid waste composting capacity (2010)
*Source:* EPA Report, Office of Resource Conservation and Recovery, November 2011

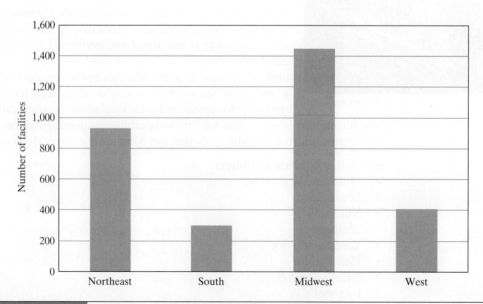

**FIGURE 11.15**  Yard trimming composting facilities (2010)
*Source:* EPA Report, Office of Resource Conservation and Recovery, November 2011

Ulrich Mueller / Shutterstock.com

## Waste-to-Energy

As shown in Table 11.7 and Figure 11.16, there are 86 **waste-to-energy facilities** in the United States (2010 data) that convert waste to electricity. These facilities use different technologies to produce energy from waste: *Mass–burn plants* are designed to burn waste in a single combustion chamber with excess air to ensure that all waste is burned; *Modular systems* are smaller in size and moveable; and *Refuse-derived fuel systems* first shred and separate noncombustible from combustible materials. The system then uses the combustible materials as supplemental fuel. A schematic of a waste-to-energy plant based on a refuse-derived fuel system is shown in Figure 11.17; note the pollution control system consists of stages where nitrogen oxide, mercury, acid gases, and particulate matter are removed and tested.

| TABLE 11.7 | Municipal Waste-to-Energy Plants (2010) | |
|---|---|---|
| **Region** | **Number Operational** | **Design Capacity (tpd)** |
| **NORTHEAST** | 40 | 46,704 |
| **SOUTH** | 22 | 31,896 |
| **MIDWEST** | 16 | 11,393 |
| **WEST** | 8 | 6,171 |
| **U.S. Total[1]** | **86** | **96,164** |

[1] Projects on hold or inactive were not included. WTE includes mass burn, modular, and refuse-derived fuel combustion facilities.

*Source:* Based on "The 2010 ERC Directory of Waste-to-Energy Plants." Energy Recovery Council (ERC). December 2010

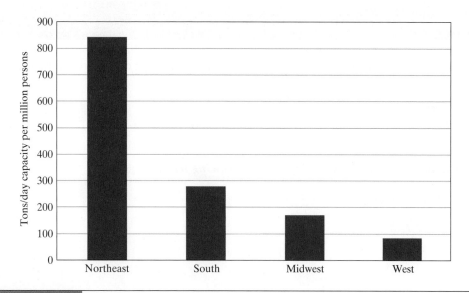

| FIGURE 11.16 | Municipal waste-to-energy capacity (2010)
*Source:* U.S. Census Bureau, Energy Recovery Council (ERC), December 2010

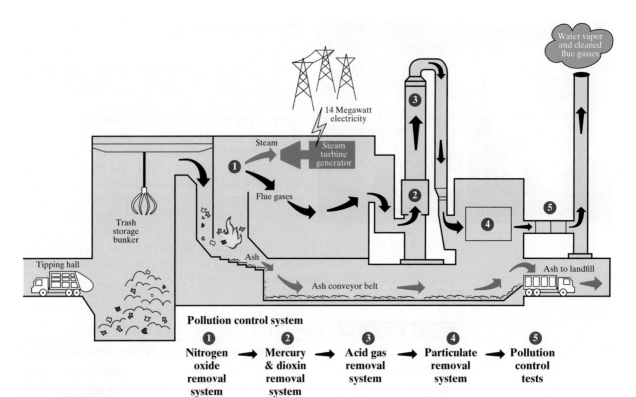

**Pollution control system**

❶ Nitrogen oxide removal system → ❷ Mercury & dioxin removal system → ❸ Acid gas removal system → ❹ Particulate removal system → ❺ Pollution control tests

**Waste-to-energy**
• 90% reduction of trash volume
• Power generation
• Pollution control

**FIGURE 11.17**    A waste-to-energy plant
*Source:* Based on www.ecomaine.org

# Before You Go On

Answer the following questions to test your understanding of the preceding section:

1. What do we mean by composting?

2. Name at least four items that can be composted.

**Vocabulary—***State the meaning of the following term:*

Waste-to-energy

# SUMMARY

## LO¹ Municipal Waste

By now you should know how we generate waste and be mindful of the entire life cycle of trash. Before you throw away something, you should think about the natural resources that were used to make the item and how much energy it took to produce, process, transport, and eventually dispose of it. Waste is classified into two broad categories: municipal and industrial. Municipal waste is the trash that we throw away every day, and it consists of items such as food scraps, packaging materials, bottles, cans, and so on.

You should also be familiar with the concepts of durable and nondurable goods. The EPA reports the amount of waste generation and recovery using the following categories: durable goods, nondurable goods, containers and packing materials, and plastic packing. Durable goods include products such as major and small appliances, furniture, carpets and rugs, rubber tires, and lead-acid batteries. By nondurable goods, the EPA refers to products such as office paper, newspaper, books, magazines, paper plates and cups, tissue paper and paper towels, disposable diapers, plastic plates and cups, trash bags, clothing and footwear, towels, sheets, and so on.

## LO² Industrial Waste

Industrial waste makes up a significant portion of solid waste in the United States. It consists of construction, renovation, and demolition materials. It also represents medical waste and the waste that is created during exploration, development, and production of fossil fuels. You also should be familiar with the basic components and design of landfills.

Hazardous waste refers to waste that, if improperly disposed, could be harmful to human health and the environment. Household examples of hazardous waste include most cleansers, paint, pesticides, batteries, and used car oil. These products may contain toxic or corrosive ingredients.

## LO³ Recycling and Composting

We can manage waste by *reducing*, *recycling*, and *composting*. When we recycle, we limit landfill use, we consume less energy, and we use our natural resources more effectively. Each year we throw away large quantities of food, fruit and vegetable skins, and so on. This category makes up nearly 30 percent of our trash. Composting offers an environmentally friendly alternative to throwing food and yard trimmings in landfills. Composting refers to the biological decomposition or decay of food wastes, yard trimmings, and other organic materials. There are also facilities that convert waste to electricity. These facilities use available technologies to produce energy from waste.

# KEY TERMS

Composting 353
Durable Goods 333
Hazardous Waste 345

Industrial Waste 332
Leachate 346
Municipal Waste 332

Nondurable Goods 333
Waste-to-Energy Facility 355

# Apply What You Have Learned

This project will help you determine how much trash you generate each year. Maintain a daily logbook to keep track of what you throw away each day for at least a week. Project your findings into monthly and annual amounts, compile your findings into a single report, and present it to the class. State all your assumptions and suggest ways to reduce waste and increase recycling.

Artieskg / Shutterstock.com

# PROBLEMS

*Problems that promote life-long learning are denoted by* 🔑

**11.1** During the holiday seasons, we tend to produce more waste. For example, we discard more wrapping paper, cards, and shopping bags. Moreover, according to the EPA, each year over 30 million live Christmas trees are sold in North America. Suggest at least five ways to reduce waste during the holidays.

**11.2** Consumer electronics and their components (e.g., batteries, printer ink cartridges, CDs) are becoming increasingly a major portion of our trash. Suggest at least three ways to reduce waste caused by electronics and their components.

**11.3** 🔑 Products containing mercury pose health hazards, and as a result, they must be disposed of properly. Some clocks, mirrors, button cell batteries, switches, relays, and thermostats are among products that may contain mercury. Identify at least five products in your daily life containing mercury that you were not aware of and identify alternatives to these products that do not use mercury.

**11.4** Use the tables given in this chapter to look up the following information: total plastics, glass, paper and paperboard generation in nondurable goods.

**11.5** 🔑 There are many facilities across the United States that deal with energy recovery from waste. Investigate how much waste material was converted into electricity in the most recent year in a mass-burn facility in your region. Discuss your findings in a brief report.

**11.6** 🔑 As we mentioned in Chapter 8, methane contributes to global warming. To reduce methane emissions from landfills and to promote its recovery and use as an energy source, the EPA offers a program entitled the Landfill Methane Outreach Program. Investigate how landfill gas is converted to energy and the types of collection systems in use today. Discuss your findings in a brief report.

**11.7** Use Figure 11.5 to determine the amount of municipal solid waste that was recycled in 1960, 1970, 1980, 1990, 2000, and 2010.

**11.8** Estimate the amount of plastic in the water bottles that you throw away each year. State your assumptions.

**11.9** There are some organic materials that you should not compost. For example, meat and fish bones could attract flies and rodents.

Give examples of other organic materials that should not be composted.

**11.10** According to the EPA, low concentration levels of prescription and over-the-counter drugs and chemicals used to make cosmetics and personal care products now are found in our water supply. Investigate how these substances find their way into our water supply and how they should be disposed of properly. Discuss your findings in a brief report.

**11.11** According to the United States Census Bureau data, each year hundreds of thousands of residential and nonresidential buildings are demolished, resulting in millions of tons of debris. Investigate what is being done to recycle some of the recovered items such as doors, windows, and beams. Discuss your findings in a brief report.

**11.12** Investigate the role and mission of the U.S. Green Building Council. Write an executive summary discussing your findings.

**11.13** Each year, we discard hundreds of millions of tires that find their way into stockpiles and landfills. Investigate the current state of scrap-tire management in your region. Discuss your findings in a brief report.

**11.14** Each year, mining activities generate large quantities of waste. Investigate this area of industrial waste, and report your findings to your instructor.

**11.15** Each compact fluorescent light bulb, on average, contains 4 milligrams of mercury. Investigate how these types of light bulbs should be disposed of properly. Write a one-page summary of your findings.

**11.16** Investigate the recycling activities in your community. Prepare a colorful and informative brochure to increase awareness of these activities in your community.

**11.17** Investigate current practices dealing with recycling of used motor oil. Discuss your findings in a brief report.

**11.18** Recylemania is a competition designed to promote recycling among colleges in the United States. Visit the Recylemania website to find ways to participate actively in the challenge. Use the knowledge you have gained from this course.

**11.19** Each year, millions of people stop at rest areas along U.S. Interstates. Investigate recycling activities at rest stops in your state. Prepare a one-page summary sheet showing your findings.

**11.20** Each year, millions of people attend professional and collegiate sporting events. Investigate recycling activities at sporting facilities on your campus, and write a brief report discussing your findings.

*"The greatest injustices proceed from those who pursue excess, not by those who are driven by necessity."*

— ARISTOTLE (384–322 BC)

# Sustainability

## CIRCULAR ECONOMY

RESOURCES

MANUFACTURING

RECYCLING

CONSUMPTION & USE

WASTE

In Part Four, we introduce you to the Earth Charter which eloquently puts into words an ethical guideline for building a sustainable, just, and peaceful global society in the 21st century. We also introduce you to key sustainability concepts, methods, and tools. Every good global citizen must develop a keen understanding of the Earth's finite resources, environmental and socioeconomic issues related to sustainability, ethical aspects of sustainability, and the necessity for sustainable development. You will learn about life-cycle analysis, resource and waste management, and environmental impact analysis, and will become familiar with sustainable-development indicators such as the U.S. Green Building Council (USGBC) and Leadership in Energy and Environmental Design (LEED) rating systems.

# CHAPTER
# 12

# Sustainability

## LEARNING OBJECTIVES

**LO¹**   The Earth Charter: describe the Earth
         Charter

**LO²**   Sustainability Concepts, Assessments,
         and Tools: explain the key sustainability
         concepts, assessments, and tools such as
         life-cycle analysis

**LO³**   Apply What You Have Learned—Knowledge
         Is Power: make specific plans to make the
         world a better place

# Discussion Starter

I n the discussion starter of Chapter 1, we said that we all want to make the world a better place: "How do we do it, and where do we start?" We also quoted Leo Tolstoy, who said:

*"Everyone thinks of changing the world, but no one thinks of changing oneself."*

The knowledge that you have gained from studying the previous eleven chapters allows you to understand your daily environmental impact by estimating how much water, food, and material you consume and discard, how much energy you

Gustavo Frazao / Shutterstock.com

expend, and how much emissions are actually due to your lifestyle and habits. Consider the saying, "knowledge is power."

**To the Students:** The question is then, "How are you planning to use this power and apply what you have learned to change yourself to make the world a better place?" What are your specific plans?

## LO¹ 12.1 The Earth Charter

We start this chapter by reminding you that in order to address our energy needs, clean air and water requirements, and food supply intelligently, *we need to work together*. Unfortunately, today there is international competition for the Earth's finite resources because each nation works hard to address its own energy, water, and food needs. Recall the example of the human body, which is made of many interacting parts that work well together and share resources effectively; when any part of our body—as small as our little finger—is in pain, the body as a whole is uncomfortable until the pain is gone. We should develop a similar, holistic view of our societies—one that increases commonality of human purpose and gives a greater meaning to life beyond the walls of our homes, the boundaries of our cities and our own countries.

To emphasize the sense of global interdependence and shared responsibility for the well-being of the entire human family, the *Earth Charter* was introduced by an independent international commission on June 29, 2000, in The Hague, Netherlands. The Earth Charter eloquently puts into words an ethical guideline for building a sustainable, just, and peaceful global society in the 21st century. The Earth Charter, which is presented here, is intended as both a vision of hope and a call to action.

# Earth Charter

## Preamble

We stand at a critical moment in Earth's history, a time when humanity must choose its future. As the world becomes increasingly interdependent and fragile, the future at once holds great peril and great promise. To move forward we must recognize that in the midst of a magnificent diversity of cultures and life forms we are one human family and one Earth community with a common destiny. We must join together to bring forth a sustainable global society founded on respect for nature, universal human rights, economic justice, and a culture of peace. Towards this end, it is imperative that we, the peoples of Earth, declare our responsibility to one another, to the greater community of life, and to future generations.

## Earth, Our Home

Humanity is part of a vast evolving universe. Earth, our home, is alive with a unique community of life. The forces of nature make existence a demanding and uncertain adventure, but Earth has provided the conditions essential to life's evolution. The resilience of the community of life and the well-being of humanity depend upon preserving a healthy biosphere with all its ecological systems, a rich variety of plants and animals, fertile soils, pure waters, and clean air. The global environment with its finite resources is a common concern of all peoples. The protection of Earth's vitality, diversity, and beauty is a sacred trust.

## The Global Situation

The dominant patterns of production and consumption are causing environmental devastation, the depletion of resources, and a massive extinction of species. Communities are being undermined. The benefits of development are not shared equitably and the gap between rich and poor is widening. Injustice, poverty, ignorance, and violent conflict are widespread and the cause of great suffering. An unprecedented rise in human population has overburdened ecological and social systems. The foundations of global security are threatened. These trends are perilous—but not inevitable.

## The Challenges Ahead

The choice is ours: form a global partnership to care for Earth and one another or risk the destruction of ourselves and the diversity of life. Fundamental changes are needed in our values, institutions, and ways of living. We must realize that when basic needs have been met, human development is primarily about being more, not having more. We have the knowledge and technology to provide for all and to reduce our impacts on the environment. The emergence of a global civil society is creating new opportunities to build a democratic and humane world. Our environmental, economic, political, social, and spiritual challenges are interconnected, and together we can forge inclusive solutions.

## Universal Responsibility

To realize these aspirations, we must decide to live with a sense of universal responsibility, identifying ourselves with the whole Earth community as well as our local communities. We are at once citizens of different nations and of one world in which the local and global are linked. Everyone shares responsibility for the present and future well-being of the human family and the larger living world. The spirit of human solidarity and kinship with all life is strengthened when we live with reverence for the mystery of being, gratitude for the gift of life, and humility regarding the human place in nature.

We urgently need a shared vision of basic values to provide an ethical foundation for the emerging world community. Therefore, together in hope we affirm the following interdependent principles for a sustainable way of life as a common standard by which the conduct of all individuals, organizations, businesses, governments, and transnational institutions is to be guided and assessed.

The basic principles of the Earth Charter include *respect and care for the community of life, ecological integrity, social and economic justice,* and *democracy, nonviolence, and peace.* The Earth Charter in its entirety, with a complete list of principle descriptions, is given in Appendix D. Take some time to read it carefully and contemplate how you can answer the call to action.

## LO² 12.2 Key Sustainability Concepts, Assessment, and Tools

As we explained in Chapter 1, there is no universal definition for sustainability. It means different things to different professions. However, one of the generally accepted definitions for **sustainability** is

> *"design and development that meets the needs of the present without compromising the ability of future generations to meet their own needs."*

As a society, we are expected to design and provide goods and services that increase the standard of living and advance health care while addressing serious environmental and sustainability concerns. As you know by now, we must consider the link among Earth's finite resources and related environmental, social, ethical, technical, and economical factors. It cannot be emphasized enough that we need global citizens who can come up with solutions that address infrastructure, energy, water, and food needs while simultaneously addressing sustainability issues. The potential shortage of citizens with a grasp of the concept of sustainability—people who can apply the sustainability concepts, methods, and tools to their problem-solving and decision-making processes—could have serious consequences for our future.

In an article published in the *American Society of Civil Engineers News* titled "Board of Direction Views Sustainability Strategy as Key Priority," William Wallace, the author of "Becoming Part of the Solution," offers five issues that must be understood:

1. The world's current economic development is not sustainable—the world population already uses approximately 20 percent more of the world's resources than the planet can sustain. (*UN Millennium Ecosystem Assessment Synthesis Report*, 2005)

2. The effects of outpacing the Earth's carrying capacity have now reached crisis proportions—extreme weather events causing huge losses, such as the prospect of rising sea levels threatening coastal cities. Global population increase outstrips the capacity of institutions to address it.

3. An enormous amount of work will be required if the world is to shift to sustainable development—a complete overhaul of the world's processes, systems, and infrastructure will be needed.

4. The engineering community should be leading the way toward sustainable development, but it has not yet assumed that responsibility. Civil engineers have few incentives to change. Most civil engineers deliver conventional engineering designs that meet building codes and protect the status quo.

5. People outside the engineering community are capitalizing on new opportunities—accounting firms and architects are examples cited by Wallace. The architects bring their practices into conformity with the U.S. Green Building Council's Leadership in Energy and Environmental Design (LEED) Green Building Rating System.

You can learn about LEED by visiting www.usgbc.org/LEED. As stated on their website,

> *"LEED is an internationally recognized green building certification system, providing third-party verification that a building or community was designed and built using strategies aimed at improving performance across all the metrics that matter most: energy savings, water efficiency, $CO_2$ emissions reduction, improved indoor environmental quality, and stewardship of resources and sensitivity to their impacts. Developed by the U.S. Green Building Council (USGBC), LEED provides building owners and operators a concise framework for identifying and implementing practical and measurable green building design, construction, operations and maintenance solutions."*

In addition to LEED, there are many other organizations based on agriculture, business, community development, energy, and environmental concerns that deal with sustainability.

Throughout this book, we have introduced you to concepts that are important in understanding sustainability. By now, you should have a good understanding of basic human needs and the demands of a growing world population as given here.

- As a society, we create and consume many different products and services. We need energy for building structures, growing food, and having access to clean water. We need clean air, clean water, food, and shelter.
- The energy use per capita in the world has been increasing steadily, particularly in developing countries as the economies of the world grow.

- Adding to these concerns, the expected rise in the population of the world is from over 7 billion currently to about 9.3 billion people by mid-century.
- Meeting these human needs causes pollution. Stationary, mobile, and natural sources contribute to outdoor air pollution. Human activities such as mining, construction, manufacturing goods, and agriculture contribute to water pollution.

As you can see, in order to address our needs and maintain a good standard of living, we are faced with the problems of finding energy sources and reducing pollution and waste.

*Ethics* refers to the study of morality and the moral choices that we all have to make in our lives. Professional societies, such as those in medicine and engineering, have long-established guidelines, standards, and rules that govern the conduct of their members. These rules are also used by the members of the board of ethics of professional organizations to interpret ethical dilemmas that are submitted by a complainant. We rely quite heavily on professionals to provide us with safe and reliable goods and services. There is no room for mistakes or dishonesty! An incompetent and unethical surgeon could cause the death of a person during a surgery. An unethical engineer could cause the deaths of hundreds of people at one time if—in order to save money—he or she designs a bridge or a part for an airplane that does not meet the safety requirements. Think about it; hundreds of people's lives are at risk! We don't have to be doctors or engineers to be ethical. As good global citizens, it is important to realize that we must also hold to the highest standards of honesty and integrity, particularly when it comes to the Earth—our home. Think about this carefully; an unethical surgeon can kill a patient, but uninformed and unethical citizens through their daily actions can destroy the planet Earth—our only home.

# Key Sustainability Assessment: Life-Cycle Analysis

We use a lot of products in our daily lives: clothing, running shoes, furniture, TVs, hair driers, coffee machines, computers, cars, refrigerators, ovens, dishwashers, and so on. *Life-cycle analysis* looks at the raw material and the energy consumption, emissions, and other factors related to a product over its *entire* life from its origin to its disposal: *from its inception and birth to its death.* Generally, a life-cycle analysis consists of four steps:

1. Purpose or goal definition
2. Inventory analysis
3. Impact assessment
4. Improvement

**Step 1:** The life-cycle analysis starts with the *definition of the scope* of the analysis. First, the purpose or the goal of the analysis must be clearly defined. Questions such as: "What do you want to study, and why do you want to study it? How detailed should the study be?" should be asked and answered.

**Step 2:** Next, you perform what is called an *inventory analysis.* You need to make a chart showing all of the components in a system's or a product's life cycle and show how these components are related (we discussed the definition of a system and its components in Chapter 2).

*(continued)*

*(continues)*

Pavel Shchegolev / Shutterstock.com

Then, you need to collect data for each component in terms of raw material, energy use, and emissions. For example, if you were to perform a life-cycle analysis on a system such as an automatic coffee machine, you would need to consider many different components. You can start with the coffee itself that would eventually be brewed in the machine and ask questions such as "How much energy and water did it take to grow the beans? How much energy did it take to transport the beans to a processing plant? How much energy did it take to roast and grind the beans? How much energy did it take to package and deliver the coffee to a supermarket?" You can then ask similar questions about the production and delivery of the paper coffee filters to stores.

Next, you proceed with the actual machine components and ask similar questions about the parts of the coffee machine that are made from plastics, metal, and glass. You ask about their production, how different components (parts) were assembled, and how the coffee machine was packaged and transported to the store.

Finally, you need to estimate the coffee machine's consumption in terms of water and energy. You also need to consider how the coffee, thousands of filters, and the coffee machine are disposed of. How many pounds or kilograms of coffee beans, paper filters, plastic, metal, and glass were consumed? How much energy was required to extract the raw materials to make the parts, assemble them, transport them to the manufacturing place, and eventually transport the finished product, the coffee machine, to the stores? How many kilowatt-hours (kWh) will the coffee machine consume during its life time?

How much water will be consumed? How much energy was used to deliver the water to your home or office, and so on? Don't forget about the amount of energy that was consumed in the process of extracting and delivering the fossil fuels themselves—the sources of energy— that were used during raw material extraction, manufacturing, transportation, usage, and disposal of the coffee machine.

**Step 3:** The third step in the life-cycle analysis is called *impact assessment*. For this step, you need to analyze the magnitude of environmental impact. For example, you need to consider emissions due to burning all of the fossil fuels that are associated with extraction of the raw materials, manufacturing, and transportation to the stores. You also need to consider emissions associated with the energy that is consumed during the useful life and during the disposal phase of the coffee machine. Remember, garbage and recycling trucks also burn fossil fuels which produce emissions. As you can see, a comprehensive life-cycle analysis requires a great deal of data and is very complex.

**Step 4:** The final step involves *improvement*. Based on what is learned, the designer should then modify the product to have a smaller environmental footprint. You, as a good global citizen and the consumer of the product, also have an important role to play. Even though you might not be able to conduct a thorough analysis, you can obtain useful information for your decisions using a limited life-cycle analysis, particularly when comparing different products.

## Key Sustainability Tools

The life-cycle assessment, environmental assessment, and U.S. Green Building Council Leadership in Energy and Environmental Design (LEED) rating system, each serve as sustainability tools. In addition to these tools, there are many resource, waste, and emission calculators available online through various government and international agencies.

## Before You Go On

Answer the following questions to test your understanding of the preceding section:

1.  What is the Earth Charter?

2.  What are the steps in a life-cycle analysis?

3.  Use an example from a product you use every day to explain the life-cycle analysis.

4.  In your own words, explain ethics.

*Vocabulary—State the meaning of the following words:*

Earth Charter

LEED

Ethics

Sustainability tool

## LO³ 12.3 Apply What You Have Learned—Knowledge Is Power

You have heard the old saying, "Knowledge is power." How are you planning to use your power? You may not realize it, but you now have many tools in your toolbox that you can use to analyze your daily water, material, and energy consumption, as well as your waste and emissions. Table 12.1 summarizes the focus of each chapter that you have studied and indicates how the knowledge gained from each chapter may be applied. A brief review of the topics covered in each chapter follows Table 12.1. Moreover, at the end of this chapter, there is a list of projects that you can initiate on your campus or community.

### Fundamental Dimensions and Systems of Units

You should have a good understanding of the fundamental dimensions in your everyday life. We have realized that we need only a few physical dimensions or quantities to describe our surroundings and daily events. For example, we need a length dimension to describe how tall, how long, or how wide something is. Time is another physical dimension that we need to answer questions such as "How old are you?" You should also know that today, based on what we know about our world, we need seven fundamental dimensions to describe our surroundings. They are length, mass, time, temperature, electric current, amount of substance, and luminous intensity. The other important concept is that not only do we need to define these physical dimensions to describe our surroundings, but we also need some way to scale

**TABLE 12.1    A Summary of Topics Covered in This Book and How to Apply Them**

| Chapter | Topic | Focus | How to Apply What You Have Learned |
|---|---|---|---|
| 1 | Basic human needs | We need clean air, clean water, food, shelter, and energy to address our needs. Both the population of the world and the energy use per capita in the world have been increasing steadily. We use more of the world's resources than the planet can sustain. Human activities are contributing to outdoor air and water pollution. | The solution to any problem starts with understanding the problem. Before coming up with solutions, you need to develop a clear understanding of what the problems are. |
| 2 | Fundamental dimensions and units | Certain physical quantities are used to describe and quantify our surroundings and activities. These physical quantities include length, mass, time, temperature, and electric current. These are divided into units such as meter or foot, second, kilogram or pound, degrees Celsius or Fahrenheit, and ampere. | These physical quantities allow you to understand and quantify your needs and waste. For example, you can estimate how much material went into making a product, or the relationship between the room temperature setting and the energy consumption of a heating unit in winter. |
| 3 | Energy, power, energy content of fuels, and efficiency | To quantify the requirements to move objects such as our cars, to lift things like an elevator, or to heat or cool our homes, energy is defined and classified into different categories such as kinetic, potential, and thermal energy. Power represents how fast you are expending energy. The energy content of a fuel quantifies the amount of energy that is released when a unit mass (kilogram or pound) or a unit volume (cubic meter, cubic foot, or gallon) of a fuel is burned. All machines have efficiencies that are less than 100%; in other words, they require more input than what they put out. | You can quantify how much energy it takes to address your daily needs, for example, to open a garage door, to move a car, to heat a home, to heat water for a shower, or to lift an elevator. You can also estimate how much fossil fuel it will take to perform these tasks. |
| 4 | Electric power | It is important to understand the electric power consumption requirements for typical home appliances, electronics, and lighting systems. Ampere, resistance, voltage, and electric power constitute the basic concepts of electricity. You should also be familiar with basic lighting terminology. For example, the amount of light emitted by a lamp is expressed in lumens, and efficacy is the ratio of how much light is produced by a lamp (in lumens) compared to how much energy is consumed by the lamp (in watts). | When you buy and use a product that consumes electricity, you can calculate how much energy the product will consume daily or annually and how much coal or natural gas is burned in a power plant to provide the electric power needed to operate the product. |
| 5 | Heating and cooling of buildings | Heat always flows from a high-temperature region to a low-temperature region. There are three different mechanisms by which thermal energy is transferred: conduction, convection, and radiation. The R-value of a material provides a measure of resistance to heat flow; the higher the value, the more resistance to heat flow the material offers. Daylighting refers to the use of windows and skylights to bring natural light into a building. Factors that are considered when selecting windows and skylights include the U-factor, solar heat gain coefficient, air leakage, sunlight transmittance, visible transmittance, and light-to-solar gain. A degree day is the difference between 65°F (typically) and the average temperature of the outside air during a 24-hour period. | You can calculate how much fuel it takes to keep your house warm in winter and cool in summer. You can use historical degree-day data to estimate the monthly and annual energy consumption rates. |

| Chapter | Topic | Focus | How to Apply What You Have Learned |
|---|---|---|---|
| 6 | Energy consumption and non-renewable energy sources | Your personal energy consumption depends on your standard of living, and industrial energy consumption depends on economic activities such as production, distribution, use, and trade of goods and services. Energy use also depends on the weather. Currently, petroleum, coal, and natural gas make up most of the fuel used in generating energy. You should have a good grasp of how we consume energy through industrial activities, in buildings, and by transporting people and goods. Nearly one-third of the U.S. energy consumption is spent transporting people and products; gasoline and diesel fuel account for nearly 85% of energy consumed by vehicles. | You can estimate energy consumption and set a goal to reduce it by a certain percentage each year. You can also estimate the equivalent amount of fuel that won't be burned due to the savings. |
| 7 | Renewable energy sources: solar, wind, hydro, and biomass | Currently, renewable energy accounts for a small percentage of our energy sources. Solar systems can be classified as active, passive, and photovoltaic systems. The amount of solar radiation available at a place depends on many factors, including geographical location, season, local landscape and weather, and time of day. Wind energy is a form of solar energy. The power generated by the wind is directly proportional to the speed of the wind cubed.<br><br>Hydropower represents the power of moving water that is converted into electricity using a number of techniques. Biomass refers to organic materials such as wood, plants, or algae grown specifically to be converted to produce biofuels using different processes. Ethanol refers to alcohol-based fuel that is made from sugar found in crops such as corn and sugar cane. Biodiesel refers to fuel made from vegetable oils, animal fats, or recycled restaurant grease. | When selecting a renewable energy source, such as a solar system or a wind turbine, you can first perform simple calculations and then use the results of your calculation to make informed decisions. |
| 8 | Emissions and air quality | Carbon dioxide plays an important role in sustaining plant life; however, if the atmosphere contains too much carbon dioxide, it will result in the greenhouse effect. The Environmental Protection Agency (EPA) measures the concentration levels of six major air pollutants: carbon monoxide (CO), lead (Pb), nitrogen dioxide ($NO_2$), ozone ($O_3$), sulfur dioxide ($SO_2$), and particulate matter (PM). The indoor levels of pollutants may be many times higher than outdoor levels. The World Health Organization (WHO) sets the global air quality standards and monitors and provides technical support. | You can calculate your carbon dioxide footprint due to your daily activities such as driving a car, taking a shower, watching TV, or powering your electronic devices. |

*(continued)*

**TABLE 12.1**    **A Summary of Topics Covered in This Book and How to Apply Them** *(continued)*

| Chapter | Topic | Focus | How to Apply What You Have Learned |
|---|---|---|---|
| 9 | Water resources and consumption | The total amount of water available on Earth remains constant; we don't lose or gain water. You should be familiar with how much water you consume to address your personal needs. The U.S. Geological Survey (USGS) groups major water-consuming activities into broad categories such as public, domestic, irrigation, livestock, aquaculture, industrial, mining, and thermoelectric power generation and reports the data for each category. Human activities and naturally occurring microorganisms contribute to the level of contaminants in our water supply. The EPA sets the standards for the maximum level of contaminants that can be in our drinking water and still be considered safe to drink. The WHO is responsible for setting water standards and monitoring and providing technical support globally. | You can estimate your annual water consumption and make a plan to reduce it by a certain amount. |
| 10 | Materials—what products are made from | Products are typically made from metals, plastics, glass, composites, or wood; construction materials include wood, aluminum, steel, and concrete. | You can estimate how much material you consume annually and devise a plan to reduce your consumption. |
| 11 | Waste and recycling | Before you throw away something, you should think about the natural resources that were used to make the item and how much energy it took to produce, process, transport, and eventually dispose of it. The EPA reports the amount of municipal and industrial waste. Each year, we also throw away large quantities of food, fruit and vegetable skins, and so on. Composting offers an environmentally friendly alternative to throwing food and yard trimmings in landfills. | You can estimate how many durable and nondurable goods you throw away and come up with a plan to reduce the amount by a certain percentage. You can manage waste by reducing, recycling, or composting. |
| 12 | Life-cycle assessment | Life-cycle analysis looks at the raw material and the energy consumption, emissions, and other factors related to a product over its entire life, from its origin to its disposal. | Before making decisions, you can perform a life-cycle analysis (even a limited version) to obtain useful information, particularly when comparing different products. |

or divide them into units. For example, the time dimension can be divided into both small and large portions, such as seconds, minutes, hours, days, months, years, etc. The SI is the most common system of units used in the world; you should be familiar with its units of length (meter), time (second), mass (kilogram), temperature (Kelvin or degree Celsius), electric current (ampere), amount of substance (mole), and luminous intensity (candela). The SI system also makes use of a series of prefixes and symbols of decimal multiples, such as Mega (M), Giga (G), Kilo (k), etc. You should also be familiar with the U.S. Customary units of length (feet), time (second), mass (pound-mass), temperature (degree Rankine or degree Fahrenheit), electric current (ampere), amount of substance (mole), and luminous intensity (candela). Every product that you own or will purchase some day is considered a system and is made of components.

## Energy, Power, and Efficiency

We need energy to create goods, to build shelter, to cultivate and process food, and to maintain our living places at comfortable temperatures. Energy can have different forms, and to better explain quantitatively the requirements to move objects such as our cars, to lift things like an elevator, or to heat or cool our homes, energy is defined and classified into different categories such as kinetic, potential, and thermal energy. *Kinetic energy* is a way we quantify how much energy is required to move something. The energy required to lift an object over a vertical distance is called *potential energy. Thermal energy* or heat transfer occurs whenever a temperature difference exists within an object or between a body and its surroundings. The units of kinetic and potential energy are joules (J) or pound force-feet (lbf·ft). The three units that are commonly used to quantify thermal energy are the British thermal unit (Btu), the calorie, and the joule (J). One British thermal unit (Btu) represents the amount of thermal energy needed to raise the temperature of one pound mass (lbm) of water by one degree Fahrenheit (°F). The calorie represents the amount of heat required to raise the temperature of one gram (g) of water by one degree Celsius (°C).

*Power* is the time rate of doing work or how fast you are expending energy. The value of power required to do the work (perform a task) represents how fast you want the work (task) done. The heating value or energy content of a fuel quantifies the amount of energy that is released when a unit mass (kilogram or pound) or a unit volume (cubic meter, cubic foot, or gallon) of a fuel is burned. All machines and systems require more input than what they put out. You should know the basic definition of efficiency, which is

$$\text{efficiency} = \frac{\text{what you want to get out of a system}}{\text{what you need to put into the system}}$$

## Electricity and Electric Power

As a good citizen, you need to know about your home's electric power distribution system and consumption. You should know that in the United States a typical house has a total 200 amperage rating. You should be familiar with basic principles of electricity; for example, the electric resistance is measured in units of ohms (Ω). An element with one ohm resistance allows a current flow of one ampere (A) when there exists a potential of one volt (V) across the element. The flow of electric charge is called the electric current or simply current. The electric current is measured in amperes. One ampere or "amp" (A) is defined as the flow of one unit of charge per second.

*Voltage* represents the amount of work required to move charge between two points, and the amount of charge that is moving between the two points per unit time is called current. Moreover, direct current (dc) is the flow of electric charge that occurs in one direction. Batteries and photovoltaic systems create direct current. Alternating current (ac) is the flow of electric charge that periodically reverses. Alternating current is created by generators at power plants. The current drawn by various electrical devices at your home is alternating current.

At the heart of every home electrical distribution system is wire. Electrical wires are typically made of copper or aluminum. The actual size of the wires is commonly expressed in terms of gage number as denoted by the American Wire Gage (AWG). You should recall that the smaller the gage number, the bigger the wire diameter. The *National Electrical Code*, published by the Fire Protection Association, contains specific information on the type of wires used for general wiring. You also should know what the power consumptions are for typical home appliances and electronics such as refrigerators; electric stoves; microwaves; dishwashing machines; clothes washing machines; clothes dryers; heating, cooling, and ventilating units; TVs; computers; and radios.

Familiarize yourself with basic lighting terminology and know how to calculate power consumption rates for lighting systems. When purchasing light bulbs, you should look for illumination and efficacy ratings. *Illumination* refers to the distribution of light on a horizontal surface, and the amount of light emitted by a lamp is expressed in lumens. As a reference, a 100-watt incandescent lamp may emit 1,700 lumens. A common unit of illumination intensity is the footcandle, which is equal to one lumen distributed over an area of one square-foot. For example, to find your way around at night, you will need between 5 to 20 footcandles. *Efficacy* is the ratio of how much light is produced by a lamp (in lumens) compared to how much energy is consumed by the lamp (in watts). How true the colors of an object appear when illuminated by a light source is represented by the color rendition index (CRI). The color rendition index has a scale of 1 to 100 with a 100-watt incandescent light bulb having a CRI value of approximately 100. There are different types of lighting systems, including incandescent light bulbs, fluorescent lamps, compact fluorescent lamps, high-intensity discharge (HID) lamps, and light-emitting diode (LED) lights.

## Thermal Energy: Heat Loss and Gain in Buildings

Thermal energy transfer occurs whenever there exists a temperature difference; this form of energy is called heat transfer. Recall that heat always flows from a high-temperature region to a low-temperature region. The three different mechanisms by which thermal energy is transferred are referred to as the modes of heat transfer. The three modes of heat transfer are conduction, convection, and radiation.

*Conduction* refers to the mode of heat transfer that occurs when a temperature difference exists in a medium. The R-value of a material provides a measure of resistance to heat flow: the higher the value, the more resistance to heat flow the material offers.

*Convection* heat transfer occurs when a fluid (a gas or a liquid) in motion comes into contact with a solid surface whose temperature differs from the moving fluid. There are two broad areas of convection heat transfer: forced and free (natural). Forced convection refers to situations where the flow of fluid is forced by a fan or a pump. Free convection, on the other hand, refers to situations where the flow of fluid occurs naturally due to density variation in the fluid.

All matter emits thermal *radiation*. The higher the temperature of the surface of the object, the more thermal energy is emitted by the object. Unlike the conduction and convection modes, heat transfer by radiation can occur in a vacuum. *Daylighting* refers to the use of windows and skylights to bring natural light into a building. The National Fenestration Rating Council defines the ratings for window and skylight energy performance. Factors that are considered when selecting windows and skylights include the U-factor, solar heat gain coefficient, air leakage, sunlight transmittance, visible transmittance, and light-to-solar gain. A degree-day (DD) is the difference between 65°F (typically) and the average temperature of the outside air during a 24-hour period. In practice, historical degree-day values (based on the average of data over many years) are used to estimate monthly and annual energy consumptions to heat buildings.

## Energy Consumption Rates and Non-Renewable Energy Sources

Your personal energy consumption depends on your standard of living, and industrial energy consumption depends on economic activities such as production, distribution, use, and trade of goods and services. Energy use also depends on the weather. Currently, petroleum, coal, and natural gas provide over 80 percent of all the fuel used in generating energy. You should have a good understanding of how we consume energy at home and in commercial buildings. Commercial buildings include retail and service areas, malls and stores, car dealerships, dry cleaners, gas stations, professional and government offices, banks, schools and colleges, hospitals, and hotels.

*Electricity* and *natural gas* are the most common energy sources used in residential and commercial buildings. Nearly one-third of the United States energy consumption is spent transporting people and products. Most of the transportation energy is consumed by automobiles and light trucks; gasoline and diesel fuel account for nearly 85 percent of energy consumed by vehicles. Automobiles are the most common modes of transportation in the United States, with most of these vehicles fueled by either gasoline or diesel. You also should know how much gasoline is processed from a barrel of crude oil and which countries are among the world's top five oil producers. The liquid fuel consumption rates are expected to increase in the coming years. Nearly one hundred countries produce crude oil. In a refinery, from each barrel of oil— which is equal to 42 gallons—19 gallons of gasoline are made. The remaining 23 gallons are turned into diesel, heating oil, jet fuel, and other petroleum-based products.

*Diesel* fuel accounts for nearly one-fifth of the total transportation fuel consumption in the United States. *Fuel oil* is a petroleum product used to heat homes in America—especially in the northeast. Heating oil and diesel fuel are similar in composition; the main difference between the two fuels is sulfur content. Heating oil has more sulfur than diesel fuel does. It is also important to know that the natural gas transportation network in the United States is made up of nearly 1.5 million miles of mainline and secondary pipelines. You should also be able to explain the process for making and distributing *liquefied petroleum gases* (LPGs). Refineries and natural gas processing plants also make millions of barrels per day of LPGs, such as propane. A gas such as propane is referred to as liquid petroleum gas because it is stored in a tank under relatively high pressures, which makes it liquid. The LPGs become gas once released from the pressurized tank. In the northeast, a liquid petroleum gas, such as propane, is used for cooking and to heat water and homes. Propane is also used in the chemical industry to make plastics and other materials.

*Coal* is a fossil fuel, which based on its carbon and energy content, is classified into anthracite, bituminous, subbituminous, and lignite. Most of the coal mined in the United States is used for generating electricity.

Currently, *nuclear energy* represents 5 and 8 percent of the world and United States energy sources, respectively. There are two processes by which nuclear energy is harnessed: nuclear fission and nuclear fusion. Nuclear power plants use nuclear fission to heat water to create steam to turn the turbines that in turn run the generators that produce electricity. In nuclear fission, to release energy, atoms of uranium are bombarded by neutrons. This process splits the atoms of uranium and releases more neutrons and energy in the form of heat and radiation. The additional neutrons go on to bombard other uranium atoms, and the process keeps repeating itself, leading to a chain reaction.

The fuel most widely used by nuclear power plants is uranium 235 or simply U-235. U-235 is relatively rare and must be processed from uranium that is mined. After it is processed, the uranium fuel is made into ceramic pellets that are stacked end-to-end to form fuel rods. The fuel rods are then bundled together to create fuel assemblies which are then used in the reactor core of a nuclear power plant. Because the spent fuel assemblies are highly radioactive, they must be stored in pools underwater for several years and are then moved to dry cask concrete or steel storage containers that are cooled by air. Eventually, the spent fuel assemblies are moved from interim storage sites to permanent underground storage facilities.

Currently, in the United States, there are 61 nuclear power plants with 99 reactors which produce about 800 billion kilowatt-hours of electricity annually. Today, there are 30 countries in the world that have nuclear power plants, with the top 10 countries generating nearly 1,977 billion kilowatt-hours of electricity.

## Renewable Energy

Renewable energy refers to solar energy, wind energy, hydro-energy, and biomass. Solar energy starts with the sun at an average distance of approximately 150 million kilometers from the Earth. The solar radiation can be divided into three bands: ultraviolet, visible, and infrared. The *visible band* comprises about 48 percent of useful radiation for heating, and the infrared makes up the rest. The amount of radiation available at a place depends on many factors including geographical location, season, local landscape and weather, and time of day. As solar energy passes through the Earth's atmosphere, some of it is absorbed, some of it is scattered, and some of it is reflected by clouds, dust, pollutants, forest fires, and water vapor. The solar radiation that reaches the Earth's surface without being diffused is called *direct beam solar radiation*. On a clear day at noon when the sun appears at its highest point in the sky, the greatest amount of solar energy reaches a horizontal surface on the Earth. Seasonal effects are also important. During the winter, the sun's angle is lower than it is in the summer, which results in a lower amount of radiation being intercepted by a horizontal surface.

*Solar energy systems* can be categorized into active, passive, and photovoltaic systems. There are two basic types of active solar heating systems: liquid and air. The liquid systems make use of water, water-antifreeze mixture, or other liquids to collect solar energy. In such systems, the liquid is heated in a solar collector, and then heat is transferred to a storage system. In contrast, in air systems, the air is heated in "air collectors" and is transported to storage or space using blowers. The passive solar systems do not make use of any mechanical components such as collectors, pumps, blowers, or fans to collect, transport or distribute solar heat to various parts of a building. Instead, a direct passive solar system uses large glass areas on the south wall of a building and a thermal mass to collect the solar energy. The solar energy is stored in interior thick masonry walls and floors during the day and is released at night.

A *photovoltaic system* converts light energy directly into electricity. A photovoltaic (PV) cell is the backbone of any photovoltaic system. Photovoltaic cell materials include crystalline silicon, polycrystalline silicon, and amorphous silicon. The manufacturers of photovoltaic systems combine cells to form a module, and then modules are combined to form an array. A photovoltaic system consists of batteries, a charge controller, and an inverter. A charge controller protects the batteries from overcharging. An inverter is a device that converts direct current into alternating current. Photovoltaic systems are classified into stand-alone systems, hybrid systems, or grid-tied systems.

*Wind energy* is a form of solar energy. It is important to understand that the wind speed increases with vertical distance from ground, and the power generated by wind is directly proportional to the speed of the wind cubed. Not all wind power can be captured. If that were to happen, the air behind the rotor would have a zero speed, which would mean that no air is flowing over the blades. Two types of wind turbines are used to extract the energy from the wind: vertical axis and horizontal axis. Wind turbines are typically classified as small (<100 kW), intermediate (<250 kW), and large (250 kW to 8 MW).

*Hydro-energy* represents the energy of moving water that is converted into electricity using a number of techniques including impoundment, diversion, and pumped storage hydropower. The impoundment approach makes use of dams to store water. The water is guided into water turbines located in hydroelectric power plants housed within the dam to generate electricity. The diversion technology diverts part of the water running through a river through turbines. This technology does not require a large dam and makes use of the natural flow of water. The pumped storage technique pumps the water from a lower elevation and stores it in a higher elevation at night when the energy demand is low. During the daytime when the energy use is high, the water is released from the higher elevation storage to the lower elevation to produce electricity.

*Biomass* refers to organic materials such as forest and wood trimmings, plants, fast growing grasses and trees, crops, or algae grown specifically to be converted to produce biofuels using different processes. Wood is considered a biomass fuel. Throughout our history, wood has been used as fuel in stoves and fireplaces. Today, wood is still a major source of energy for people in many developing countries. In recent years, sawdust has been compressed to form pellets—commonly known as wood pellets—that are burned in heating stoves. Also, wood and paper product industries use their wood waste as fuel to generate electricity. In recent years, much attention has been focused on algae as biofuel. Algae are small aquatic organisms that convert sunlight to energy. They can be grown in algae farms, which are basically large man-made ponds. Some algae store energy in the form of oil that can be extracted by breaking down the cell structure using solvents or sound waves. The extracted oil is then refined to serve as biofuel.

*Ethanol* refers to an alcohol-based fuel that is made from sugar found in crops such as corn and sugar cane. *Biodiesel* refers to fuel made from vegetable oils, animal fats, or recycled restaurant grease. Most of the ethanol produced in the United States is made from corn and is mixed with gasoline; most of the biodiesel fuel comes from soybean oil and is mixed with diesel fuel derived from petroleum.

## Air and Air Quality Standards

It is important to know the characteristics of the atmosphere, the difference between climate and weather, and to understand the greenhouse effect. Air is a mixture of mostly nitrogen and oxygen, and small amounts of other gases such as argon, carbon dioxide, sulfur dioxide, and nitrogen oxide. The air surrounding the Earth, depending on its temperature, can be divided into four regions: troposphere, stratosphere, mesosphere, and thermosphere. *Weather* represents atmospheric conditions such as the temperature that could occur during a period of hours or days. *Climate*, on the other hand, represents the average weather conditions over a long period of time (i.e., decades or centuries). Carbon dioxide plays an important role in sustaining plant life; however, if the atmosphere contains too much carbon dioxide, it will not allow the Earth to cool down effectively by radiation, which results in the *greenhouse effect*.

The *Environmental Protection Agency* (EPA) measures the concentration level of pollutants in many urban areas and collects air quality information. The source of outdoor air pollution is classified into three broad categories: stationary, mobile, and natural sources. The EPA is responsible for setting standards for six major air pollutants: carbon monoxide (CO), lead (Pb), nitrogen dioxide ($NO_2$), ozone ($O_3$), sulfur dioxide ($SO_2$), and particulate matter (PM). The EPA also is continuously working to set standards and monitor the emission of pollutants that cause acid rain, which damages bodies of water and fish, buildings, and our national parks. The EPA works with individual states to reduce the amount of sulfur in fuels and sets more stringent emission standards for cars, buses, trucks, and power plants. Because we all contribute to air pollution, we need to be aware of the consequences of our lifestyles and find ways to reduce emissions.

According to EPA studies of human exposure to air pollutants, the indoor levels of pollutants may be many times higher than outdoor levels.

Some common health symptoms caused by poor indoor air quality are headaches, fatigue, and shortness of breath. The factors that influence indoor air quality are classified into several categories: the heating, ventilation, and air-conditioning (HVAC) system; sources of indoor air pollutants; and occupants. In order to save energy, we are building tight houses with lower ventilation rates that also have lower air infiltration compared to that of older structures. We are also using more synthetic building materials in newly built homes that can give off harmful vapors. Moreover, we are using more chemical pollutants, such as pesticides and household cleaners. There are several ways to control the level of contaminants: (1) source elimination or removal, (2) source substitution, (3) proper ventilation, (4) exposure control, and (5) air cleaning. It is very important that you bring indoor air quality issues to the attention of your friends, classmates, and family. We all need to be well-educated in this topic and try to do our part to create and maintain a healthy indoor air quality.

The *World Health Organization* (WHO) is the authority on global health matters, including air-quality-related health issues. The WHO is responsible for setting and monitoring air quality standards and for providing the technical support needed to achieve these standards. According to the WHO, air pollution is a major global environmental risk to health that causes respiratory infections, heart disease, and lung cancer. Each year, nearly 2 million premature deaths are attributed to indoor air pollution in developing countries. The latest WHO air-quality guidelines recommend limits for the concentration of selected air pollutants such as particulate matter (PM), ozone ($O_3$), nitrogen dioxide ($NO_2$), and sulfur dioxide ($SO_2$).

## Water Sources, Consumption Rates, and Quality Standards

The total amount of water available on Earth remains constant; even though water can change phase from liquid to solid (ice) or from liquid to vapor, we don't lose or gain water. You should be familiar with how much water you consume to address your personal needs. You should also realize that we use the greatest amount of water in flushing toilets, followed by washing clothes and taking showers. We should consider taking conservative measures such as installing low-flow showers and water efficient toilets in our homes.

The *U.S. Geological Survey* (USGS) has been estimating total water consumption in the United States since 1950. They report their findings every five years for both the groundwater and surface water sources. For bookkeeping purposes, the USGS groups major water-consuming activities into broad categories such as public, domestic, irrigation, livestock, aquaculture, industrial, mining, and thermoelectric power generation and reports the data for each category. A public water supply refers to water that is drawn by the government. Most of our domestic water supply is delivered by a public supplier. Public suppliers also provide water for businesses, schools, firefighting, community parks and swimming pools, and at times for commercial applications. The irrigation category refers to the amount of water that is provided by man-made systems for agricultural purposes. The amount of water that is used for activities such as dairy operations, feed lots, and providing drinking water for livestock is represented by the livestock category. Aquaculture refers to farming fish, shrimp, and other animals that live in water. The farming of plants and algae that live in water is also grouped into this category. The water that is used for industrial purposes such as making paper, chemicals, or steel is classified as industrial. The water used for the excavation of rocks and minerals is classified as mining water. Finally, the water used in power plants that generate electricity is classified as thermoelectric power. Thermoelectric power generation, irrigation, and public supply sectors are among the largest consumers of water in the United States.

The EPA sets the standards for the maximum level of contaminants that can be in our drinking water and still be considered safe to drink. Human activities and naturally occurring microorganisms contribute to the level of contaminants in our water supply. The EPA sets two standards for the level of water contaminants: (1) the maximum contaminant level goal (MCLG) and (2) the maximum contaminant level (MCL). The MCLG represents the maximum level of a given contaminant in the water that causes no known harmful health effects. On the other hand, the MCL, which may represent slightly higher levels of contaminants in the water, is the level of contaminants that are legally enforceable.

The WHO is also responsible for setting and monitoring water standards and providing technical support. According to the WHO, nearly one billion people lack access to clean drinking water; as a result, millions of deaths are attributed to unsafe water each year. Examples of diseases transmitted through water contaminated by human waste include diarrhea, cholera, dysentery, typhoid, and hepatitis. Diarrhea is largely preventable but is a major killer responsible for 1.5 million deaths every year, mostly among children five years old and younger living in developing countries. Hygiene education and the promotion of hand washing are simple, cost-effective measures that can reduce diarrhea cases by up to 45 percent. More than 50 countries still report cholera to the WHO. An estimated 2.6 billion people lack access to adequate sanitation globally. Also, according to the WHO, wastewater in agriculture is associated with serious public health risks.

## Common Materials that We Use Every Day

Products are typically made from metals, plastics, glass, composites, or wood; construction materials include wood, aluminum, copper, steel, and concrete. Aluminum, titanium, and magnesium, because of their small densities (relative to steel), are commonly referred to as lightweight metals and are used in many structural and aerospace applications.

*Aluminum* and its alloys have densities (density = mass/volume) that are approximately one-third the density of steel. Aluminum is commonly alloyed with other metals such as copper, zinc, and magnesium. Everyday examples of common aluminum products include beverage cans, household aluminum foil, staples in tea bags that do not rust, building insulation, and so on.

*Titanium* has an excellent strength-to-weight ratio. Titanium alloys are used in golf clubs, bicycle frames, and spectacle frames. *Magnesium* is another lightweight metal that looks like aluminum, but it is lighter. Magnesium is commonly alloyed with other elements such as aluminum, manganese, and zinc to improve its properties. Magnesium and its alloys are used in nuclear applications, in dry-cell batteries, and in aerospace applications.

*Copper* is a good conductor of electricity and heat, and because of these properties, it is commonly used in many electrical, heating, and cooling applications. Copper alloys are also used as tubes, pipes, and fittings in plumbing. When copper is alloyed with zinc, it is commonly called brass. Bronze is an alloy of copper and tin.

*Steel* is an alloy of iron with approximately 2 percent or less carbon. The properties of steel can be modified by adding other elements, such as chromium, nickel, manganese, silicon, and tungsten. The 18/8 stainless steels, which contain 18 percent chromium and 8 percent nickel, are commonly used for tableware and kitchenware products. Cast iron is also an alloy of iron that has 2 to 4 percent carbon. Steel is a common material that is used in the framework of buildings and bridges; the body of appliances such as refrigerators, ovens, dishwashers, clothes-washers and dryers; and for cooking utensils.

*Plastic* products include grocery and trash bags, soft drink containers, household cleaning containers, vinyl siding, polyvinyl chloride (PVC) piping, valves, and fittings. Styrofoam™ plates and cups, plastic forks, knives, spoons, and sandwich bags are other examples of plastic products that are consumed every day. Plastics are often classified into two categories: thermoplastics and thermosets. When heated to certain temperatures, thermoplastics can be molded and remolded. By contrast, thermosets cannot be remolded into other shapes by heating. Silicon is a nonmetallic chemical element that is used quite extensively in the manufacturing of transistors and various electronic and computer chips. It is found in the form of silicon dioxide in sands and rocks or combined with other elements—such as aluminum, calcium, sodium, or magnesium—in a form that is commonly referred to as silicates. Silicon, because of its atomic structure, is an excellent semiconductor, which is a material whose electrical conductivity properties can be changed to act either as a conductor of electricity or as an insulator (preventer of electricity flow).

*Glass* is commonly used in products such as windows, light bulbs, housewares (such as drinking glasses), chemical containers, beverage and beer containers, and decorative items. The composition of the glass depends on its application. The most widely used form of glass is soda–lime–silica glass. The materials used in making soda–lime–silica glass include sand (silicon dioxide), limestone (calcium carbonate), and soda ash (sodium carbonate). Other materials are added to create desired characteristics for specific applications. Silica glass fibers are commonly used today in fiber optics, which is a branch of science that deals with transmitting data, voice, and images through thin glass or plastic fibers.

*Composite materials* are found in military planes, helicopters, satellites, commercial planes, fast-food restaurant tables and chairs, and many sporting goods. In comparison to conventional materials (such as metals), composite materials can be lighter and stronger. Composite materials consist of two main ingredients: matrix materials and fibers. Fibers are embedded in matrix materials, such as aluminum or other metals, plastics, or ceramics. Glass, graphite, and silicon carbide fibers are examples of the types of fibers used in the construction of composite materials. The strength of the fibers is increased when embedded in the matrix material, and the composite material created in this manner is lighter and stronger.

Common examples of *wood* products include hardwood flooring, roof trusses, furniture frames, wall supports, doors, decorative items, window frames, kitchen cabinets, trimming in luxury cars, tongue depressors, clothes-pins, baseball bats, bowling pins, fishing rods, and wine barrels. Timber is commonly classified as softwood and hardwood. Softwood timber is made from trees that have cones (coniferous), such as pine, spruce, and Douglas fir. On the other hand, hardwood timber is made from trees that have broad leaves or flowers.

*Concrete* is used in the construction of roads, bridges, buildings, tunnels, and dams. Concrete consists of three main ingredients: aggregate, cement, and water. Aggregate refers to materials such as gravel and sand, and cement refers to the bonding material that holds the aggregate together. Concrete is usually reinforced with steel bars or steel mesh that consists of thin metal rods to increase its load-bearing capacity. Another common construction practice is the use of precast concrete. Precast concrete slabs, blocks, and structural members are fabricated in less time and at less cost in factory settings where the surrounding conditions are controlled. Because concrete has a higher compressive strength than tensile strength, it is prestressed by pouring it into forms that have steel rods or wires. The steel rods or wires are stretched so that the pre-stressed concrete then acts as a compressed spring, which will become un-compressed under the action of tensile loading.

## Municipal and Industrial Waste and Recycling

We should be mindful of how much waste we generate and consider the entire life cycle of trash. Before you throw away something, you should think about the natural resources that were used to make the item and how much energy it took to produce, process, transport, and eventually dispose of it.

*Municipal waste* is the trash that we throw away every day: food scraps, packaging materials, bottles, cans, and so on. The EPA reports the amount of waste generation and recovery using the following categories: durable goods, nondurable goods, containers and packing materials, and plastic packing. Durable goods cover products such as major and small appliances, furniture, carpets and rugs, rubber tires, and lead-acid batteries. Nondurable goods represent products such as office papers, newspapers, books, magazines, paper plates and cups, tissue paper and paper towels, disposable diapers, plastic plates and cups, trash bags, clothing and footwear, towels, sheets, and so on.

*Industrial waste* refers to rubbish that is produced in industry. A significant portion of solid waste in the United States is industrial waste. It consists of construction, renovation, and demolition materials. It also represents medical refuse and the waste that is created during the exploration, development, and production of fossil fuels.

We can manage waste by *reducing*, *recycling*, and *composting*. When we recycle, we reduce landfill use, we consume less energy, and we use our natural resources more effectively. Each year we throw away large quantities of food, fruit and vegetable skins, and so on. This category makes up nearly 30 percent of our trash. Composting offers an environmentally friendly alternative to throwing food and yard trimmings in landfills. Composting refers to the biological decomposition or decay of food wastes, yard trimmings, and other organic materials.

# SUMMARY

## LO¹ The Earth Charter

In order to address our energy, clean air and water, and food supply intelligently, *we need to work together*. To emphasize the sense of global interdependence and shared responsibility for the well-being of the entire human family, the *Earth Charter* was introduced by an independent international commission, on June 29, 2000, in The Hague, Netherlands. The Earth Charter eloquently puts into words an ethical guideline for building a sustainable, just, and peaceful global society in the 21st century. The Earth Charter is intended as both a vision of hope and a call to action. The Earth Charter consists of six sections: Preamble, Earth—Our Home, The Global Situation, The Challenges Ahead, Universal Responsibility, and Principles.

## LO² Sustainability Concepts, Assessments, and Tools

### Key Sustainability Concepts

One of the generally accepted definitions of sustainability is "*design and development that meets the needs of the present without compromising the ability of future generations to meet their own needs.*" As a society, we are expected to design and provide goods and services that increase the standard of living and advance health care while addressing serious environmental and sustainability concerns. We must consider the link among Earth's finite resources and environmental, social, ethical, technical, and economical factors.

*Ethics* refers to the study of morality and the moral choices that we all have to make in our lives. Professional societies, such as those in medicine and engineering, have long established guidelines, standards, and rules that govern the conduct of their members. These rules are also used by the members

of the board of ethics of the professional organization to interpret ethical dilemmas that are submitted by a complainant. As good global citizens, it is important to realize that we must also hold to the highest standards of honesty and integrity, particularly when it comes to the Earth—our home. Think about this carefully: an unethical surgeon can kill a patient, but uninformed and unethical citizens can destroy the planet Earth—our only home.

### Key Sustainability Assessment: Life-Cycle Analysis

*Life-cycle analysis* looks at the raw material and the energy consumption, emissions, and other factors related to a product over its entire life—from its origin to its disposal; *from its inception and birth to its death*. Generally, the life-cycle analysis consists of four steps:

1. Purpose or goal definition
2. Inventory analysis
3. Impact assessment
4. Improvement

### Key Sustainability Tools

The life-cycle assessment, environmental assessment, and U.S. Green Building Council Leadership in Energy and Environmental Design (LEED) rating system serve as sustainability tools to evaluate our environmental impact and sustainability. In addition to these tools, there are many resource, waste, and emission calculators available online through various government and international agencies.

## LO³ Apply What You Have Learned— Knowledge Is Power

You have heard the old saying: "knowledge is power." How are you planning to use your power? Well, you can start at home at your campus.

# KEY TERMS

Earth Charter 363
Ethics 367

Life-Cycle Analysis 367

Sustainability 365

# PROBLEMS

Each day, students get involved with sustainability projects in their homes, on their campuses, or for their cities. They are making their communities better places to live and work. Here are examples of projects that were initiated on campuses across the country. These projects also could be extended to other municipalities. Consider a similar project that is suitable for your campus or city. For your project, apply what you have learned by calculating how much energy savings, water savings, and reduction in waste and emissions could be expected from such project.

**Bike-Share Program** This project could be funded by the student activity fees or the city to allow students and citizens to check out bicycles for a certain period of time using their ID or credit cards. How many bicycles, helmets, and locks do you need to provide in order to have a successful program?

**Ride-Share Program** This can be a web-based system that matches participants in the same geographical area to encourage carpooling. Those who participate in these types of programs can share the cost of fuel and the parking permit.

**Photovoltaic Parking Rooftops** These systems can provide shaded parking for bicycles and cars. The photovoltaic system that generates electricity can be installed on the parking lot's rooftop. The initial investment costs for these types of projects are normally high, so they will require some form of partnerships with the city or state government and the utilities company.

**Solar Hot-Water Systems** These systems provide hot water for use on campus for different purposes such as cafeterias, labs, or swimming pools. Again, the initial investment costs for these types of projects are normally high, so they will require some form of partnerships with the city or state government and the utilities company.

**Campus Vegetable Gardens** These types of projects will promote growing some of the vegetables used in the cafeterias on campus. If land is not available,

you may want to consider creating green roofs, which are roofs of building that are partially or completely covered with vegetation. In addition to serving as a garden space, green roofs also provide energy savings for the buildings; the green roofs reduce heat loss through the roof of the building during winter and reduce heat gain during summer. The campus gardens could also have composting components where student volunteers take food waste to composting bins located in the campus gardens.

**Heating of Buildings** Students can work with facility engineers to develop policies for classroom and office temperatures to be set slightly lower during occupied hours. You can promote wearing extra layers of clothing to students and faculty.

**Water Collection Systems** Rainwater collection systems can provide water for your campus gardens or flushing toilets.

**Biodiesel Production** Student volunteers can collect cafeteria waste cooking oil on your campus to be converted into biodiesels.

**Awareness Programs** Focus on the first-year students to promote saving energy and water and encouraging waste reduction. For example, you can arrange to have dormitories go dark for a short period of time to bring attention to energy consumption and conservation. Think of other means to educate the new students on campus.

**Recycling** Develop activities during the sporting events on your campus to reduce the amount of trash and increase recycling. You can also develop an end-of-semester recycling drive where you collect and donate clothing, old electronics, and food that normally end up in the trash.

You can come up with many more similar projects on your own for home, work, and your community! The following is a list of universities and institutes that offer useful information related to sustainability. Don't forget, we are all in it together!

Carleton College: http://apps.carleton.edu/campus/sustainability/

Cornell University: http://www.sustainablecampus.cornell.edu/

Emory University: http://www.emory.edu/sustainability.cfm

George Washington University: http://sustainability.gwu.edu/

Princeton University: http://www.princeton.edu/sustainability/

Tufts University: http://www.tufts.edu/tie/tci/TuftsSustainability.htm

University of Arkansas: http://sustainability.uark.edu/

University of California, Berkeley: http://sustainability.berkeley.edu/

University of California, Los Angeles: http://www.sustain.ucla.edu/

University of Florida: http://www.sustainable.ufl.edu/

University of New Hampshire: http://www.sustainableunh.unh.edu/

University of North Carolina: http://sustainability.unc.edu/

University of Oregon: http://sustainability.uoregon.edu/

University of Tennessee, Knoxville: http://environment.utk.edu/

University of Vermont: http://www.uvm.edu/greening/

University of Washington: http://www.washington.edu/about/environmentalstewardship/

University of Virginia: http://www.virginia.edu/sustainability/

Virginia Tech: http://www.facilities.vt.edu/sustainability/

University of Wisconsin-Madison: http://www.sage.wisc.edu/

Yale University: http://www.yale.edu/sustainability/

American College and University Presidents Climate Commitment: http://www.presidentsclimatecommitment.org/

Guide to Developing a Sustainable Food Purchasing Policy: http://www.aashe.org/resources/pdf/food_policy_guide.pdf

Sustainability Endowments Institute: http://www.endowmentinstitute.org/

Photo by Boyer/Roger Viollet/Getty Images

*"The great aim of education is not knowledge but action."*

— HERBERT SPENCER (1820–1903)

# Appendix A

## A Summary of Formulas Discussed in this Book

weight = (mass)(acceleration due to gravity)

$$\text{temperature}\,(°C) = \frac{5}{9}[\text{temperature}\,(°F) - 32]$$

$$\text{temperature}\,(°F) = \frac{9}{5}\text{temperature}\,(°C) + 32$$

work = (force)(distance)

$$\text{kinetic energy} = \left(\frac{1}{2}\right)(\text{mass})(\text{speed})^2 = \left(\frac{1}{2}\right)mV^2$$

work = change in kinetic energy of the object

$$= (\text{force})(\text{distance}) = \left[\left(\frac{1}{2}\right)(\text{mass})(\text{speed})^2\right]_{\text{final}} - \left[\left(\frac{1}{2}\right)(\text{mass})(\text{speed})^2\right]_{\text{initial}}$$

change in potential energy = (weight of the object)(change in elevation)

$$= \overbrace{(\text{mass of the object})(\text{acceleration due to gravity})}^{\text{weight of the object}}(\text{change in elevation})$$

$$\text{power} = \frac{\text{work}}{\text{time}} = \frac{\text{energy}}{\text{time}}$$

$$\text{efficiency} = \frac{\text{desired output}}{\text{required input}}$$

voltage = (resistance)(current) = $V = (R)(I)$

electric power = (voltage)(current) = $P = (V)(I)$

$$\text{efficacy} = \frac{\text{light produced (lumens)}}{\text{energy consumed by the lamp (W)}}$$

power plant efficiency $= \dfrac{\text{net energy generated}}{\text{energy input from fuel}}$

Fourier's law for conduction heat transfer, $q = kA\left(\dfrac{T_1 - T_2}{L}\right)$

heat flow $= \dfrac{\text{temperature difference}}{\text{thermal resistance}}$

Newton's law of cooling for convection heat transfer, $q = hA(T_s - T_f)$

Radiant energy emitted, $q = \varepsilon \sigma A T_s^4$

$Q_{DD} = \dfrac{\text{building heat loss}\left(\dfrac{\text{Btu}}{\text{h}}\right) \times 24\,\text{hrs}}{\text{design temperature difference}\,(°F)}$

$Q_{monthly} = (Q_{DD})(\text{monthly degree days})$

$Q_{yearly} = (Q_{DD})(\text{yearly degree days})$

density $= \dfrac{\text{mass}}{\text{volume}}$

volume flow rate $= \dfrac{\text{volume}}{\text{time}}$

mass flow rate $= \dfrac{\text{mass}}{\text{time}}$

mass flow rate $= (\text{density})(\text{volume flow rate})$

wind power $= (\text{efficiency})\left(\dfrac{1}{2}\right)(\text{air density})(\text{sweep area})(\text{speed})^3$

hydro power $= (\text{overall efficiency})(\text{mass flow rate})(\text{acceleration due to gravity})(\text{change in elevation})$

# Appendix B

## Conversion Factors

| Quantity | From SI to U.S. Customary | From U.S. Customary to SI |
|---|---|---|
| Length | 1 m = 39.37 in. <br> 1 m = 3.28 ft <br> 1 km = 0.62 miles | 1 in. = 0.0254 m <br> 1 ft = 0.3048 m <br> 1 mile = 1.61 km |
| Volume* | 1 liter = 0.264 gallons | 1 gallon = 3.785 liters |
| Mass | 1 kg = 2.20 lbm | 1 lbm = 0.45 kg |
| Density | 1 kg/m$^3$ = 0.0624 lbm/ft$^3$ | 1 lbm/ft$^3$ = 16.025 kg/m$^3$ |
| Force, Weight | 1 N = 0.225 lbf | 1 lbf = 4.45 N |
| Pressure | 1 Pa = 1.45 $\times$ 10$^{-4}$ lbf/in.$^2$ | 1 lbf/in.$^2$ = 6895 Pa |
| Work, Energy | 1 J = 0.737 ft·lbf | 1 ft·lbf = 1.356 J |
| Power | 1 kW = 1.341 hp | 1 hp = 0.746 kW |
| Temperature | $^{\circ}C = \dfrac{5}{9}(^{\circ}F - 32)$ | $^{\circ}F = \dfrac{9}{5}\,^{\circ}C + 32$ |

* 1 ft$^3$ = 7.48 gallons
1 m$^3$ = 1000 liters

# Some Useful Data

density of water $= 1,000$ kg/m$^3$ $= 62.4$ lbm/ft$^3$

density of air $= 1.2$ kg/m$^3$ $= 0.076$ lbm/ft$^3$

## The Energy Content of Common Fuels

| Fuel | Quantity | Average Energy Content |
|------|----------|------------------------|
| Coal | One pound | 10,000 Btu ($10.5 \times 10^6$ J) |
| Diesel | One gallon | 139,000 Btu ($146.6 \times 10^6$ J) |
| Gasoline | One gallon | 124,000 Btu ($130.8 \times 10^6$ J) |
| Fuel oil (home heating oil) | One gallon | 139,000 Btu ($146.6 \times 10^6$ J) |
| Natural Gas | One cubic foot | 1,000 Btu ($1.05 \times 10^6$ J) |
| Wood | One cord (128 ft$^3$) (4 feet by 4 feet by 8 feet pile of wood stacked neatly) | 20,000,000 Btu ($21.1 \times 10^9$ J) |

## Examples of American Wire Gage (AWG) for Solid Copper Wire

| American Wire Gage (AWG) Number | Diameter (mils) | Current | Common Use |
|---|---|---|---|
| 00 | 365.0 | 200 A | Service entrance |
| 0 | | 150 A | |
| 1 | 289.0 | | |
| 2 | 258.0 | 100 A | Service panels |
| 5 | 182.0 | | |
| 6 | 162.0 | 60 A | Electric furnaces |
| 7 | 144.0 | 40 A | Kitchen appliances, receptacles, light fixtures |
| 10 | 91.0 | 30 A | |
| 12 | 81.0 | 20 A | Residential wiring |
| 14 | 64.0 | 15 A | Lamps, light fixtures |
| 16 | 51.0 | | |
| 18 | 40.0 | | |
| 20 | 32.0 | | |

## Thermal Conductivity of Some Common Materials

| Material | Thermal Conductivity (W/m · °C) |
| --- | --- |
| Air (at atmospheric pressure) | 0.0263 |
| Aluminum (pure) | 237 |
| Aluminum alloy (4.5% copper, 1.5% magnesium, 0.6% manganese) | 177 |
| Asphalt | 0.062 |
| Bronze (90% copper, 10% aluminium) | 52 |
| Brass (70% copper, 30% zinc) | 110 |
| Brick | 1.0 |
| Concrete | 1.4 |
| Copper (pure) | 401 |
| Glass | 1.4 |
| Gold | 317 |
| Human fat layer | 0.2 |
| Human muscle | 0.41 |
| Human skin | 0.37 |
| Iron (pure) | 80.2 |
| Stainless steels | 13.4 to 15.1 |
| Lead | 35.3 |
| Paper | 0.18 |
| Platinum (pure) | 71.6 |
| Sand | 0.27 |
| Silicon | 148 |
| Silver | 429 |
| Zinc | 116 |
| Water (liquid) | 0.61 |

## Typical Values of Heat Transfer Coefficients

| Convection Type | Heat Transfer Coefficient, $h$ (W / m² · °C) | Heat Transfer Coefficient, $h$ (Btu/h · ft² · °F) |
|---|---|---|
| *Free Convection* | | |
| Gases | 2 to 25 | 0.35 to 4.4 |
| Liquids | 50 to 1000 | 8.8 to 175 |
| *Forced Convection* | | |
| Gases | 25 to 250 | 4.4 to 44 |
| Liquids | 100 to 20,000 | 17.6 to 3500 |

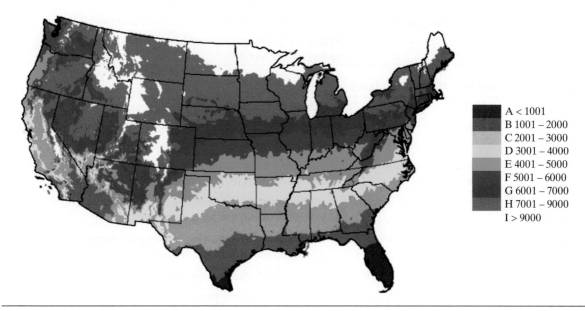

A < 1001
B 1001 – 2000
C 2001 – 3000
D 3001 – 4000
E 4001 – 5000
F 5001 – 6000
G 6001 – 7000
H 7001 – 9000
I > 9000

The annual degree-days for the United States
*Source:* NOAA

## The $CO_2$ Emission by the Source of Energy

| Source of Energy | Pounds of $CO_2$ Emission (average values) |
|---|---|
| 1 kWh | 1.7 |
| 1,000 ft³ of natural gas | 120 |
| 1 gallon of gasoline | 20 |
| 1 gallon of diesel | 22 |
| 1 gallon of fuel oil | 22.5 |
| 1 gallon of propane | 12.4 |

## The Outdoor and Indoor Pollutants, the WHO Recommended Limits, and Their Health Effects (µg = microgram = $10^{-6}$ gram)

| Pollutant | Limits | Causes/Health Effects |
|---|---|---|
| $PM_{2.5}$ | 10 µg/m³ annual mean<br>25 µg/m³ 24-hour mean | Indoor open fire and leaky stoves in developing countries<br>Industrial activities/vehicles/ power production |
| $PM_{10}$ | 20 µg/m³ annual mean<br>50 µg/m³ 24-hour mean | Cardiovascular and respiratory diseases, lung cancer |
| $O_3$ | 100 µg/m³ 8-hour mean | Vehicles/industrial activities<br>Breathing problems, asthma, reduced lung function, lung diseases, and heart disease |
| $NO_2$ | 40 µg/m³ annual mean<br>200 µg/m³ 1-hour mean | Industrial activities/vehicles/power production/human activities<br>Reduced lung function |
| $SO_2$ | 20 µg/m³ 24-hour mean<br>500 µg/m³ 10-minute mean | Industrial/power production/human activities<br>When combined with water, it produces acid rain<br>Respiratory problems, reduced lung function, infections of the respiratory tracts |

## Examples of Drinking Water Standards

| Contaminant | MCGL | MCL | Source of Contaminant by Industries |
|---|---|---|---|
| Antimony | 6 ppb | 6 ppb | copper smelting, refining, porcelain plumbing fixtures, petroleum refining, plastics, resins, storage batteries |
| Asbestos | 7 M.L. (million fibers per liter) | 7 M.L. | asbestos products, chlorine, asphalt felts and coating, auto parts, petroleum refining, plastic pipes |
| Barium | 2 ppm | 2 ppm | copper smelting, car parts, inorganic pigments, gray ductile iron, steel works, furnaces, paper mills |
| Beryllium | 4 ppb | 4 ppb | copper rolling and drawing, nonferrous metal smelting, aluminum foundries, blast furnaces, petroleum refining |
| Cadmium | 5 ppb | 5 ppb | zinc and lead smelting, copper smelting, inorganic pigments |
| Chromium | 0.1 ppm | 0.1 ppm | pulp mills, inorganic pigments, copper smelting, steel works |
| Copper | 1.3 ppm | 1.3 ppm | primary copper smelting, plastic material, poultry slaughtering, prepared feeds |
| Cyanide | 0.2 ppm | 0.2 ppm | metal heat treating, plating, and polishing |
| Lead | zero | 15 ppb | lead smelting, steel works and blast furnaces, storage batteries, china plumbing fixtures |
| Mercury | 2 ppb | 2 ppb | electric lamps, paper mills |
| Nickel | 0.1 ppm | 0.1 ppm | petroleum refining, gray iron foundries, primary copper, blast furnaces, steel |
| Nitrate | 10 ppm | 10 ppm | nitrogenous fertilizer, fertilizing mixing, paper mills, canned foods, phosphate fertilizers |
| Nitrite | 1 ppm | 1 ppm | |
| Selenium | 0.05 ppm | 0.05 ppm | metal coating, petroleum refining |
| Thallium | 0.5 ppb | 2 ppb | primary copper smelting, petroleum refining, steel works, blast furnaces |

## The Approximate Mass for Each Layer of the Earth

|  |  | Approximate Mass (pounds) | Approximate Mass (kg) | Percentage of the Earth's Total Mass |
|---|---|---|---|---|
| Atmosphere |  | $1.12 \times 10^{19}$ | $5.1 \times 10^{18}$ | 0.000086 |
| Oceans |  | $3.08 \times 10^{21}$ | $1.4 \times 10^{21}$ | 0.024 |
| Crust |  | $5.73 \times 10^{22}$ | $2.6 \times 10^{22}$ | 0.44 |
| Mantle |  | $8.9 \times 10^{24}$ | $4.04 \times 10^{24}$ | 68.47 |
| Core | Outer core | $4.03 \times 10^{24}$ | $1.83 \times 10^{24}$ | 31.01 |
|  | Inner core | $2.12 \times 10^{23}$ | $9.65 \times 10^{22}$ | 1.63 |

The chemical elements to date (2016)

# The Earth Charter

## Preamble

We stand at a critical moment in Earth's history, a time when humanity must choose its future. As the world becomes increasingly interdependent and fragile, the future at once holds great peril and great promise. To move forward we must recognize that in the midst of a magnificent diversity of cultures and life forms we are one human family and one Earth community with a common destiny. We must join together to bring forth a sustainable global society founded on respect for nature, universal human rights, economic justice, and a culture of peace. Towards this end, it is imperative that we, the peoples of Earth, declare our responsibility to one another, to the greater community of life, and to future generations.

## Earth, Our Home

Humanity is part of a vast evolving universe. Earth, our home, is alive with a unique community of life. The forces of nature make existence a demanding and uncertain adventure, but Earth has provided the conditions essential to life's evolution. The resilience of the community of life and the well-being of humanity depend upon preserving a healthy biosphere with all its ecological systems, a rich variety of plants and animals, fertile soils, pure waters, and clean air. The global environment with its finite resources is a common concern of all peoples. The protection of Earth's vitality, diversity, and beauty is a sacred trust.

## The Global Situation

The dominant patterns of production and consumption are causing environmental devastation, the depletion of resources, and a massive extinction of species. Communities are being undermined. The benefits of development are not shared equitably and the gap between rich and poor is widening. Injustice, poverty, ignorance, and violent conflict are widespread and the cause of great suffering. An unprecedented rise in human population has overburdened ecological and social systems. The foundations of global security are threatened. These trends are perilous—but not inevitable.

## The Challenges Ahead

The choice is ours: form a global partnership to care for Earth and one another or risk the destruction of ourselves and the diversity of life. Fundamental changes are needed in our values, institutions, and ways of living. We must realize that when basic needs have been met, human development is primarily about being more, not having more. We have the knowledge and technology to provide for all and to reduce our impacts on the environment. The emergence of a global civil society is creating new opportunities to build a democratic and humane world. Our environmental, economic, political, social, and spiritual challenges are interconnected, and together we can forge inclusive solutions.

## Universal Responsibility

To realize these aspirations, we must decide to live with a sense of universal responsibility, identifying ourselves with the whole Earth community as well as our local communities. We are at once citizens of different nations and of one world in which the local and global are linked. Everyone shares responsibility for the present and future well-being of the human family and the larger living world. The spirit of human solidarity and kinship with all life is strengthened when we live with reverence for the mystery of being, gratitude for the gift of life, and humility regarding the human place in nature.

We urgently need a shared vision of basic values to provide an ethical foundation for the emerging world community. Therefore, together in hope we affirm the following interdependent principles for a sustainable way of life as a common standard by which the conduct of all individuals, organizations, businesses, governments, and transnational institutions is to be guided and assessed.

## Principles

### I. RESPECT AND CARE FOR THE COMMUNITY OF LIFE

**1. Respect Earth and life in all its diversity.**

 a. Recognize that all beings are interdependent and every form of life has value regardless of its worth to human beings.

 b. Affirm faith in the inherent dignity of all human beings and in the intellectual, artistic, ethical, and spiritual potential of humanity.

**2. Care for the community of life with understanding, compassion, and love.**

 a. Accept that with the right to own, manage, and use natural resources comes the duty to prevent environmental harm and to protect the rights of people.

 b. Affirm that with increased freedom, knowledge, and power comes increased responsibility to promote the common good.

**3. Build democratic societies that are just, participatory, sustainable, and peaceful.**

 a. Ensure that communities at all levels guarantee human rights and fundamental freedoms and provide everyone an opportunity to realize his or her full potential.

 b. Promote social and economic justice, enabling all to achieve a secure and meaningful livelihood that is ecologically responsible.

**4. Secure Earth's bounty and beauty for present and future generations.**

 a. Recognize that the freedom of action of each generation is qualified by the needs of future generations.

 b. Transmit to future generations values, traditions, and institutions that support the long-term flourishing of Earth's human and ecological communities.

### II. ECOLOGICAL INTEGRITY

**5. Protect and restore the integrity of Earth's ecological systems, with special concern for biological diversity and the natural processes that sustain life.**

 a. Adopt at all levels sustainable development plans and regulations that make environmental conservation and rehabilitation integral to all development initiatives.

 b. Establish and safeguard viable nature and biosphere reserves, including wild lands and marine areas, to protect Earth's life support systems, maintain biodiversity, and preserve our natural heritage.

 c. Promote the recovery of endangered species and ecosystems.

 d. Control and eradicate non-native or genetically modified organisms harmful to native species and the environment, and prevent introduction of such harmful organisms.

 e. Manage the use of renewable resources such as water, soil, forest products, and marine life in ways that do not exceed rates of regeneration and that protect the health of ecosystems.

 f. Manage the extraction and use of non-renewable resources such as minerals and fossil fuels in ways that minimize depletion and cause no serious environmental damage.

**6. Prevent harm as the best method of environmental protection and, when knowledge is limited, apply a precautionary approach.**

 a. Take action to avoid the possibility of serious or irreversible environmental harm even when scientific knowledge is incomplete or inconclusive.

 b. Place the burden of proof on those who argue that a proposed activity will not cause significant harm, and make the responsible parties liable for environmental harm.

 c. Ensure that decision making addresses the cumulative, long-term, indirect, long distance, and global consequences of human activities.

 d. Prevent pollution of any part of the environment and allow no build-up of radioactive, toxic, or other hazardous substances.

 e. Avoid military activities damaging to the environment.

**7. Adopt patterns of production, consumption, and reproduction that safeguard Earth's regenerative capacities, human rights, and community well-being.**

 a. Reduce, reuse, and recycle the materials used in production and consumption systems, and ensure that residual waste can be assimilated by ecological systems.

b. Act with restraint and efficiency when using energy, and rely increasingly on renewable energy sources such as solar and wind.

c. Promote the development, adoption, and equitable transfer of environmentally sound technologies.

d. Internalize the full environmental and social costs of goods and services in the selling price, and enable consumers to identify products that meet the highest social and environmental standards.

e. Ensure universal access to health care that fosters reproductive health and responsible reproduction.

f. Adopt lifestyles that emphasize the quality of life and material sufficiency in a finite world.

8. **Advance the study of ecological sustainability and promote the open exchange and wide application of the knowledge acquired.**

a. Support international scientific and technical cooperation on sustainability, with special attention to the needs of developing nations.

b. Recognize and preserve the traditional knowledge and spiritual wisdom in all cultures that contribute to environmental protection and human well-being.

c. Ensure that information of vital importance to human health and environmental protection, including genetic information, remains available in the public domain.

## III. SOCIAL AND ECONOMIC JUSTICE

9. **Eradicate poverty as an ethical, social, and environmental imperative.**

a. Guarantee the right to potable water, clean air, food security, uncontaminated soil, shelter, and safe sanitation, allocating the national and international resources required.

b. Empower every human being with the education and resources to secure a sustainable livelihood, and provide social security and safety nets for those who are unable to support themselves.

c. Recognize the ignored, protect the vulnerable, serve those who suffer, and enable them to develop their capacities and to pursue their aspirations.

10. **Ensure that economic activities and institutions at all levels promote human development in an equitable and sustainable manner.**

a. Promote the equitable distribution of wealth within nations and among nations.

b. Enhance the intellectual, financial, technical, and social resources of developing nations, and relieve them of onerous international debt.

c. Ensure that all trade supports sustainable resource use, environmental protection, and progressive labor standards.

d. Require multinational corporations and international financial organizations to act transparently in the public good, and hold them accountable for the consequences of their activities.

11. **Affirm gender equality and equity as prerequisites to sustainable development and ensure universal access to education, health care, and economic opportunity.**

a. Secure the human rights of women and girls and end all violence against them.

b. Promote the active participation of women in all aspects of economic, political, civil, social, and cultural life as full and equal partners, decision makers, leaders, and beneficiaries.

c. Strengthen families and ensure the safety and loving nurture of all family members.

12. **Uphold the right of all, without discrimination, to a natural and social environment supportive of human dignity, bodily health, and spiritual well-being, with special attention to the rights of indigenous peoples and minorities.**

a. Eliminate discrimination in all its forms, such as that based on race, color, sex, sexual orientation, religion, language, and national, ethnic or social origin.

b. Affirm the right of indigenous peoples to their spirituality, knowledge, lands and resources and to their related practice of sustainable livelihoods.

c. Honor and support the young people of our communities, enabling them to fulfill their essential role in creating sustainable societies.

d. Protect and restore outstanding places of cultural and spiritual significance.

## IV. DEMOCRACY, NONVIOLENCE, AND PEACE

**13. Strengthen democratic institutions at all levels, and provide transparency and accountability in governance, inclusive participation in decision making, and access to justice.**

a.  Uphold the right of everyone to receive clear and timely information on environmental matters and all development plans and activities which are likely to affect them or in which they have an interest.

b.  Support local, regional and global civil society, and promote the meaningful participation of all interested individuals and organizations in decision making.

c.  Protect the rights to freedom of opinion, expression, peaceful assembly, association, and dissent.

d.  Institute effective and efficient access to administrative and independent judicial procedures, including remedies and redress for environmental harm and the threat of such harm.

e.  Eliminate corruption in all public and private institutions.

f.  Strengthen local communities, enabling them to care for their environments, and assign environmental responsibilities to the levels of government where they can be carried out most effectively.

**14. Integrate into formal education and life-long learning the knowledge, values, and skills needed for a sustainable way of life.**

a.  Provide all, especially children and youth, with educational opportunities that empower them to contribute actively to sustainable development.

b.  Promote the contribution of the arts and humanities as well as the sciences in sustainability education.

c.  Enhance the role of the mass media in raising awareness of ecological and social challenges.

d.  Recognize the importance of moral and spiritual education for sustainable living.

**15. Treat all living beings with respect and consideration.**

a.  Prevent cruelty to animals kept in human societies and protect them from suffering.

b.  Protect wild animals from methods of hunting, trapping, and fishing that cause extreme, prolonged, or avoidable suffering.

c.  Avoid or eliminate to the full extent possible the taking or destruction of non-targeted species.

**16. Promote a culture of tolerance, nonviolence, and peace.**

a.  Encourage and support mutual understanding, solidarity, and cooperation among all peoples and within and among nations.

b.  Implement comprehensive strategies to prevent violent conflict and use collaborative problem solving to manage and resolve environmental conflicts and other disputes.

c.  Demilitarize national security systems to the level of a non-provocative defense posture, and convert military resources to peaceful purposes, including ecological restoration.

d.  Eliminate nuclear, biological, and toxic weapons and other weapons of mass destruction.

e.  Ensure that the use of orbital and outer space supports environmental protection and peace.

f.  Recognize that peace is the wholeness created by right relationships with oneself, other persons, other cultures, other life, Earth, and the larger whole of which all are a part.

# Index